AI and IoT-Driven Social Sensing for Smart City Innovation

Hari Mohan Rai
Nazarbayev University, Astana, Kazakhstan

Abdul Razaque
Satbayev University, Kazakhstan

Syed Atif Moqurrab
University of Bedfordshire, UK

IGI Global
Scientific Publishing
Publishing Tomorrow's Research Today

Vice President of Editorial — Melissa Wagner
Director of Acquisitions — Mikaela Felty
Director of Book Development — Jocelynn Hessler
Production Manager — Mike Brehm
Cover Design — Jose Rosado

Published in the United States of America by
IGI Global Scientific Publishing
701 East Chocolate Avenue
Hershey, PA, 17033, USA
Tel: 717-533-8845 | Fax: 717-533-7115
Website: https://www.igi-global.com E-mail: cust@igi-global.com

Library of Congress Cataloging-in-Publication Data

Names: Rai, Hari Mohan, 1986- editor | Razaque, Abdul, 1974- editor | Moqurrab, Syed Atif, 1988- editor
Title: AI and IoT-driven social sensing for smart city innovation / edited by: Hari Mohan Rai, Abdul Razaque, and Syed Atif Moqurrab.
Description: Hershey PA : IGI Global Scientific Publishing, [2027] | Includes bibliographical references. | Summary: "This book explores the evolving field of social sensing within smart cities, presenting its potential, challenges, and future directions"-- Provided by publisher.
Identifiers: LCCN 2025028106 (print) | LCCN 2025028107 (ebook) | ISBN 9798337349626 hardcover | ISBN 9798337349633 paperback | ISBN 9798337349640 ebook
Subjects: LCSH: Smart cities | Artificial intelligence--Industrial applications | Internet of things--Industrial applications
Classification: LCC TD159.4 (print) | LCC TD159.4 (ebook)
LC record available at https://lccn.loc.gov/2025028106
LC ebook record available at https://lccn.loc.gov/2025028107

British Cataloguing in Publication Data
A Cataloguing in Publication record for this book is available from the British Library.

Table of Contents

Detailed Table of Contents

Zurich, Switzerland is recognized as the world's leading smart city as per the Smart City Index 2024. It leverages advanced technologies to drive sustainable innovation. The study focuses on the role of advanced and emerging technologies in transforming Zurich into a model smart city. The study explores how AI-powered solutions contribute to urban efficiency and sustainability. The study evaluates the impact of smart city initiatives on 15 key sustainability indicators important to a smart city. The data is collected from government reports, industry publications, academic research etc. The findings highlight the Zurich strategic approach to integrating AI in urban planning, offering valuable lessons for cities throughout the world to achieve sustainability through digital transformation.

The rapid development of smart cities can be explained by the combination of artificial intelligence and the IoT. Implementation of the IoT infrastructures, which includes smartphones, wearable devices, surveillance cameras, environmental monitors and numerous other devices, to a municipality gives its access to vast amounts of data on real-time. A combination of machine learning, deep learning, and natural language processing allows AI to reduce such data sets into patterns, anomalies, and trend predictions by numerous aspects of cities. The discussion will focus on AI models of IoT data processing, such as human activity recognition, traffic-flow forecasting, and social-media sentiment analysis, and analyze the use of edge AI and federated learning to reduce the latency and still preserve the data privacy all of which is extremely important in metropolitan areas. The corresponding applications of disaster management, safety, health-related monitoring, and optimization of energy

The rapid advancement of technology has led to the development of innovative solutions in childcare, addressing the challenges faced by modern parents. This paper presents the design and implementation of an IoT-based Smart Baby Cradle that enhances infant safety and comfort through real-time monitoring and automated soothing mechanisms. The system integrates various sensors to detect temperature, humidity, sound, motion, and health parameters. It employs a microcontroller to process sensor data and trigger appropriate actions, such as gentle rocking or playing lullabies, in response to the baby's needs. Parents receive real-time alerts and can remotely monitor and control the cradle using a mobile application. The proposed system not only reduces caregiver stress but also ensures the well-being of the infant, representing a significant advancement in smart parenting solutions.

The infrastructure that promises urban revolution of innovations, such as real-time traffic optimization, community surveillance of health metrics, and data-intensive urban governance, does not come without significant risks, including mass surveillance, cybersecurity threats, and discrimination via algorithms. The chapter evaluates the challenges of privacy, security, and ethics, shedding light on governance gaps in legal regulations such as the GDPR and the NIST frameworks by detailing these gaps using examples from a ransomware case in Atlanta and in another case, the racial bias in predictive policing. To address these challenging gaps, the implementation of privacy-by-design system architecture, decentralised data governance systems, and participatory policy systems that bring technological evolution in line with human rights protection and democratic principles was proposed. The conclusion affirms the urgent need to have interdisciplinary cooperation that can promote the production of smart cities that are profitable and fair by developing a thorough culture of trust.

sensing enhances situational awareness, reduces response times, and fosters citizen engagement. We also address the challenges related to data privacy, inclusivity, and digital equity. The chapter concludes with a forward-looking perspective on how combining AI, federated learning, and decentralized systems such as blockchain can ensure resilient, ethical, and adaptive smart zone ecosystems for the future.

Mariem Smaoui, National Engineering School of Sfax, Tunisia
Mouna Rekik, National Engineering School of Sfax, Tunisia
Lotfi Krichen, National Engineering School of Sfax, Tunisia

Waste management is a major environmental challenge, driven by urban growth and modern consumption, which surpasses the capacity of traditional sorting methods. To address this challenge, integrating Artificial Intelligence (AI) and the Internet of Things (IoT) into waste management offers an innovative solution. This chapter presents an autonomous and intelligent system capable of automatically identifying and sorting waste (plastic, metal, glass, cardboard, organic) using smart cameras and specialized sensors. A central unit equipped with a camera and AI models detects plastic and cardboard in real time, while other sensors identify the characteristics of glass, metals, and organic waste. Mechanical actuators then direct the waste to the appropriate bins. Level sensors prevent overflows and optimize collection, with data accessible via a mobile application. A functional prototype was tested, confirming technical feasibility, effective integration, and reliable automated sorting, thus laying the groundwork for industrial or large-scale deployment.

Valisher Sapayev, Mamun University, Khiva, Uzbekistan
S. Aarthi, Marwadi University, Rajkot, India
Turobova Sevara, Termez University of Economics and Service, Termez, Uzbekistan
R. N. Ravikumar, Marwadi University, Rajkot, India
Khayrulla Urozboev, Alfraganus University, Tashkent, Uzbekistan
Rasulov Avazbek, Tashkent University of Information Technologies, Tashkent, Uzbekistan

The rapid growth of smart cities has led to extensive use of social sensing technologies, generating vast amounts of citizen and environmental data. While this data enables real-time insights for traffic management, public health, and energy optimization, it also raises significant privacy concerns. This chapter explores the application of

Federated Learning (FL) as a privacy-preserving, decentralized approach to social sensing in urban environments. FL allows local model training on edge devices, sharing only encrypted updates, thereby reducing risks of data leakage. The chapter details FL principles, architecture, privacy mechanisms, and case studies on traffic prediction and public health monitoring, demonstrating improvements in accuracy, efficiency, and trust. It also addresses challenges such as heterogeneous data, limited device resources, and communication bottlenecks, offering governance strategies for ethical deployment. A roadmap for expanding FL applications in smart cities is presented.

Social sensing offers a transformative approach to urban intelligence by leveraging citizen-generated data from mobile devices, social media, and participatory platforms. Unlike traditional sensors, social sensing captures real-time, context-rich insights into urban dynamics such as traffic, public sentiment, safety, and infrastructure conditions. This chapter explores how the integration of social sensing with AI and IoT technologies enables responsive, inclusive, and adaptive city systems. It presents technical frameworks, real-world applications, and policy considerations while addressing key challenges such as data quality, bias, and privacy. Ultimately, it advocates for collaborative sensing ecosystems where citizens are active participants in shaping smarter, more resilient cities.

Preface

Cities are no longer merely physical environments of streets, buildings, and infrastructure; they have become dense, continuously generating fields of data. Every smartphone in a pocket, every connected vehicle, every environmental sensor, and every social media post contributes to an unprecedented digital portrait of urban life as it unfolds. The central premise of this book, AI and IoT-Driven Social Sensing for Smart City Innovation, is that the convergence of Artificial Intelligence (AI) and the Internet of Things (IoT) has given rise to a new paradigm for understanding and governing cities, one in which the experiences, movements, and expressions of citizens themselves become a primary source of urban intelligence.

Social sensing refers to the practice of treating people, and the devices they carry and interact with, as distributed sensors that report on the state of their environment. In contrast to traditional infrastructure-bound monitoring, social sensing fuses citizen-generated information from mobile devices, social media, and crowdsourcing with conventional IoT measurements to capture the human pulse of the city in real time. When this rich but noisy stream of data is processed using modern AI techniques, including machine learning, deep learning, edge computing, and federated learning, it can be transformed into actionable insight for transportation, public safety, utilities, energy, waste management, environmental monitoring, and public health. This book brings together a diverse, international community of researchers and practitioners to examine how that transformation is being designed, implemented, and governed.

THE TOPIC IN THE WORLD TODAY

The timeliness of this subject is difficult to overstate. The accelerating pace of urbanization in the twenty-first century has placed enormous pressure on city services and infrastructure, while simultaneously raising citizen expectations for responsiveness, sustainability, and inclusiveness. Around the world, municipalities are investing heavily in smart city initiatives, and global indices now rank cities on the maturity of their digital transformation. At the same time, the technological foundations for social sensing have matured rapidly: IoT devices are inexpensive and ubiquitous, edge computing has made low-latency intelligence feasible, and advances in AI have made it possible to extract meaning from heterogeneous, large-scale, and unstructured data.

Yet the same technologies that promise more efficient, livable, and sustainable cities also raise serious concerns. Mass data collection invites risks of surveillance, cybersecurity breaches, algorithmic bias, and the erosion of privacy. The digital divide threatens to exclude the very citizens whose participation social sensing depends upon. Consequently, the field today sits at a crucial intersection of engineering capability and societal responsibility. This book deliberately reflects that duality: it celebrates the technical ingenuity that makes social sensing possible while insisting that trust, ethics, and governance are not afterthoughts but design requirements.

TARGET AUDIENCE

This book is intended for a broad and interdisciplinary readership. Researchers and graduate students in computer science, electronics and communication, data science, and urban informatics will find rigorous treatments of architectures, algorithms, and emerging research directions. Practitioners, including system engineers, IoT developers, and urban technologists, will benefit from concrete system designs, prototypes, and case studies that illustrate how concepts translate into deployable solutions. Equally, urban planners, public administrators, and policymakers will find accessible discussions of the opportunities, limitations, and ethical implications of social sensing that can inform real-world decisions. By spanning foundational concepts, technical implementations, real-city case studies, and governance frameworks, the volume aims to serve as both a reference and a bridge across these communities.

ORGANIZATION OF THE BOOK AND CHAPTER CONTRIBUTIONS

The book comprises nineteen chapters contributed by authors from numerous countries and institutions, reflecting the genuinely global nature of the smart city endeavor. Although the chapters can be read independently, they collectively progress from foundational reviews and integration frameworks, through domain-specific applications and system implementations, to the critical concerns of privacy, security, and ethics. The following paragraphs introduce each contribution.

Chapter 1, "A Case Study on Zurich Smart City Initiatives: Advancing Sustainability Innovation and Urban Development," by Rajni Bala, grounds the discussion in a real-world exemplar. Taking Zurich, ranked as the world's leading smart city in the Smart City Index 2024, as its subject, the chapter evaluates how AI-powered solutions advance urban efficiency and sustainability across fifteen key indicators. Drawing on government reports, industry publications, and academic research, it distills strategic lessons that other cities can adapt in pursuit of sustainable digital transformation.

Chapter 2, "AI and IoT Integration for Social Sensing Optimization in Smart Cities," by Premavathi T and Damodharan Palaniappan, examines how the combination of AI and IoT infrastructure can turn vast real-time data streams into urban insight. The chapter discusses AI models for human activity recognition, traffic-flow forecasting, and social-media sentiment analysis, and considers how edge AI and federated learning can reduce latency while preserving privacy. Applications in disaster management, safety, health monitoring, and energy optimization are presented with concrete examples.

Chapter 3, "Crowd-AI for Smart Cities: Leveraging Social Sensing and Edge AI for Real-Time Urban Intelligence," by Darshan Mansukhbhai Tank, proposes Crowd-AI, a framework that synergizes social sensing, edge AI, and IoT to deliver human-centric urban intelligence. By processing social signals from social media, wearables, and IoT infrastructure at the edge, the approach reduces latency, safeguards privacy, and enables hyper-local decisions. Case studies in traffic management, public sentiment monitoring, and disaster response illustrate the framework, which the chapter extends toward digital twins, federated learning, and citizen co-creation.

Chapter 4, "Energy and Safety Optimization in Smart Cities through Edge-AI-Driven Adaptive Street Lighting with Ambient Sensing and Traffic Pattern Prediction," by Viji Gripsy J, Praneesh M, and John T. Mesia Dhas, focuses on one of the largest contributors to urban energy consumption: street lighting. The chapter demonstrates how edge-AI-driven adaptive lighting, informed by ambient sensing and real-time traffic prediction, can optimize energy use while improving safety. It details system architecture, sensing modalities, and predictive algorithms, and confronts practical challenges of privacy, cost-efficiency, and scalability.

Chapter 5, "Enhancing Urban Living with AI and IoT-Based Social Sensing Systems," by Raj Kumar S and Kavitha K, offers a broad survey of how social sensing systems make cities more responsive to citizens' needs. By collecting real-time data from sensors, mobile devices, and social media, and analyzing it with AI to predict trends and detect anomalies, such systems improve operational efficiency and citizen engagement. The chapter surveys applications ranging from smart transportation and environmental monitoring to crowd management and public health tracking.

Chapter 6, "Integration of AI and IoT for the Optimization of Social Sensing in Smart City," by Monika Kumari and Nikhil Kumar Goyal, presents architectural layouts and integration schemes for AI-IoT social sensing systems. The chapter explains how machine learning, deep learning, and reinforcement learning can improve sensing precision and reduce data redundancy, with illustrative applications in traffic management, environmental monitoring, citizen engagement, and public safety. It also addresses privacy, security, and ethics, and points toward edge computing, human-centric AI, and 6G integration.

Chapter 7, "IoT-Based Smart Cradle Baby Monitoring," by Theivanathan G, Akash S, Arunvel R, and Akash N, applies social sensing principles to childcare. The chapter describes a smart baby cradle that integrates sensors for temperature, humidity, sound, motion, and health parameters, using a microcontroller to trigger soothing actions such as gentle rocking or lullabies. Parents receive real-time alerts and can monitor and control the cradle remotely through a mobile application, reducing caregiver stress while safeguarding infant well-being.

Chapter 8, "Privacy, Security, and Ethical Challenges in AI and IoT Social Sensing Systems," by Nathaniel Adeniyi Akande, confronts the risks that accompany pervasive urban sensing, including mass surveillance, cybersecurity threats, and algorithmic discrimination. Drawing on real incidents and on regulatory frameworks such as GDPR and NIST, the chapter exposes governance gaps and proposes privacy-by-design architectures, decentralized data governance, and participatory policy mechanisms. It makes a compelling case for interdisciplinary cooperation to ensure that smart cities are not only efficient but also fair and trustworthy.

Chapter 9, "Smartphone Detection with Social Sensor Systems," by Binastya Anggara Sekti, examines the smartphone as a powerful social sensor that gathers data through GPS, cameras, microphones, and connectivity modules. The chapter shows how smartphone-based detection integrates with AI and IoT frameworks to provide real-time insight into mobility, public safety, and environmental quality, emphasizing participatory and opportunistic sensing and cloud-edge integration. Case studies on crowd-density monitoring and smart tourism illustrate the transformative, citizen-centric potential of these ubiquitous devices.

Chapter 10, "Social Sensing as a Catalyst for Effective Public Utilities in Urban Areas," by Lokesh Jain, explores how social sensing reinvents urban governance by combining human-generated information with conventional IoT systems. The chapter shows how real-time citizen insight can support intelligent transport, water and energy management, waste management, and emergency response. While acknowledging challenges of reliability, privacy, and data integration, it argues that emerging tools such as AI, blockchain, and interoperability standards can enhance trust, inclusivity, and resilience in citizen-centric cities.

Chapter 11, "Social Sensing for Effective Public Utilities in Smart Zones," also by Binastya Anggara Sekti, focuses specifically on public utility management within smart zones. By leveraging both participatory data, such as mobile app reports and social media posts, and opportunistic data from IoT devices, cities can detect issues in water, electricity, sanitation, and waste in near real time. The chapter presents architectures and case studies, addresses data privacy and digital equity, and looks toward AI, federated learning, and blockchain for resilient and ethical smart zone ecosystems.

Chapter 12, "Towards Smart and Sustainable Waste Management through IoT Integration for Automated Sorting," by Mariem Smaoui, Mouna Rekik, and Lotfi Krichen, presents an autonomous system for identifying and sorting waste, including plastic, metal, glass, cardboard, and organic materials, using smart cameras and specialized sensors. A central AI-equipped unit detects materials in real time while mechanical actuators direct waste to appropriate bins, and level sensors prevent overflows. A tested functional prototype confirms technical feasibility, laying the groundwork for large-scale deployment in sustainable urban waste management.

Chapter 13, "Trust-Centric and Privacy-Preserving Federated Learning Architectures for Ethical Data-Driven Smart City Services," by Valisher Sapayev, Aarthi S, Turobova Sevara, Ravikumar R N, Khayrulla Urozboev, and Rasulov Avazbek, advances Federated Learning as a privacy-preserving, decentralized approach to social sensing. By training models locally on edge devices and sharing only encrypted updates, the architecture reduces the risk of data leakage. Through case studies in traffic prediction and public health monitoring, the chapter demonstrates gains in accuracy, efficiency, and trust, and offers governance strategies and a roadmap for ethical deployment in smart cities.

Chapter 14, "Urban Intelligence through Social Sensing: Monitoring the Pulse of Smart Cities," by Aruna M, Arthi B, Padmapriya G, and Sridhar S, articulates how citizen-generated data captures real-time, context-rich insight into urban dynamics such as traffic, public sentiment, safety, and infrastructure. The chapter presents technical frameworks, real-world applications, and policy considerations, while addressing challenges of data quality, bias, and privacy. It advocates for collaborative sensing ecosystems in which citizens are active participants in shaping smarter, more resilient cities.

Chapter 15, "What are the Key E-Readiness Factors for Smart City? A Qualitative Study of Malang City," by Ali Roziqin, Muhammad Kamil, Shannaz Mutiara Deniar, Yusril Saputra, and Rifki Muhammad Bintang, closes the volume with a qualitative case study of Malang City, Indonesia. Through interviews and document analysis, the chapter identifies adaptive regulation, implementer competency, inter-agency coordination, and digital infrastructure as vital enablers of smart city development, while examining barriers such as the digital divide, budget limitations, and resistance to change. Its practical insights on collaboration and capacity-building offer valuable guidance for policymakers.

CONCLUSION

Taken together, these nineteen contributions provide a comprehensive and balanced portrait of a field in rapid evolution. They demonstrate that the value of social sensing lies not in any single technology but in the thoughtful integration of many: IoT devices that gather data, AI that interprets it, edge and federated architectures that make it timely and private, and governance frameworks that keep it trustworthy. They also show that the smart city is fundamentally a human project. The recurring emphasis on citizen participation, ethics, privacy, and inclusiveness across these chapters reflects a maturing recognition that technical performance and social responsibility must advance together.

It is our hope, as editors, that this volume will serve as a catalyst for further research and responsible innovation. By assembling foundational reviews, novel frameworks, working prototypes, and real-city case studies within a single resource, we aim to help researchers, practitioners, and policymakers build smart cities that are not only more efficient and sustainable, but also more equitable, resilient, and genuinely responsive to the people who live in them. We thank the contributing authors and reviewers whose dedication made this book possible, and we invite our readers to engage critically with the ideas presented here as they shape the urban futures to come.

Acknowledgment

The editors wish to express their sincere gratitude to all those whose efforts and support made this book possible.

First and foremost, we thank the contributing authors. Their expertise, dedication, and willingness to share their research form the very substance of this volume. Across many countries and institutions, they responded to our queries and revision requests with patience and professionalism, and we are grateful for their collaboration throughout the editorial process.

We are equally indebted to the reviewers, whose careful and constructive evaluations were essential to maintaining the scholarly quality of each chapter. Their thoughtful feedback strengthened the contributions and helped the authors refine and clarify their work. In particular, we wish to acknowledge Ms. Maria Gul of Nazarbayev University and Dr. Aditya Pal of Dongguk University, South Korea, who provided valuable assistance throughout this book project, contributing to the finalization of chapters and to the review process alongside our other reviewers.

We also extend our heartfelt thanks to our fellow co-editors. The completion of this volume is the result of a truly collaborative effort, and we are grateful for the shared commitment, coordination, and mutual support that guided the project from conception to publication.

Our thanks go to the editorial and production team at IGI Global for their professional guidance and consistent support at every stage, from the initial proposal through to final publication. Their responsiveness and attention to detail were invaluable.

We gratefully acknowledge the support of Nazarbayev University, which provided the environment and resources that enabled this work.

On a personal note, I am deeply grateful to my wife, Mahima Rai, and to my sons, Heyansh Rai and Vivaansh Rai, for their patience, encouragement, and unwavering support during the many hours devoted to bringing this book to completion. Their understanding made this work possible.

Chapter 1
A Case Study on Zurich Smart City Initiatives:
Advancing Sustainability Innovation and Urban Development

Rajni Bala
http://orcid.org/0000-0002-1225-939X
Chitkara Business School, Chitkara University, India

ABSTRACT

Zurich, Switzerland is recognized as the world's leading smart city as per the Smart City Index 2024. It leverages advanced technologies to drive sustainable innovation. The study focuses on the role of advanced and emerging technologies in transforming Zurich into a model smart city. The study explores how AI-powered solutions contribute to urban efficiency and sustainability. The study evaluates the impact of smart city initiatives on 15 key sustainability indicators important to a smart city. The data is collected from government reports, industry publications, academic research etc. The findings highlight the Zurich strategic approach to integrating AI in urban planning, offering valuable lessons for cities throughout the world to achieve sustainability through digital transformation.

INTRODUCTION

The 'Smart City' concept was developed to enhance urban living by providing efficient public services through digital technology. The aim is to provide an efficient and sustainable environment to improve well-being and quality of life. Today around

DOI: 10.4018/979-8-3373-4962-6.ch001

55% of the world's population lives in urban areas. As per a UN report, around 68% of people are expected to live in urban areas by 2050 (United Nations, 2018). This means that two out of every three people will be living in urban areas. As we see, the population in urban areas is increasing day by day. So, problems like environmental pollution, insufficient physical infrastructure, energy requirements, traffic congestion etc. are also increasing. This is the reason that cities want to develop and progress through innovative and smart technologies. The adoption of these technologies not only will help to solve above problems but also improve quality of life by being more sustainable, more energy efficient, and more environmentally friendly. That's why the concept of smart cities came which focuses on the use of digital technologies for modern urban development.

The concept of a smart city originated in the United States in 1992. The term was first used in the book 'The Technopolis Phenomenon', which discussed the growth of information technology and its use in modern infrastructure. The smart city concept is also known as city 4.0. It is a digital version of a physical city. As a newly developed idea, the term is still evolving and being revised frequently. Today we see a smart city as the relationship between physical infrastructure and digital technologies to solve urban problems, quality of life, create a safe environment, and be more sustainable. A smart sustainable city is an innovative city that uses information and communication technologies (ICTs) and other means to improve quality of life, efficiency of urban operations and services, and competitiveness while ensuring that it meets the needs of present and future generations concerning economic, social and environmental aspects. It is difficult for a city to be smart in all components. The extent of smart features that a city carries depends upon the budget and technology availability.

DEFINITIONS

"A city that monitors and integrates conditions of all of its critical infrastructures, including roads, bridges, tunnels, rails, subways, airports, seaports, communications, water, power, even major buildings, can better optimize its resources, plan its preventive maintenance activities, and monitor security aspects while maximizing services to its citizens." (Hall, R. E., 2000).

"The use of Smart Computing technologies to make the critical infrastructure components and services of a city which include city administration, education, healthcare, public safety, real estate, transportation, and utilities—more intelligent, interconnected, and efficient" (Washburn, D., Sindhu, U., Balaouras, S., Dines, R. A., Hayes, N. M., & Nelson, L. E., 2010).

A city "combining ICT and Web 2.0 technology with other organizational, design and planning efforts to de- materialize and speed up bureaucratic processes and help to identify new, innovative solutions to city management complexity, in order to improve sustainability and livability." (Toppeta, D., 2010).

ZURICH: THE WORLD SMARTEST CITY, 2024

Zurich is Switzerland's largest city and a major economic, cultural, and social center. It's located in north-central Switzerland on the northwest tip of Lake Zurich. It is a continuously growing metropolitan area that extends into central, northern, and eastern Switzerland. It is the hub of industrial, financial, and cultural developments. Zurich is the largest city in Switzerland and the capital of the eponymous Canton, with a population of 430,000 inhabitants (over one million with the greater metropolis area included). It's also the country's economic and financial capital. In recent years, the city has boomed, becoming an internationally renowned finance center. It's very attractive, and its population is growing rapidly.

The journey of Zurich as a smart city began with Switzerland's forward-thinking ambition of creating cities that combine technology innovation, environmental sustainability, and excellent living standard. The concept began when the country realized that increasing urban population, mounting climate concerns, rising energy demands required novel solutions. Swiss lawmakers and city planners recognized that depending solely on traditional infrastructure and services would not be sufficient to assure long term success. As a result, Zurich was chosen as a focal point to pioneer the smart city model due to its strong economy, cultural significance and history of successful urban planning.

Initially, the government, commercial firms, and research institutions collaborated to develop plans for digital transformation, energy efficiency and long-term growth. The objective was not only to modernize city services but also to maintain Zurich's global leadership in technology, innovation and finance. Heavy investments were made in, digital platforms, public transport systems and renewable energy to cut carbon emissions and improve daily life. Switzerland's dedication to sustainability, efficiency, long-term planning served as major motivators.

The idea of becoming a smart city was also influenced by Zurich's emphasis on citizen well-being. The country believed that smart solutions such as renewable energy, efficient mobility, and e-governance could directly benefits its citizens. Citizens' participation was encouraged from the beginning, making them crucial to the project's success. Zurich expanded its projects to include digital education, waste reduction systems, and smart healthcare. Gradually, these initiatives established the city as a pioneer. Today, Switzerland's ambition to make Zurich a city where innovation, technology, and sustainability coexist has successfully turned into reality. Zurich not only leads worldwide as a smart city, but it also acts as an inspiration for other countries aspiring for future-ready urbanization.

The journey of Zurich as a smart city started in 2017. It began with a smart streetlight that illuminates only when required. It collects environmental data, provides power to electric vehicles, acts as a smart WiFi antenna, and measures traffic flow. Zurich ranked 2nd in the 2017 Sustainable Cities Mobility Index of Arcadis, a company that performs analysis and consulting on urban issues. It was ranked 1st in the European mobility ranking of the Omio travel site in 2019, 5th in the Urban Mobility Readiness Index and 2nd in the Public Transport category in 2022, and 1st as the world's smartest city by smart city index 2024. The index is developed by the collaboration between IMD World Competitiveness Center's Smart City Observatory and the World Smart Sustainable Cities Organization (WeGO). Around 142 cities throughout the world participated in Smart City Index 2024.

Zurich remained at the top for the fifth consecutive year, recognized for its excellent digital infrastructure, eco-friendly mobility systems, and inclusive citizen programs that enhance quality of life. The index evaluates cities based on resident feedback focusing on infrastructure quality and the effective use of technology to address urban challenges such as health, mobility, governance, and environmental sustainability. European and Asian cities dominate the top ranks, with Zurich leading, followed by Oslo and Canberra. The report highlights the increasing importance of trust, good governance, AI integration, and resilience in shaping future-ready cities. Notably, North American cities experienced a decline in rankings, reflecting challenges in infrastructure and public safety.

Table 1. Top 20 Smart Cities as per IMD Smart City index, 2024 (IMD World Competitiveness Center, 2024)

Rank	Top Cities	Rank	Top Cities	Rank	Top Cities	Rank	Top Cities
1	Zurich, Switzerland	6	Copenhagen, Denmark	11	Stockholm, Sweden	16	Taipei City, China
2	Oslo, Norway	7	Lausanne, Switzerland	12	Dubai, United Arab Emirates	17	Seoul, South Korea
3	Canberra, Australia	8	London, England	13	Beijing, China	18	Amsterdam, Netherland
4	Geneva, Switzerland	9	Helsinki, Finland	14	Hamburg, Germany	19	Shanghai, China
5	Singapore	10	Abu Dhabi, United Arab Emirates	15	Prague, Czech Republic	20	Hong Kong, China

Table 1 presents the list of top 20 smart cities worldwide as per IMD smart city index, 2024. Overall, the Smart City Index 2024 showcases cities that successfully blend innovation with sustainability and resident engagement to create smarter urban environments.

Smart City Index, 2024 utilizes 15 indicators, out of which respondents were asked to select any 5 indicators which they think have grate priority for the city. An indicator that has a higher response means has top concern for the city.

Table 2. Percentage of Zurich respondents chose priority areas out of 15 indicators (IMD World Competitiveness Center, 2024)

No of Areas	Priority Areas	% of Respondents chose Priority Areas
1	Affordable Housing	76.4%
2	Road Congestion	58.8%
3	Air Pollution	32.2%
4	Unemployment	29.4%
5	Security	28.2%
6	Green Spaces	27.8%
7	Fulfilling Employment	25.9%
8	Health Services	24.0%
9	Public Transport	22.3%
10	Social Mobility	21.0%
11	Recycling	20.1%
12	Citizen Engagement	13.7%
13	Basic Amenities (Water & Waste Management)	11.8%
14	Corruption/Transparency	11.2%
15	School Education	11.1%

Table 2 describes 15 priority indicators used in the Smart City Index, 2024, where respondents were asked to select any five indicators, they consider most important for a city. An indicator receiving a higher number of responses indicates it is a top priority concern for the city.

Table 3. IMD Smart City Ranking, 2024 (IMD World Competitiveness Center, 2024)

City	Smart City Rank 2024	Smart City Rating 2024	Structure	Technology	Smart City Rank 2023
Zurich	1	AAA	AAA	AA	1

Table 3 shows that Zurich ranked top city with AAA ranking as per IMD Smart city index 2024. It achieved higher ratings in both structure and technology.

Zurich is one of the top cities in the world that sets high-quality life standards. It is the city that is always at the forefront of the smart city movement. In 2000, the city started to implement digital infrastructure projects and focused on sustainable urban mobility. In 2012, Zurich began its Smart City Zurich Initiative, an effort to integrate technology, sustainability, and innovation into its urban development. Since then, Zurich has continuously expanded its smart city projects, focusing on integrating cutting-edge technologies like AI, IoT, and 5G to enhance quality of life, sustainability, and urban efficiency.

Figure 1. Big Tech & AI Companies in Greater Zurich (Greater Zurich Area, 2024).

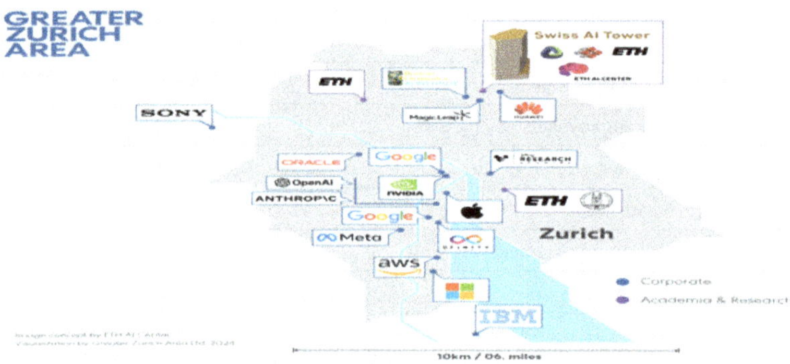

Figure 1 shows that Zurich is having world's leading technical institutes like Swiss AI Lab IDSIA and ETH Zurich. It is at the top list to attract the world's best companies like Microsoft, Google, Meta, OpenAI, IBM, NVIDIA, and many more (Greater Zurich Area, 2024).

Figure 2. Sustainable Investment in Switzerland (Zurich Financial Centre, 2023)

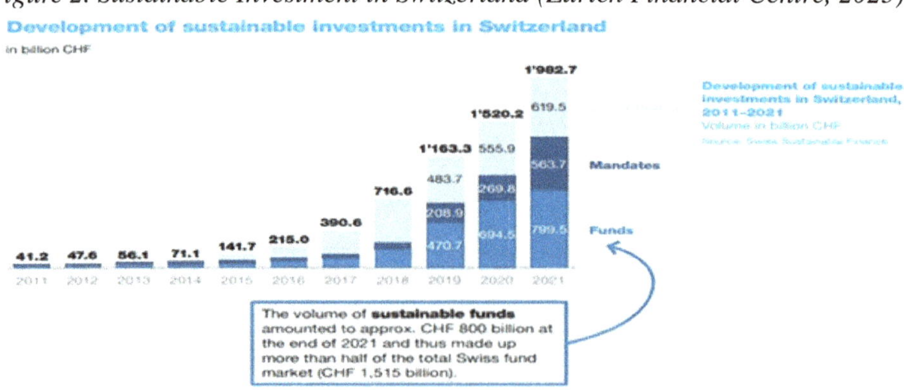

Figure 2 illustrates the growth of sustainable investments in Switzerland over the period 2011–2021, measured in billion CHF. It categorizes sustainable investments into three segments: Funds, Mandates, and Asset Owners.

Figure 3. Sustainable Investment in Zurich Region (Zurich Financial Centre, 2023)

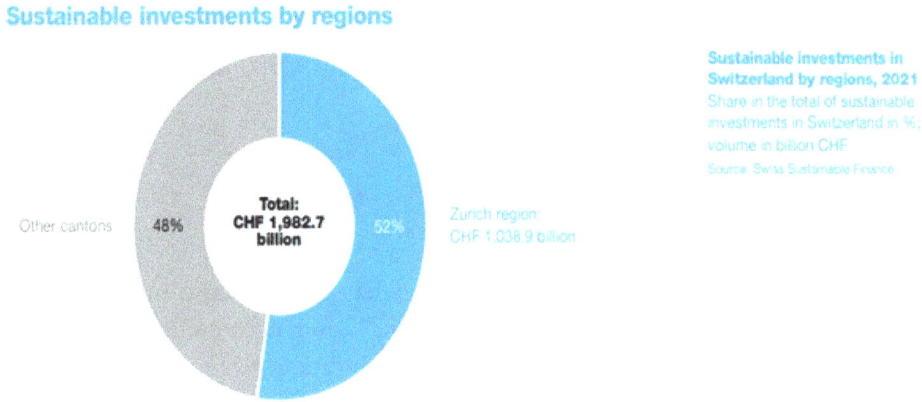

Figure 3 illustrates the distribution of sustainable investments across different regions in Switzerland for the year 2021. The total sustainable investment in Switzerland amounted to CHF 1,982.7 billion.

The city has developed 'Zurich 2040 Strategies for the challenges that it is expecting to face in the future. Out of these, Smart City Strategy', Net Zero 2035/2040', and 'Urban Space & Mobility, 2040' are topic-specific strategies (Mayors User, 2024).

NEED OF THE STUDY

Zurich is a model for sustainability and innovation in urban development. The city has successfully integrated advanced technologies like 5G, AI, and IOT to enhance quality of life while addressing urban challenges. The city focuses on renewable energy, smart grids, and eco-friendly initiatives that align with the Global sustainability goals (SDGs 3, 4, 6, 7, 9 & 11). The study shows how cities can better address challenges like energy consumption, waste management, and traffic congestion with AI while maintaining people's quality of life. The study discusses the solutions to minimize environmental impact and better utilization of resources. In this way, other cities can learn how they can adopt innovative and sustainable practices tailored to their unique needs. Zurich's smart city framework can provide urban planners and policymakers with evidence-based solutions to design smarter and more resilient cities.

OBJECTIVES

1. To analyze the role of technology in transforming Zurich's world smartest city
2. To find the impact of smart city initiatives on sustainability and urban living standards

METHODOLOGY

The study is secondary research based. The data is collected from government reports, industry publications, academic research, various case studies etc. to ensure comprehensive analysis. To make study more reliable, data was collected from only credible resources and only updated information was considered. Multiple resources were cross verified. The use of diverse data gives a holistic perspective to the subject matter. Not only that, insights from benchmark reports and best global practices were reviewed to strength the depth of analysis.

SMART CITY INITIATIVES BY ZURICH BASED ON 15 PRIORITY INDICATORS

Figure 4.

1. Affordable Housing (76.4%)

- Affordable Housing Projects
- Cooperative Housing Models
- Innovative Housing Design
- Smart Grid Technology

Zurich government is investing a lot in making affordable houses available for residents. To acquire land, renovate existing buildings, and construct new apartments, the government has spent 300 million Swiss Francs (De Boer J., 2023). The architectural firm, Baechi Foundation along with Gus Wüstemann Architects has developed projects that focus on high-quality living with effective cost. The apartments are designed in such a way that maximizes the flow of light and air, creates a spacious environment, and ensures privacy within budget limits (Archello, 2019). The Zurich government focuses on cooperative housing projects, where people can collectively maintain their living space. (Kalagas A., 2018). This promotes community engagement and offers more affordable housing options to residents. The government's focus is on energy-efficient buildings to reduce energy consumption and greenhouse gas emissions. This is the reason Zurich promotes energy-efficient buildings that meet the "2000-watt society" standard by 2050. This means that daily a zurichian can consume only 2 kilos watt energy per hour (Hubner, K. (2009). This action by the government is not only promoting the sustainability concept but also reducing living costs. Where many countries are facing the issue of high energy consumption and greenhouse gas emissions, Zurich has a continuous reduction in it. The per capita energy consumption in Zurich has been reduced to 4000 watts from 6000 watts and an approximate 50% reduction in greenhouse gases. (Wikipedia Contributors, 2025). Zurich has deployed smart grids that manage and distribute energy across residential and commercial buildings, optimizing the city's overall energy consumption. This system makes efficient use of energy and aligns with the city's sustainability goals (Engeler, S., Auer, M., & La Fauci, R., 2022).

Figure 5.

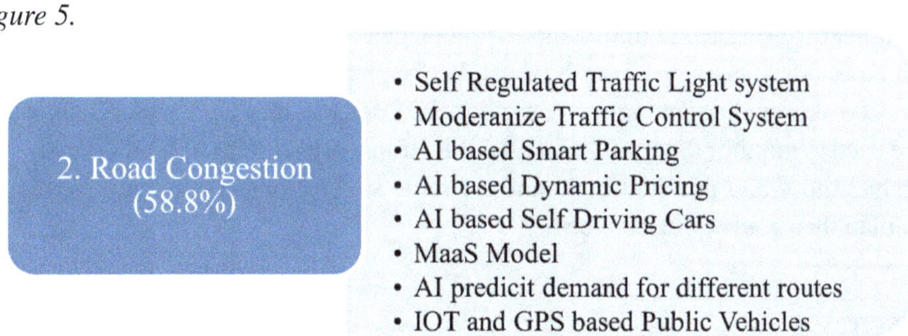

In the last five years, Zurich has modernized more than 200 traffic signal systems. These systems are equipped with Yutraffic sX controllers and 40-volt LED technology. This ensures the smooth operation of traffic lights across the city. It has increased the reliability and energy efficiency of traffic lights. (Yunex Traffic, 2024). The Traffic Engineering group in Zurich provides up-to-date traffic information to transport Management systems. ZüriTraffic, an adaptive access control system monitors traffic states in busy streets within in inner city. When congestion reaches predefined thresholds, traffic signals are automatically adjusted to regulate the number of vehicles entering the city center, thereby reducing overall congestion. Like in morning rush hours, traffic lights are automatically adjusted and thus reduce congestion. The traffic Engineering group collects data from 3500 loop detectors for effective monitoring and provides it with a transport management system to make effective traffic control light strategies. (Wissenstransfer, 2012). AI utilizes neural networks, deep learning, and machine learning to analyze large sets of structured and unstructured data. This allows parking authorities to respond quickly to changes in parking (PARQEX, 2025). Multiple ticketing platforms give the facility to passengers to use multiple modes of transportation like trams, buses, and trains with a single ticket. AI based dynamic pricing method is used to control parking demand, encourage the use of alternative transportation modes, and increase off-peak travel. Like parking fees are adjusted as per location, time of day, demand etc. The demand for automated vehicles is very high in Zurich. It helps to control the traffic system in the city. The use of self-driving cars does not require any driving license. These cars can operate without passengers thus opening new opportunities for taxi services and car sharing. AI-based self-driving technology helps ensure that autonomous vehicles can navigate the city's streets safely (Gebert C., 2018). The city uses a Mobility as a Service (MaaS) platform that aggregates services such as trains, buses, trams, car-sharing, and bike-sharing into a single app, providing users with flexible and personalized travel options. The system is based on AI model to provide multi transport options to citizens. AI systems make use of historical data, weather patterns, and public events to predict demand for different routes at different time periods. This helps Zurich adjust the frequency of buses, trams, and trains to avoid overcrowding and improve service. IOT sensors and GPS with AI algorithm are used to monitor the location of all public transport vehicles. It predicts estimated arrival times and provides personalized route suggestions. These help passengers to plan their journey effectively

Figure 6.

As per Climate and Clean Air Coalition, Switzerland was among the first members of the Climate and Clean Air Coalition (CCAC) since 2013. It served as Co-Chair of its Working Group and Steering Committee from October 2018 to June 2021. Switzerland has helped to reduce national emissions. It has also contributed to addressing climate change and air pollution at the international level. In the last 25 years, Air quality in Switzerland has improved significantly. Zurich is at the top among European cities fighting air pollution. Switzerland has retrofitted vehicles and machines over more than 20,000 help to reduce diesel soot and black carbon emissions. Due to smart agriculture policies, Zurich reduced 16.7% of methane emissions from 1990 to 2015 and Black carbon emissions by 70% from 2000 to 2018(Climate and Clean Air Coalition, 2021). Zurich has reduced the vehicle's speed limit from 50km/h to 30km/h on parts of its street network. This benefited 28% of residents from harmful noise levels in the daytime and 17% protected in nighttime. The aim is not only to reduce noise levels but lower traffic-related emissions and improve residents' quality of life. (European Environmental Agency, 2022). Zurich Airport measures air quality by monitoring nitrogen oxides, particulate matter, and ultra-fine particulates. This data helps in assessing and managing air pollution levels around the airport area (Flughafen Zürich AG., 2025). The Chair of Air Quality and Particle Technology at ETH Zurich conducts research and offers courses on air quality monitoring and control, aerosol measurement, and characterization. This academic focus contributes to developing effective strategies for air pollution management (ETH Zurich, 2025).

Figure 7.

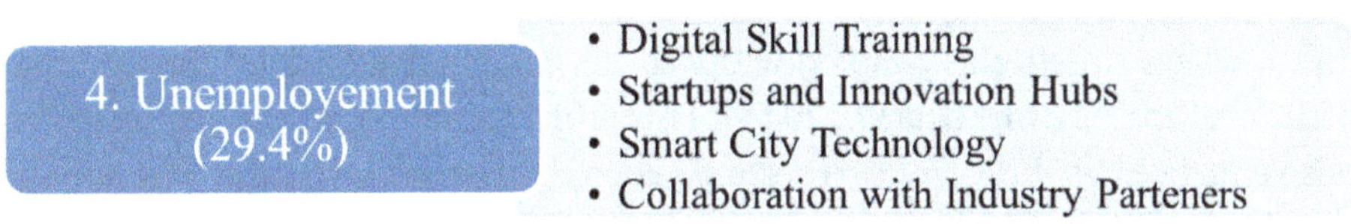

Zurich focuses on digital skill development among its citizens to prepare them for the job market. The government has partnerships with various educational institutions to provide training on data analysis, coding, digital marketing, blockchain, AI etc. This initiative aims to equip citizens with digital skills necessary to get technology sector employment. Zurich supports various innovation hubs and incubators to provide mentorship, resources, and networking opportunities to entrepreneurs, create new job opportunities etc. (Business Environment News, 2024). The city uses IOT solutions in job creation in different sectors like data analysis, digital marketing and urban planning etc. This attracts businesses to contribute to employment growth (Matuszak J., 2023). The city collaborates with business organizations to align educational courses as per market needs. It ensures that the workforce should have current job skills. This not only bridges the education and employment gap but facilitates smoother transitions from job seekers into the workforce.

Figure 8.

5. Security (28.2%)

- Integrated Public Safety
- Integration of IOT and LRWAN Technology
- Smart Street Lightning
- Hybrid Cloud Infrstructure
- Data Driven Decision Making

Over the last decade, Zurich police have partnered with Hexagon Solutions so that a technology-oriented police force can be created to enhance privacy, mobility, and collaboration with law enforcement. The agency enables police, railways, Ambulance, Fire, and other organizations to work together without interfering with their day-to-day work (Hexagon, 2022). This all results in efficient resource management and enhancing public safety. The city uses a Long-range wide area network (LRWAN) to connect all IoT devices to measure traffic conditions and manage parking, water levels, and air quality. This contributes to environmental scanning and increases people's safety and security (Matuszak J., 2023). Recently, Zurich started a streetlight project, where streetlights adapted with sensors can control their brightness as per traffic level and daytime (Lia O., 2024). This project resulted in 70% energy saving. It also enhances visibility for drivers and pedestrians. Zurich IT Infrastructure uses a hybrid cloud approach that enables on-premises applications and cloud services integration. This keeps the data secure and contributes to urban digital transformation efforts. (Hirt R., 2023). Zurich also uses a data analytics approach in security measures and urban planning. It collects data from various resources, analyzes it and that helps to address security issues and make the best resource allocation.

Figure 9.

6. Green Spaces (27.8%)

- Green Roofs and Green Infrastructure
- IOT for Environment Monitoring
- AR for Urban Planing

Since 1991, Zurich has made it mandatory that all flat rooftops be greened to enhance biodiversity, and rainfall storage and to contribute to city green infrastructure (Wikipedia Contributors, n.d.). Zurich uses IOT technology to manage and monitor green space. It makes use of LRWAN technology to optimize green space use. So that they can follow environmental regulations and better serve the society. (Matuszak J., 2023). Zurich uses AR technology to imagine planned green spaces before its construction. Residents are allowed to experience and give their feedback on the proposed project. This not only increases the participation of local residents but also generates a sense of community development among them (Ming En Liew, 2022). Zurich uses data analytics in the maintenance of parks and recreational areas so that it can meet residents' needs and contribute to environmental sustainability.

Figure 10.

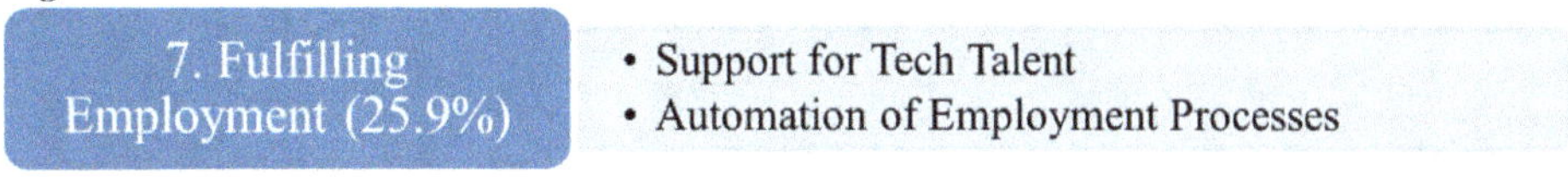

Switzerland's tech industry is experiencing significant growth, with high demand for roles such as Integration Engineers, Cybersecurity Specialists, and Cloud Engineers. The country offers steady work visa quotas, creating opportunities for skilled tech professionals to grow their careers in a globally recognized innovation hub (Kumari R., 2024). Companies like Service Hunter AG have developed platforms such as quitt, which automate the registration, insurance, and payroll accounting for private employees. This innovation streamlines employment processes, reducing administrative burdens and enhancing efficiency (Wikipedia Contributors, n.d.).

Figure 11.

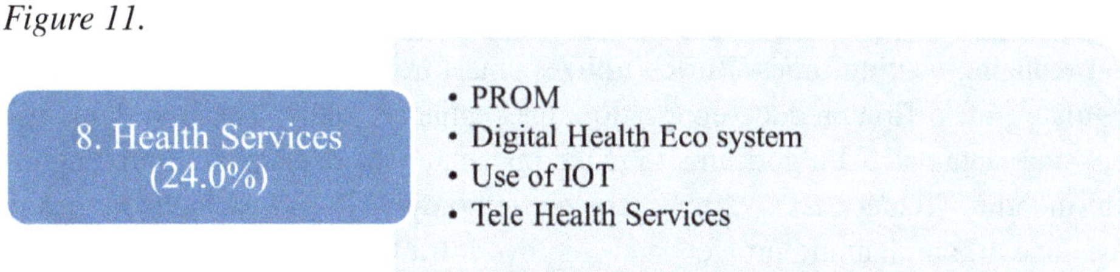

Digital Health Zurich's focus on to implement digital health solutions in hospitals. It has developed the Patient Reported Outcome Measure (PROMs) platform which is used to collect and analyze patient data. This platform incorporates digital infrastructure and the latest technology (Digital Health Zurich, 2022). Zurich is a hub of digital health companies. Around 121 digital health companies were active in Zurich till 2024. Digital health centers offer products and services for doctors, hospitals, patients, and digital health companies. The city has its 'Digital Health Map 2024' which describes the city's role in encouraging digital health solutions. (Mitteilung, 2024). The city uses IOT technology to improve health care services. Wearable devices are used to monitor patient health data and enable real-time data collection and analysis. (Abhilash, 2024). The University Hospital of Zurich provides telehealth services to provide online medical advice to citizens. A team of doctors consists of six physicians specialists in clinical telemedicine who take queries and answer them within 24-48 hours. Till now 59360 medical queries have been handled. Most users are female having an average age group of 38. But with time, many men and older people have also begun to consult online about health issues. (Kessler, S. H., & Schmidt-Weitmann, S., 2021). Zurich provides key services accessible within 15 minutes reach for over. 95% of its residents. (Niranjan A., 2024)

Figure 12.

9. Public Transport (22.3%)

- Autonomous Vehicles
- Demand Responsive Transport
- Smart Traffic Management
- Intergration of AI and AR
- Sustainable Mobility Solutions

Zurich is piloting autonomous public transport vehicles to address mobility challenges, particularly in rural areas. This project aims to provide efficient and sustainable transportation options (Smart Cities World, 2025). The city has implemented on-demand public transport services powered by Via Transportation. This system allows passengers to book rides via a mobile app, optimizing routes and reducing waiting times. Zurich utilizes smart traffic management systems to regulate traffic flow, reduce congestion, and enhance safety. These systems use real-time data and AI algorithms to make instantaneous decisions, improving urban mobility (Banerjee, A., 2023). The city employs AR and AI technologies to visualize urban planning projects and improve infrastructure management. These technologies assist in planning and maintaining transportation infrastructure more effectively (Liew M.E., 2022). Zurich promotes sustainable transportation options, including electric vehicles and bike-sharing programs, to reduce carbon emissions and enhance urban livability.

Figure 13.

10. Social Mobility (21.0%)

- Smart Mobility Solutions
- Digital Platform for Public Service
- Smart Infrastructure Development

Zurich has developed smart mobility strategy to make transportation system more efficient and accessible. To achieve it, the city uses various modes of transportation like public transit, cycling, shared vehicles etc. to provide residents with affordable mobility options (EBP, 2020). The city uses various digital platforms to provide access to public services like education, housing, healthcare etc. Residents can easily access information from these platforms. These reduce barriers to essential services and promote social inclusion. Zurich makes use of digitalization to enhance social mobility. It focuses on digitalization of transportation systems to provide all residents with more equitable access to mobility (University of Zurich, n.d.).

Figure 14.

11. Recycling (20.1%)

- Plastic Recycling
- AI Powered Waste Analysis
- Waste to Energy Facilities

Zurich makes use of chemical recycling methods to produce high quality products from plastic waste. Under this method, long chain plastic molecules break down into building blocks that can be reassembled into high quality plastic. It is a better way to recycle plastic as compared to mechanically. Not only that, through chemical recycling waste plastic is converted into molecules and then turned up into fuel (Petrol, jet fuel), Lubricants (Engine oil) etc. This gives a strong base to world researchers to develop more efficient ways to recycle plastic waste (Bergamin F., 2024). Zurich based startup Up Circle makes use of waste analysis platform that is based on artificial Intelligence system. It allows the collection and processing of waste to be analyzed in real time. This helps recyclers and waste handlers to estimate the value of waste materials and to make decision that which material they need to send to recycling plants (Greater Zurich Area, 2023). Zurich-based Hitachi Zosen Inova (HZI) converts non recycled waste into energy like electricity and heat for local residents and business. The aim is to reduce carbon footprint and enhance energy supply. This project treats local waste, create jobs and expand the district heating grid (ESWET, 2023)

Figure 15.

12. Citizen Engagement (13.7%)

- E-Participation Portal
- SMart City Hubs (AI powered Kisok)
- Data Dashboards
- VR

Zurich uses digital platforms to facilitate communication between citizens and city planners. Digital platforms like *Züri wie neu* enable citizens to propose ideas for urban development projects or vote on municipal decisions, fostering inclusive decision-making. It allows citizens to report issues such as damaged infrastructure or environmental concerns. (Liew M. E., 2022). AI-powered kiosks placed in public spaces provide real-time updates on city projects, gather citizen feedback, and offer localized services. Applications like *MyZurich* allow residents to access information, provide feedback, and participate in local events. (Liew M. E., 2022). Zurich employs data dashboards to provide information on city projects. These display live data on air quality, traffic, public transport, and more, empowering residents to make informed decisions. In 2008, Zurich residents endorsed the 2000-Watt Society initiative to reduce energy consumption to 2000 watts and annual green house gas emissions to one tone per person by 2050. This shows the citizen engagement in making environment more sustainable (Wikipedia Contributors, n.d.). Zurich employs VR tools to engage citizens in urban planning and infrastructure projects. VR systems integrated with AI allow citizens to simulate and analyze the impact of proposed changes in real-time.

Figure 16.

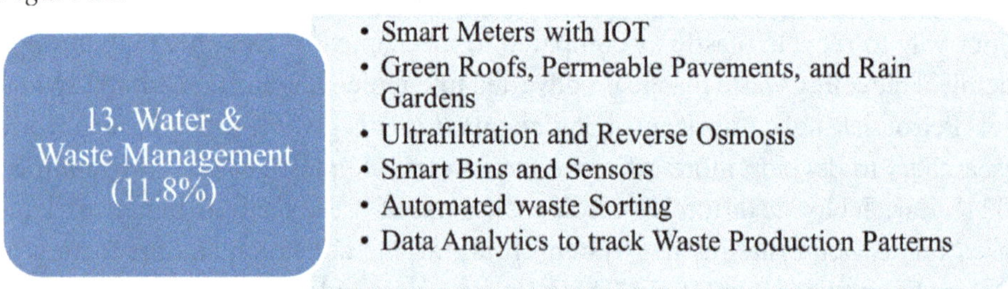

Zurich is using advanced technologies to enhance water management system. The city is having collaboration with INTEGRA metering, one of leading providers of smart meter solutions. The city is rolling out Switzerland's largest network of smart meters. These smart water meters with IOT provide accurate water consumption data. These monitors water usage pattern, leak detection, optimize water distribution. Over the next 12 years, Zurich is planning to replace 32,500 mechanical meters with ultrasonic smart meters. This will enable the city to achieve accuracy and efficiency in water monitoring (INTEGRA, 2023). In the city, water suppliers collaborate with energy and transportation sectors to develop integrated smart solutions. Like smart metering and smart grids are contributing to city sustainable goal. (TEKTELIC, 2023). The government in collaboration with swiss company Ayyeka embedded water pipes with IOT sensors to monitor minute by minute water quality. These sensors provide real time data on various parameters like ORP (Oxidation-reduction potential or Redox), temperature, turbidity, and pH of well water. (Mulhern, J. L., 2023). Zurich uses green roofs, permeable pavements, and rain gardens, to manage stormwater. These systems help rainwater to be absorbed into the ground that reduces the flood risk and improve water retention in urban areas. It uses wastewater for irrigation and industrial use and ultra filtration and reverse osmosis to ensure that drinking water remains safe and clean (Monteiro, C. M., Mendes, A. M., & Santos, C., 2023). Zurich is implementing advanced technologies to enhance waste management. The city uses smart bins embedded with sensors to monitor waste level. Waste management system can communicate with these smart bins with the help of sensors. Based on information gathered, it can schedule its visit to smart bind for waste collection. That will reduce fuel consumption and improve efficiency. (Civil D., 2023). Zurich uses automated sorting technologies to separate materials at recycling centers. This technology helps to segregate recyclable materials from waste. Organic waste is sent to biogas plant to produce methane which is used to generate energy. (Mcmahon J., 2020). Zurich make use of ERZ (Entsorgung + Recycling Zürich) app which provide information to residents about waste collection schedules, waste sorting reminders, and find recycling points locations (W4 Marketing AG., 2018). Zurich uses data analytics to track waste production patterns and optimize collection logistics. The city analyzes data on waste generation to design more efficient waste management systems, such as adjusting pick-up times or routes (Beath, 2024).

Figure 17. Disposal Routes in Zurich

Figure 18. Waste Groups in Zurich

Figure 19.

14. Corruption/Transparency (11.2%)

- Anti Corruption Framework
- Zero Tolerance Corruption Poilcy
- Open Data Portal
- Blockchain Technology
- AI for Fraud Detection
- E-procurement Systems
- Digital Twin Technology
- Public Transparency Dash Boards
- Ethical Training

Zurich established Anti corruption strategy for 2021-2024 to monitor and reduce corruption in government and business organizations. An interdepartmental working committee on combating corruption was created to ensure the implementation of strategy. The committee was responsible for handling issues like anti-corruption strategy, lobbying, conflict of interest etc. (OECD, 2024). Zurich Insurance group has zero tolerance policy towards corruption. The group ensures that business activities must be fair and honest (Zurich Insurance Group, n.d.). It uses AI algorithms to analyze claims data, identifying patterns and anomalies that may indicate fraudulent activities. (Ratcliffe, A., 2024). Zurich uses open data portals to publish datasets related to public expenditures, urban development, procurement, and more. Citizens, businesses, and researchers can access this data for accountability and analysis (Stadt Zurich, n.d.). It uses block chain technology to ensure the integrity of transactions and prevent tampering with records. Block chain creates immutable records of government contracts, land registries, and public procurement processes, making them tamper-proof and auditable. (World Economic forum, 2020). Blockchain is used for secure and transparent digital voting processes, ensuring accountability in electoral practices. (Swiss Information, 2018). Zurich uses E-procurement systems that automate and document every step of the bidding process, reducing opportunities for favoritism or manipulation. Zurich uses digital twins to simulate and monitor urban governance processes, ensuring transparency in city planning and resource allocation. Citizens can view digital replicas of ongoing projects, allowing them to track progress and ensure adherence to plans. Zurich provides real-time dashboards to keep citizens informed about government activities, spending, and project statuses. Dashboards display detailed breakdowns of government budgets, expenditures, and allocations. Zurich incorporates AI-driven ethics training for public officials to raise awareness about corruption risks. AI simulates real-life scenarios involving ethical dilemmas, helping officials recognize and avoid corrupt practices.

Figure 20.

15. School Education (11.1%)

- Integration of AI and AR
- Digital Education Platform
- Smart Classroom
- R&D in Education Technology

Schools and colleges in Zurich make use of AI and AR technology to enrich teaching methods and teaching content. AR is used to visualize complex subjects. AI is used to cater to the needs of individual students and provide personalized learning experience. (Liew M. E., 2022). The city has implemented digital platforms that facilitate access to educational resources and administrative services. These platforms enable students and educators to interact, share materials, and manage schedules efficiently, fostering a more connected and streamlined educational environment. Zurich schools are equipped with smart classroom technologies, including interactive whiteboards and digital learning tools. These technologies support collaborative learning and provide students with immediate access to information, enhancing the overall educational experience (Gong, H.; Wen, Z.; Wu, Z.Y.; Yu, M.H., 2014). Institutions like ETH Zurich are at the forefront of researching the application of AI in education. They promote responsible usage of generative AI, encouraging innovative approaches while prioritizing ethical considerations (Daniel Flück & Karin Brown (2024).

STATISTICALLY IMPACT OF SMART CITY INITIATIVES ON SUSTAINABILITY AND URBAN LIVING STANDARDS

Despite a global increase in energy consumption and green house gas emissions, Switzerland observed notable reductions. Between 2000 and 2020, The Swiss Federal Office of Energy highlights that there was a reduction in per capita energy from 6,000 watts to 4,000 watts and almost 50% reduction in greenhouse gas emissions (Wikipedia Contributors, n.a.).

Over the past 25 years, Air quality has improved a lot in Switzerland. Zurich is on the top cities as per list of European cities fighting air pollution that reduces climate change. By adopting smart agriculture policies like manure management and efficient livestock production, Zurich was able to reduce 16.7% methane emissions between 1990 to 2015. Black carbon emissions are also reduced by 70% between 2000 & 2018 (CCAC Partner, n.d.).

Zurich was ranked 16th most congested city in Europe in 2012. In 2020, It ranked 46th position in most congested cities in Europe. This shows Zurich significant efforts to reduce congestion level in city (Menendez, M., & Ambühl, L., 2022). The city has efficient public transport system with 98% utilization during peak times. This shows that Zurich has reduced dependence on private vechicles.

Zurich has made significant efforts in improving its green space. Zurich has dedicatedly assigned 33.83% of area to green spaces including parks, forests and recreational areas (The Telegraph, 2021).

Zurich hopes to achieve climate neutrality by 2040 with the help of smart grides and energy efficient buildings. Zurich is using more than 50% of its energy from renewable sources (European Commission, 2024).

Zurich's startup ecosystem has experienced significant growth, with over 876 active startups as of 2023. In 2023, these startups attracted approximately CHF 872 million (over $1 billion) in venture capital funding. Many of these emerging companies focus on sectors such as artificial intelligence, financial technology, and urban solutions (Christian, A., 2024)

Zurich economy is growing at annual growth rate of 1.8% in real times after adopting smart city initiatives (Canton of Zurich, Department for Economic Affairs, Office for Economy, & Walker Späh, C., 2024).

Zurich stands out as a model for walkability in terms of urban accessibility. As per study by Nature Cities explained that almost 99.2% of Zurich residents live within a 15-minute walk of essential services. This high level of accessibility will help to reduce dependence on automobiles. It will also promote healthier lifestyle and fosters community involvement (Niranjan A., 2024).

As of the third quarter of 2024, Switzerland's employment rate stood at 80.20%. Historically, the employment rate has averaged 78.98% from 1996 to 2024. (Trading Economics, 2024).

In terms of accessibility, around 314,225 residents have access to leisure green spaces within a 10-minute walk, underscoring the city's commitment to integrating nature into urban living (Burkhard C., Burren A. L., Gretler C., Ruegg L., Stutz M., Tharmalingam E., 2021)

Zurich is ranked among the top cities globally highlighting its safety, infrastructure, healthcare and public services as per Mercer’s quality of living index (Quality of Living City Ranking, 2024).

CONCLUSION

Zurich has established itself as global leader in smart city development by integrating emerging technologies to enhance urban living, sustainability and economic growth. The city is continuously on the top due to high tech digital framework, well organized city planning and commitment to sustainability. Zurich leadership in financial and industrial sectors strengthens its position as a dynamic and metropolitan centre. Zurich is expert in balancing growth while maintaining high quality living standards. Zurich continues to set new standards by adopting data driven policies, citizen engagement and digital transformation in urban living. It serves as a model for other cities aiming to develop smart, sustainable and future ready urban environment. The city actively promotes green mobility through electric public transport, cycling infrastructure and pedestrian-friendly urban spaces. Zurich also invests heavily in renewable energy, smart grids, and energy-efficient buildings to ensure a low-carbon future. Its strong focus on research and innovation creates opportunities for start-ups and technology-driven enterprises to thrive. With effective waste management systems and water conservation strategies, the city protects natural resources while accommodating urban expansion. Advanced digital platforms make it easier for citizens to access government services, healthcare, and education in a transparent and efficient manner. Zurich's long-term vision emphasizes resilience, making it well-prepared to face global challenges such as climate change and rapid urbanization. By fostering inclusivity and citizen participation in decision-making, the city ensures a sense of community and shared responsibility. International collaborations and partnerships further enhance Zurich's reputation as a forward-thinking global hub. With its blend of tradition, innovation and sustainability, Zurich stands out as a beacon of modern urban excellence that inspires cities worldwide.

REFERENCES

Abhilash (2024). Empowering Zurich: How IOT solutions are driving smart city success. https://www.intertoons.ch/post/empowering-zurich-how-iot-solutions-are-driving-smart-city-success?

Archello. (n.d.). Affordable housing in Zurich. https://archello.com/project/affordable-housing-in-zurich

Banerjee, A. (2023). Smart Cities spur new thinking on technology risks. https://www.zurich.com/commercial-insurance/sustainability-and-insights/commercial-insurance-risk-insights/smart-cities-spur-new-thinking-on-technologys-risks

Beath. (2024). Swiss cities revolutionize public services. https://icumonitoring.ch/real-time-data-revolution-enhancing-public-services.html?

Bergamin, F. (2024). Chemical recycling plastic is ready to go. https://ethz.ch/en/news-and-events/eth-news/news/2024/08/chemical-plastics-recycling-is-ready-to-go.html?

Burkhard, C., Burren, A. L., Gretler, C., Ruegg, L., Stutz, M., & Tharmalingam, E. (2021). Urban Green Space Indicator. Usability and Accessability of UGS in the city of Zurich and Uster. https://storymaps.arcgis.com/stories/591c0eb844814673bc603296e209ffb8?

Business Environment News. (2024). Zurich again crowned as the world's smartest city. https://www.greaterzuricharea.com/en/news/zurich-again-crowned-worlds-smartest-city

Canton of Zurich, Department for Economic Affairs, Office for Economy, & Walker Späh, C. (2024). Attractiveness of the Canton of Zurich for business and talent – A comparison2024. https://www.zh.ch/content/dam/zhweb/bilder-dokumente/themen/wirtschaft-arbeit/wirtschaftszahlen/standortattraktivitaet-im-vergleich_EN.pdf

Christian, A. (2024). The hottest startups in Zurich in 2024. *WIRED*. https://www.wired.com/story/the-hottest-startups-in-zurich-in-2024/?

Civil, D. (2023). Top 10 Smart cities in the world. https://buildmytalent.com/2023/09/16/top-10-smart-cities-in-the-world

Climate and Clean Air Coalition. (2021). *Switzerland.* https://www.ccacoalition.org/partners/switzerland

Daniel Flück & Karin Brown. (2024). AI in teaching and Learning, https://ethz.ch/en/the-eth-zurich/education/ai-in-education.html?

De Boer, J. (2023). Zurich to invest 300 million francs into making housing more affordable. *iamexpat.* https://www.iamexpat.ch/housing/real-estate-news/zurich-invest-300-million-francs-making-housing-more-affordable

Digital Health Zurich. (2022). A Practice lab for patient centered clinical innovation. https://www.dizh.uzh.ch/en/2022/07/07/zurich-applied-digital-health-center-2

EBP. (2020). Smart Mobility Strategy for the city of Zurich, https://www.ebp.global/xx-en/projekte/smart-mobility-fuer-die-stadt-zuerich?

Egridere, S. (2024). Transforming Urban Living - Insights from the IMD Smart City Index 2024 on Europe's Leading and Lagging Cities, https://mayorsofeurope.eu/science-and-the-city/transforming-urban-living-insights-from-the-imd-smart-city-index-2024-on-europes-leading-and-lagging-cities/

Engeler, S., Auer, M., & La Fauci, R. (2022). Smart grid rollout in der Stadt Zürich. *Bulletin.*https://www.bulletin.ch/de/news-detail/smart-grid-rollout-in-der-stadt-zuerich.html

ESWET. (2023). A waste treatement expansion in zurich: A sustainable energy solution by Hitachi Zosen INOVA. https://eswet.eu/a-waste-treatment-expansion-in-zurich-a-sustainable-energy-solution/?

European Commission. (2024). How Zurich is becoming net zero by 2040. https://build-up.ec.europa.eu/en/news-and-events/news/how-zurich-becoming-net-zero-2040?

European Environmental Agency. (2022). Reduction to 30km/h speed limits in Zurich, Switzerland, https://www.eea.europa.eu/publications/outlook-to-2030/reduction-to-30-km-h

Flughafen Zürich, A. G. (2025). Local air quality is closely monitored, https://www.flughafen-zuerich.ch/en/company/responsibility/environmental-protection/air

Gebert, C. (2018). Induced Demand: Modelling the potential impacts of automated vehicles in Zurich. *NSL – Netzwerk Stadt Und Landschaft.* https://www.nsl.ethz.ch/induced-demand-modelling-the-potential-impacts-of-automated-vehicles-in-zurich

Gong, H., Wen, Z., Wu, Z. Y., & Yu, M. H. (2014). Design of a smart classroom broadcasting system based on WSN. In *Advanced Materials Research* (Vol. 850, pp. 520–524). Trans. Tech. Publ., DOI: 10.4028/www.scientific.net/AMR.850-851.520

Greater Zurich Area. (2023). Upcircle digitalizing waste for the circular economy, https://www.greaterzuricharea.com/en/news/upcirclc-digitalizing-waste-circular-economy?

Greater Zurich Area. (2024). Artificial intelligence in the Greater Zurich Area, https://www.greaterzuricharea.com/en/artificial-intelligence-greater-zurich-area

Hall, R. E. (2000). The vision of a smart city. In Proceedings of the 2nd International Life Extension Technology Workshop, Paris, France, September 28, http://www.osti.gov/bridge/servlets/purl/773961- oyxp82/webviewable/773961.pdf

Hexagon (2022). Protecting citizens with intearated Public Safety Solutions. https://futurescot.com/protecting-citizens-with-integrated-public-safety-solutions

Hirt, R. (2023). Smart City Zurich: How Anthos is driving the city's digital transformation, https://cloud.google.com/blog/topics/hybrid-cloud/city-of-zurich-builds-a-hybrid-cloud-with-anthos?

Hubner, K. (2009). 2000 watt society. https://ourworld.unu.edu/en/2000-watt-society

IMD World Competitiveness Center. (2024). IMD Smart City Index 2024. IMD World Competitiveness Center. https://www.coit.es/sites/default/files/imd_-smartcityindex-2024-full-report.pdf

INTEGRA. (2023). Invigorating the water supply of the city of Zurich, https://integra-metering.com/invigorating-the-water-supply-of-the-city-of-zurich/?

Kalagas, A. (2018). Co-op City: Zurich's experiment with nonprofit housing. *Medium.* https://medium.com/@alexiskalagas/coop-city-z%C3%BCrichs-experiment-with-non-profit-housing-64207f60cafc

Kessler, S. H., & Schmidt-Weitmann, S. (2021). Diseases and Emotions: An Automated Content Analysis of Health Narratives in Inquiries to an Online Health Consultation Service. *HealthCommunication.* 36 (2*): 226-235.* https://doi.org/. PMID 36112901. S2CID 210568504.DOI: 10.1080/10410236.2019.1673950

Kumari, R. (2024). Why Switzerland is next big opportunity for skilled tech talent in 2025? https://content.techgig.com/career-advice/why-switzerland-is-next-big-opportunity-for-skilled-tech-talent-in-2025/articleshow/115947673.cms

Lia, O. (2024). Top 7 smart cities in the world in 2024. https://earth.org/top-7-smart-cities-in-the-world/?

Liew, M. E. (2022). How AR and AI are making Zurich a smarter and safer city. https://govinsider.asia/intl-en/article/how-ar-and-ai-are-making-zurich-a-smarter-and-safer-city-david-weber-zurich?

Matuszak, J. (2023). Swiss cities pioneering smart tech. https://knowhow.distrelec.com/internet-of-things/swiss-cities-pioneering-smart-tech

Mayors_User. (2024). *What makes Zurich a Smart City role model: five questions for Mr David Weber, Head of Smart City Zurich. Mayors of Europe.* https://mayorsofeurope.eu/news/what-makes-zurich-a-smart-city-role-model-five-questions-for-mr-david-weber-head-of-smart-city-zurich

Mcmahon, J. (2020). From open air composting to efficient Anaerobic Digestion, https://biomassmagazine.com/articles/from-open-air-composting-to-efficient-anaerobic-digestion-17001?

Menendez, M., & Ambühl, L. (2022). Implementing Design and Operational Measures for Sustainable Mobility: Lessons from Zurich. *Sustainability (Basel)*, *14*(2), 625. DOI: 10.3390/su14020625

Mitteilung (2024). Zurich: Growth Engine for Digital Health in Switzerland, https://www.zh.ch/de/news-uebersicht/mitteilungen/2024/wirtschaft-arbeit/growth-engnge-digital-health.html

Monteiro, C. M., Mendes, A. M., & Santos, C. (2023). Green roofs as an urban NbS strategy for rainwater retention: Influencing factors—a review. *Water (Basel)*, *15*(15), 2787. DOI: 10.3390/w15152787

Mulhern, J. L. (2023). Safe Drinking water: Monitoring Quality control in Schwyz, Switzerland. https://www.ayyeka.com/blog/ensuring-safe-drinking-water-ayyekas-solution-for-water-quality-monitoring-in-schwyz-switzerland?

Niranjan, A. (2024). Europe beats the US for walkable, livable cities, study shows, https://www.theguardian.com/environment/2024/sep/16/europe-beats-the-us-for-walkable-livable-cities-study-shows?

OECD. (2024). Anti-Corruption and Integrity Outlook: Country Fact Sheet 2024, Anti-Corruption and Integrity Outlook 2024 – Country Notes: Switzerland | OECD.

PARQEX. (2025). Dynamic and predictive AI pricing for parking. https://parqex.com/dynamic-and-predictive-ai-pricing-for-parking

Partner, C. C. A. C. (n.d.). https://www.ccacoalition.org/partners/switzerland?

Quality of Living City Ranking. (2024). https://www.mercer.com/insights/total-rewards/talent-mobility-insights/quality-of-living-city-ranking/?

Ratcliffe, A. (2024). How technology, including AI, gives claims investigators an edge, https://www.zurich.com/media/magazine/2024/how-technology-including-ai-gives-claims-investigators-an-edge?

Smart Cities World. (2025). Switzerland to launch public transport autonomous driving project,https://www.smartcitiesworld.net/connected-and-autonomous-vehicles/switzerland-to-launch-public-transport-autonomous-driving-project-11081

Stadt Zurich. (n.d.). Open Government Data, https://www.stadt-zuerich.ch/opendata

Swiss information. (2018). Switzerland's first municipal blockchain vote hailed a success, https://www.swissinfo.ch/eng/business/crypto-valley-_-switzerland-s-first-municipal-blockchain-vote-hailed-a-success/44230928?

TEKTELIC. (2023). Top 3 smart cities in 2021: How IOT technologies help building smart cities. https://tektelic.com/expertise/best-smart-cities/?

The Telegraph. (2021). From Auckland to Zurich: The cities with the most green space. https://www.telegraph.co.uk/travel/galleries/cities-most-green-space/?

Toppeta, D. (2010). *The Smart City Vision: How Innovation and ICT Can Build Smart, "Livable"*. Sustainable Cities. The Innovation Knowledge Foundation.

Trading Economics. (2024). Switzerland employment rate, https://tradingeconomics.com/switzerland/employment-rate

United Nations. (2018). 68% of world population is projected to line in Urban areas by 2050, https://www.un.org/uk/desa/68-world-population-projected-live-urban-areas-2050-says-un

University of Zurich. (n.d.). The Digital Society Initiative Community Mobility, https://mobility.dsi.uzh.ch/about/?

W4 Marketing AG. (2018). *ERZ Mobile App development Clean Zurich.* https://www.w-4.ch/en/about-us/references/erz-app?

Washburn, D., Sindhu, U., Balaouras, S., Dines, R. A., Hayes, N. M., & Nelson, L. E. (2010). *Helping CIOs Understand "Smart City" Initiatives: Defining the nSmart City, Its Drivers, and the Role of the CIO.* Forrester Research, Inc.

Wikipedia Contributors. (n.d.). Green infrastructure. https://en.wikipedia.org/wiki/Green_infrastructure

Wikipedia contributors. (n.d.). 2000-watt society. Wikipedia. https://en.wikipedia.org/wiki/2000-watt_society

Wikipedia contributors. (n.d.). ServiceHunter. https://en.wikipedia.org/wiki/ServiceHunter

Wissenstransfer. (2012). Traffic management in the inner city of Zurich. https://www.nsl.ethz.ch/en/traffic-management-in-the-inner-city-of-zurich

World Economic forum (2020). Exploring Blockchain Technology for Government Transparency: Blockchain-Based Public Procurement to Reduce Corruption. https://www3.weforum.org/docs/WEF_Blockchain_Government_Transparency_Report_Supplementary%20Research.pdf?

Yunex Traffic. (2024). Large scale project successfully completed in Zurich: Yunex Traffic modernizes over 200 traffic signals. https://www.yunextraffic.com/newsroom/large-scale-project-zuerich

Zurich, E. T. H. (2025). The Chair of Air Quality and Practice Technology. https://ie.ifu.ethz.ch

Zurich Financial Centre. (2023). Zurich Financial Centre 2023/2024. https://www.zh.ch/content/dam/zhweb/bilder-dokumente/themen/wirtschaft-arbeit/wirtschaftsstandort/dokumente/englisch/zurich_financial_centre_2023-24_executive_summary.pdf

Zurich Insurance Group. (n.d.). We care about business integrity. *Zurich Insurance Group*. https://www.zurich.com/about-us/corporate-governance/code-of-conduct/we-care-about-business-integrity

Chapter 2
AI and IoT Integration for Social Sensing Optimization in Smart Cities

T. Premavathi
http://orcid.org/0009-0003-0172-2021
Marwadi University, India

Damodharan Palaniappan
http://orcid.org/0009-0003-0721-3068
Marwadi University, India

ABSTRACT

The rapid development of smart cities can be explained by the combination of artificial intelligence and the IoT. Implementation of the IoT infrastructures, which includes smartphones, wearable devices, surveillance cameras, environmental monitors and numerous other devices, to a municipality gives its access to vast amounts of data on real-time. A combination of machine learning, deep learning, and natural language processing allows AI to reduce such data sets into patterns, anomalies, and trend predictions by numerous aspects of cities. The discussion will focus on AI models of IoT data processing, such as human activity recognition, traffic-flow forecasting, and social-media sentiment analysis, and analyze the use of edge AI and federated learning to reduce the latency and still preserve the data privacy all of which is extremely important in metropolitan areas. The corresponding applications of disaster management, safety, health-related monitoring, and optimization of energy are then outlined, with the examples and results being empirical, accompanied by concrete examples.

DOI: 10.4018/979-8-3373-4962-6.ch002

1. INTRODUCTION

Smart cities are a new way of city development combination with the consideration of technology means to make their work more efficient and the life of citizens better. They use Information and Communication Technologies in bringing up the standards of the lives of people, enhancing sustainable development, and encouraging innovations of management and other general services (Yarubi et al., 2025). Such transformative process is centered on emerging technologies, especially digital tools, the Internet of Things devices, and artificial intelligence. An increase in urban infrastructure, in turn, implies the enhancement of such-like energy, water, transportation, and waste systems, making urban infrastructures more resilient and efficient (Abouhassan et al., 2024). Another important element of smart cities is social sensing that is the process of gathering and using information about city dwellers and built-in infrastructure in order to understand and interact in cities in a much-improved manner. Artificial intelligence and the Internet of Things allow processing the data in real-time using diverse data sources, such as smartphones, sensors, surveillance devices, and the like. Such algorithms like machine learning and natural language processing are used to detect trends and behavior in the city environment (Kostadimas et al., 2025). This chapter discusses the benefits of the AI and IoT combination in social sensing in smart cities. It looks at AI-based application models, including the recognition of human activity, traffic forecasting, sentiment classification. Moreover, it also speaks about the advantages of edge AI and federated learning in improving data processing and confidentiality. Applications in the real world such as disaster response, public safety, healthcare and energy management were reviewed with important ethical and technical challenges. The chapter ends by suggesting a framework of the future IoT intelligent city infrastructure.

2. UNDERSTANDING SOCIAL SENSING IN URBAN ENVIRONMENTS

One of the most important elements of smart cities is social sensing; a process that aims towards the collection and processing of data concerning human activities, interactions, and condition of the environment within cities. The proposed solution involves the use of collective intelligence of both citizens and machines to obtain new insights about the dynamics of urban areas and enhance the decision-making process (Rashid et al., 2023).

2.1 An Overview of Social Sensing Systems

Social sensing was the act of gathering and analyzing data that had been generated by a lot of different avenues, including people and their gadgets as a means of acquiring knowledge about the social problems and urban issues. It implies technologies being used to monitor and examine human activities, opinions, and sentiments and environmental conditions that might influence city life (Rashid et al., 2023). The use of ubiquitous connectivity and massive data gathering provides the opportunity to combine social sensing into a complete awareness of the social, physical and environmental environment in a city (Murroni et al., 2023). A social sensing system usually comprises multiple main elements, such as sensors, data transmittance, as well as storage and processing network, and analysis tools. The sensors may be physical, that is, smartphones, wearable fashionable items, environmental sensors, or camera-based sensors or virtual including social media pages and online discussion forums. These sensors sense the information about human activities, including its localization, mobility, communications, and physiological parameters along with the information about the environmental conditions, including the temperature, air purity, noise, and traffic (Adibi et al., 2024).

2.2 Traditional vs. Technology-Enhanced Sensing

In the conventional approach, the city management and planning has been based on data that is obtained manually with the use of surveys, censuses, and direct observations. Nevertheless, these avenues are quite lengthy, expensive, and narrow. The more efficient and dynamic sensing, with the help of technology, provides more possibilities, including the use of different digital tools and platforms, involving a large number of sources to collect hour-by-hour information in real-time (Murroni et al., 2023). As compared to conventional processes, technology-enhanced sensing utilizes the recent developments of digital technologies to enable data collection and analysis in real-time so that the process can be more flexible and scalable to better comprehend urban processes. A Comparison of Traditional and Technology-Enhanced Sensing Methods is depicted in Figure. 1. Traditional Sensing: The method of data collection is based on the manual data such as surveys, direct observations, and census data. It is always restricted in scope because of the resource limitation and logistics challenges that influence the comprehensiveness of captured data. Data analysis could be categorized as qualitative, where statistical analysis and descriptive analysis have been adopted to get interpretation of the findings. The reaction time in this method is very low as there are going to be major delays in data collection process, processing and making decisions do not get to be made in time to cause a quick and adequate action to be taken. Manual surveys are direct questionings of people in the society to obtain data concerning demographics, opinions, and behaviors. Censuses give an all-inclusive listing to a population and give detailed demographic and socioeconomic information. Direct observation involves a systematic observation of urban setting leading to collection of data on traffic, land use, and environment.

Figure 1. Comparison of Traditional and Technology-Enhanced Sensing Methods

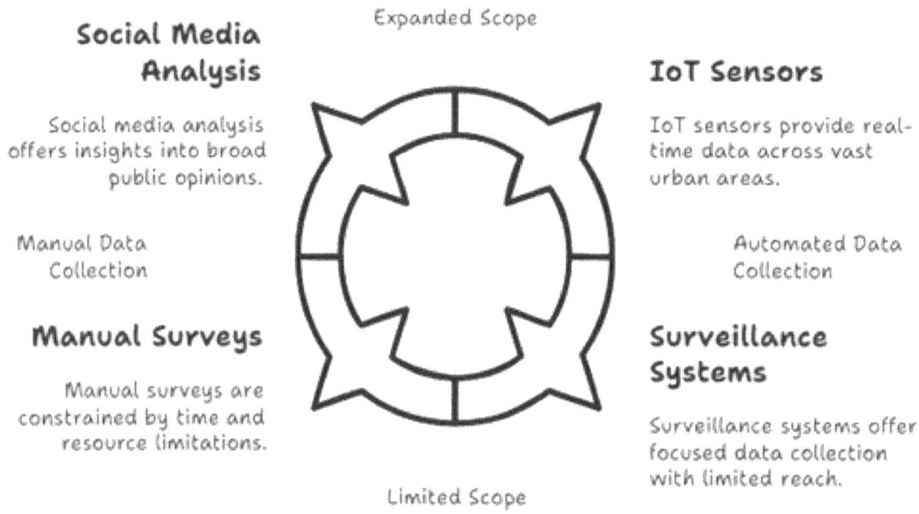

Technology-Enhanced Sensing: All this information was gathered with the help of an extensive set of digital tools, such as IoT devices, mobile phones, social media, and surveillance infrastructure. The boundaries considerably extend due to the use of the omnipresent character of digital technologies because of the possibility to continually follow the life and state of urban areas. The data analysis is quantitative and represented with the help of advanced algorithms, machine learning tools, and data analytics to draw insights, based on large volumes of data (Murroni et al., 2023). This enables the cities to base their decisions on data so that they enhance their efficiency and sustainability. Technology-enabled sensing is automation of data acquisition where sensors, IoT devices and digital platforms are used to transmit data in real-time.

2.3 Sources of Data in Smart Cities

Smart cities conduct lots of activities and activities gathered through various subjects hence allowing social sensing applications. These data sources could be classified as follows in a broad manner. Smartphones are everywhere small computers containing a number of sensors such as GPS, accelerometers, microphones, and cameras (Murroni et al., 2023). They are capable of offering crucial data, as in the following: Location and Mobility: Monitoring movements, traffic congestion and commuting trends. Communication Patterns: Tracking call histories, messaging activities and social media usage in order to get an insight into social networks and communication patterns. App Usage: Tracking the application to determine the interest, behavior and needs of the user. Wearables: The data on physiological parameters, provided by wearable devices like smartwatches and fitness trackers, especially heart rate and sleep and physical activity, can be obtained. One can use these data in the following manner:

Survey Health and Wellness: Recognizing health risks, monitoring chronic illnesses and facilitating healthy living. Evaluate Exposure to Environment: This is determining exposure to environmental pollution, noise, and other forms of environmental hazards. Improve Workplace Safety: The work fatigue, stress level, and hazard compliance of workers can be tracked in hazardous places.

Environmental Sensors: The parameters, which may be measured with the help of environmental sensors installed in the city, include: Air Quality: Air quality measured by levels of pollutants e.g. particulate matter, nitrogen dioxide and ozone. Noise Levels: An evaluation of the degree of noise pollution and the effects thereof on the health and the well-being of the populace. Weather Conditions: Monitoring of temperature, humidity, rainfall and other weather conditions. Surveillance Systems: The surveillance systems, such as the CCTV cameras and license plate recognition technologies, may present the following information. Traffic Surveillance: Monitoring the movements of the vehicles, localization of traffic snarls and improved traffic optimization. Public Safety: Surveillance of criminal operations, tracking of public areas, and optimized emergency help (Grêt-Regamey et al., 2021).

2.4 Challenges in Raw Data Acquisition and Processing

Although it is of potential benefits, there are a number of challenges of social sensing in the view of raw data due to raw data acquisition and raw processing. Data Quality and Reliability: Assuring the correctness, completeness and consistency of data derived with varying sources. Data Privacy and Security: Safeguards sensitive personally identifiable information and guard against access or use (Calzada, 2022). Interoperability and Data Integration: The integration of information between data sources and data formats to allow a combined picture of information within the urban environment. Data Bias and Fairness: How can we solve any possible biases in data collection and analysis leading to fair and equitable results? Scalability and Real-Time Processing: the ability to scale up and address the huge amount of data produced by smart cities and to process them in real time to enable wise decision-making on a timely basis. Ethical Considerations: Discussion regarding the ethical aspect of social sensing such as surveillance, privacy, and autonomy. These issues demand a complex strategy comprising the technological innovation, policy and people role. Social sensing helps achieve such responsive, efficient and sustainable environments in a city by triumphing over these obstacles.

3. INTERNET OF THINGS INFRASTRUCTURE FOR SOCIAL SENSING

Social sensing in smart cities relies on the IoT which has sufficient infrastructure that can be used to gather, analyze, and share real-time data.The IoT is characterized by a system of interconnected objects, vehicles, buildings, and other physical objects which are embedded with sensors, software, and network connectivity to allow the exchange and collection of information (Khalil et al., 2022).

3.1 Overview of IoT Architecture in Smart Cities

Smart cities can be viewed as the IoT architecture which contains a number of different layers with their particular functions and roles (Khalil et al., 2022): Sensing Layer: This is composed of a number of sensors and other gadgets that gather information about the physical domain. Some of these devices are environmental sensors, traffic sensors, biomedical, and surveillance sensors. Network Layer: the network layer enables the sensing layer to communicate with data-processing centers. It incorporates a range of communication technology and protocols that comprise Wi-Fi, cellular network (4G, 5G), LoRaWAN, and MQTT (Qin et al., 2019). Data Processing Layer: It processes and analyses the type of data retrieved on the sensing layer. It engages data storage, mining and analysis methods to draw interpretable data and trends. Application Layer: The application layer is the frontal area for providing services and applications to the end users depending on the examined information. Such apps are study of traffic-management systems, environment-observing systems, health-checking monitoring systems, and civil readiness apps.

3.2 Types of IoT Devices Used for Sensing

A large diversity of IoT devices employed in smart cities is oriented on social sensing, and each of them has different data received by it. The Figure.2 describes the IoT Devices and Data Flow in Smart Cities.

Figure 2. IoT Devices and Data Flow in Smart Cities

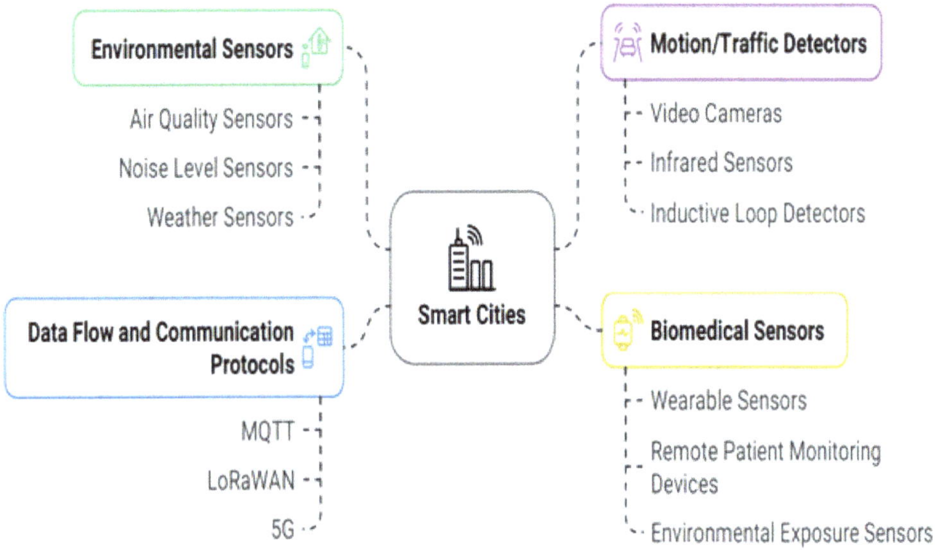

Environmental sensor A sensor is a primitive computer part that, in its simplest form, converts analog signals to digital (although digital sensors are also found). Environmental sensors detect environmental data and thus are basically any sensor that monitors the environment or activity in it. The environmental parameters are observed with the assistance of environmental sensors that help deliver vital information in determining the state of air, sounds, and weather. Sensors Air Quality: Determine the level of harmful substances in the air, including particulate matter (PM 2.5, PM 10), nitrogen dioxide, ozone (O 3), carbon monoxide.

Noise level sensors: To measure the noise pollution at cities, noise hotspots and determine the state of noise pollution on human health. Weather Sensors: Measure temperature, humidity, rain, and wind speed among other weather conditions in order to make weather predictions and monitoring of climate. Under traffic detectors, motion detectors and traffic detectors ensure the flow of traffic and people, and the data of the traffic is used to control the traffic and protect the lives of people and to plan the city (Murroni et al., 2023). Video Cameras: These record video and can be used to monitor the movement of people within the streets, public areas, and also the traffic flow in an area which makes it possible to monitor the area in real time. Infrared Sensors: Help in sensing the presence and the movement of vehicles and people even in low-light situations. Inductive Loop Detectors: Duct located on road surfaces, in this method, detection of vehicle is determined by the variation in the magnetic field. Biomedical sensors are used to detect aspects about physiological parameters and offer data to health surveillance, physical exercise following and wellness checking out.

3.3 Data Flow and Communication Protocols

In smart cities, IoT devices create constantly growing patterns of information that should be sent to data-processing hubs in a smooth and stable way. There are a few communication protocols which are usually applied in IoT data transmissions. MQTT: A low-latency messaging protocol that utilizes low bandwidth innovation networks. One of the areas in which MQTT is being employed extensively in IoT solutions is due to its simplicity and efficiency. LoRaWAN: It is a Long-range networking protocol that uses low power and wide area networking. LoRaWAN is an appropriate solution to long-range interconnection applications and battery-powered devices. 5G: The newest cellular technology promises high bandwidth, low latency and connectivity of numerous devices. 5G will support new applications of the IoT which rely on the real-time data transfer and data throughput level. Communication between IoT devices is a multi-stage process. A device gathers the data, usually having sensors installed, and transmits the data to a central point of collection.

3.4 Real-Time Data Generation and Scalability Concerns

This has resulted in the continually increasing amounts of real-time generated data thus the proliferation of IoT devices in smart cities (Khalil et al., 2022). This becomes a major problem as far as storage, processing and scalability is concerned regarding data. Data Storage: The sheer size of data that will be collected by the IoT devices necessitates large capacity and low-cost storage solutions. The cloud storage services are also potential sources of information with large datasets. Data Processing: Time-sensitive data feeds call for effective data-processing mechanisms and facilities. It can be reduced by moving data processing to the edge (to a location near the data source), so that they and completion of the process can come faster.

Scalability: The essence of the smart city IoT infrastructure should be scaled to accommodate more connected devices. A scalable network structure, data-processing systems and application services will be needed to handle future growth. To battle these challenges, there must be an all-dimensional solution by indulging in technological innovation, infrastructure development, and policy initiatives.

4. ROLE OF ARTIFICIAL INTELLIGENCE IN SOCIAL SENSING

The role of Artificial Intelligence is greatly important in converting the raw social sensing information to information that can be acted upon to provide smart cities with the ability to learn and act in reaction to the complex urban dynamics (Acemoğlu et al., 2022). Using AI, cities are able to make better decisions, manage resources better and generally raise the living standards of its residents.

4.1 Importance of AI in Interpreting IoT Data

That wealth of information that exists in smart cities through the IoT can be intense. AI will present software to handle, study and comprehend the information effectively and precisely. The capability of the AI algorithms to identify patterns, anomalies, and trends in the data otherwise unrecognizable and hard to detect manually is an important category. This will enable urban leaders to have a better insight on dynamics of the city including traffic movements, energy use and any threat to safety. The ability to draw insights out of data and connected network of devices makes AI-driven smart city applications capable of avoiding downtime, efficiency and effectiveness of operations, and delivery of improved services and services (Kuguoglu et al., 2021). The use and application of big data in urban planning and building smart cities cannot be used without big data (Son et al., 2023). The AI algorithms would provide the possibility to process the data produced through the IoT (Khan et al., 2022).

4.2 Machine Learning and Deep Learning for Pattern Recognition

Some of the powerful AI methods that have application in pattern recognition in the social sensing data are; machine learning and deep learning. Transformations of ML algorithms can be learned through data without explicit programming and, thus, be able to recognize complex relationships and patterns (Zatsu et al., 2024). DL is a branch of ML, which employs the use of an artificial neural network and extracts features on a higher level of the data using multiple layers (Acemoğlu et al., 2022). Applications - Predictive Maintenance: ML can be used to assess the infrastructure sensor information (e.g., bridges and roads) to schedule the maintenance before incurring the expenses of failing. Traffic Management: DL algorithms can study traffic camera recording to obtain the best traffic flow and minimize congestion. Energy Consumption Forecasting: A more efficient distribution of energy could be achieved by the ML model, which forecasts the energy consumption dependent on weather information and occupancy of stores and other object types. Other uses of AI algorithms include real-time processing and analysis of data, vital in most of the social sensing applications.

4.3 Natural Language Processing for Sentiment and Trend Analysis

A branch of AI is Natural Language Processing which allows computers to understand, interpret, and generate human language. Social sensing is the application of NLP to examine text data found in social media, online forums, and other sources to retrieve sentiments and emerging trends (Acemoğlu et al., 2022). Applications - Public Sentiment Analytics: NLP algorithms might be used to read social media posts to assess the opinions of people on the different topics: new policies, the construction of new infrastructure, etc. Trend Identification: NLP has the ability to analyze emerging trends in the discourse by the populace, e.g. issues of air pollution or safety on the road. Disaster Response: NLP may work with the social media data during any natural disasters to find out the affected zone and provide the rescue operations to be addressed on priority.

4.4 Examples of AI in Citizen Behavior Prediction and Anomaly Detection

AI can bridge this gap by predicting citizen behaviors and, on the same front, it can detect any unusual activity in cities. Among other things, this information can be utilized to enhance the level of citizen safety, optimize resource distribution, and improve the quality of life of citizens (Acemoğlu et al., 2022). Applications - Crime Prediction: AI can be able to analyze previous crime data, population, and other aspects to make predictions of places with high risk of crime. Anomaly Detection: AI is able to look in sensor measurements to figure out any unusual activities like a sudden spike in energy consumption or unusual traffic patterns, which could say there was an issue. Public Health Monitoring: AI can be used to analyze the data collected by wearable sensor devices and other available data to diagnose the affected population by the outbreak of the infectious diseases. Through AI, smart cities will have the capacity of converting raw data into actionable data and will thus be able to make better decisions on how to allocate resources as well as how to enhance the quality of life to their citizens (Hsu et al., 2022). AI will have a key role in social sensing as that invention develops.

5. AI MODELS AND TECHNIQUES FOR OPTIMIZING SOCIAL SENSING

When it comes to social sensing applications, artificial intelligence models and techniques will be crucial to the desired value extraction of the vastly produced data (Acemoğlu et al., 2022).

5.1 Human Activity Recognition

Human Activity Recognition refers to persistent analysis on AI that intends to recognize human behavior mechanically based on sensor data or videos (Tee et al., 2022). HAR may be either vision-grounded (through cameras) or sensor-grounded (with the help of mobile devices or body accessories) (Tee et al., 2022). Figure.3 indicates the transformation of the human activity recognition models that shift between simple and complex range.

Figure 3. Human Activity Recognition Models Range from Simple to Complex

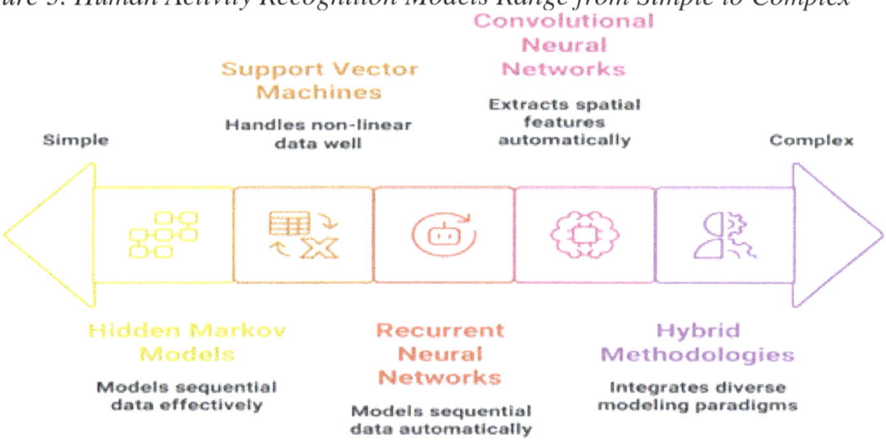

Recognition through vision in this method, computer vision is applied on video data and actions are labelled. Sensor-based recognition applies wearable sensor data i.e. accelerometers and gyroscopes. The sensor types of systems are less demanding in terms of energy, and can also be applied in more locations, since they are less bulky and consume less energy. The elderly care setting, security, and smart household (Atadoga et al., 2024) are possible applications of HAR. The Models and Techniques are mentioned below: Deep Learning Models: They usually use Convolutional Neural Networks and Recurrent Neural Networks which are often used to capture spatial and sequential data respectively. These models are automated and can learn even intricate patterns when presented with large amounts of data thus these are specifically good to use with tough activity recognition tasks. Conventional Machine Learning: Hidden Markov models and support vector machines have as well been used in HAR. SVMs work well in high-dimensional spaces and can work with non-linear data, but HMMs are appropriate in sequential modeling, hence useful in activities that occur over time. Hybrid approaches which feature an interplay of different modeling paradigms has demonstrated possibilities of improving both accuracy and robustness of human activity recognition systems. Hybrid approaches have the ability to overcome the shortcomings of individual models as they utilize their strengths and reach a top-of-the-notch performance.

5.2 Traffic Flow Prediction

Traffic flow forecasting is the process of predicting future traffic by historical and real-time sensor records. This assists in the optimum choice of traffic signals and avoids traffic congestion and enhances efficiency in transportation. Among the techniques are time-series analysis whereby historical values are used to predict the traffic through models known as the ARIMA and Kalman filters. In some of these models, neural network, SVMs, and decision tree machine learning models, train the relationships between the variables of traffic to future traffic flow (Jacyna-Golda et al., 2025). Machine learning when combined with time-series analysis helps in enhancing the level of accuracy that is involved in prediction. The predictions have also been improved with the use of ensemble learning in which predictions of a number of models are used (Jacyna-Go stays et al., 2025). The popularity of deep learning is due to the ability to find a complex pattern automatically using traffic data and enhance the prediction (Abduljabbar et al., 2021). The Long Short-Term Memory networks excel at modeling the temporal dependencies (Jia et al., 2021). Nevertheless, several models need the use of high-quality traffic data which is not readily available in low-detector areas. In order to do that, information on road networks, points of interest and weather can be combined and used to correct the performance in data-sparse environments.

5.3 Social Media Sentiment Analysis

Natural Language Processing Social media sentiment analysis is a technique applied to social media, mainly by using Natural Language Processing in order to know the feeling of the post. This interpretation measures the opinion of the people, the new trends and brand reputations. In another study, an Android-based application, Citizen Sense, was employed to collect data; whereby, users shared textual and image material and emotion/moods of 25 emotion types. Lexicon-based

Approaches: Pre-defined dictionaries of words and sentence score are applied in the process of discovering the feelings of the text. Naive Bayes, SVM and the deep learning models of Machine Learning models are used to classify the sentiments of social media posts. Deep Learning Models, like CNNs and RNNs are used to consider the contextual and enhance the ACC of the sentiment classification (Limami et al., 2024). The Linear Support Vector Machine method had a high accuracy in sentiment classification of COVID-19 tweets but the Stochastic Gradient Descent algorithm was more efficient in terms of the parameters it used in terms of computation.

5.4 Object Detection and Image Classification

Object detection and image classification deal with detection and the classification of objects within images and videos. Applications such as video surveillance, autonomous driving and industrial automation are involved. Convolution Neural Networks prevail, and its types, like Faster R-CNN, YOLO, and SSD can be found to detect and classify an arbitrary number of objects in the image. There is also common usage of Transfer Learning, where pre-trained models of CNN are refined by applying them on special datasets and enhance them to function better and with shorter training period (Ayturan et al., 2025). The deep learning approaches have then developed to present powerful methods of automatically inducing features in raw data and have also realized a significant rise in object recognition. So many computer vision applications on object detection are in operation nowadays such as surveillance and security, autonomous automobile, anomaly detection, pedestrian detection, and robot vision. Frames such as TensorFlow and PyTorch and programming languages, including Python and Anaconda environments, offer a good framework where these object detection models could be deployed and implemented (Abba et al., 2024).

6. EMERGING TRENDS: EDGE AI AND FEDERATED LEARNING

Edge AI and federated learning are new trends that can solve the most important issues concerning latency, bandwidth, and data privacy in AI-IoT application within smart cities. The use of AI models at the end points of devices like smartphones, IoT and edge servers, instead of pushing the data to central server to be processed, is known as Edge AI (Yurdem et al., 2024).

6.1 Concept of Edge AI in Reducing Latency and Bandwidth

Edge AI also diligently processes data on the location where they generate data and therefore greatly decreases latency and bandwidth consumption. Since Edge AI runs the calculations locally, it helps avoid the necessity to transfer massive data to the cloud and, consequently, improves speed and network congestion. Edge Computing expands the storage and processing capacities of cloud servers to the edge of mobile networks and close to the end-users and when coupled with AI and ML schemes can be called Edge AI (Pajooh et al., 2022). Today, it is possible to combine minuscule machine learning innovations that have allowed scholars to develop ML-centered models that are characterized by augmented energy efficiency and innovative data-centric intelligence (Khalil et al., 2022). Edge intelligence demonstrates a lot of promise in the context of real-time tracking and alerting environmental problems (Shah et al., 2021). AI algorithms play a critical role in a lot of applications, such as speech processing, image classification, and medication discovery, and specifically, in the reduction of latency, bandwidth consumption, and privacy concerns (Shi et al., 2020).

6.2 Federated Learning for Privacy-Preserving Data Training

The purpose of federated learning is to keep data confidential because the training of the model can be done by means of the device, and only the update of the model is transferred to the centralized server. In this way, sensitive data need not be forwarded to a central point, dealing with both privacy issues and regulatory obligations. In 2016, Google came up with the concept of FL. The IoTs in FL will train their models with the available local data, and the transmission of sensitive data does not have to be conveyed to a central cloud site (Alghamedy et al., 2024). The technique plays a very critical role in healthcare, finance and personal data analytics as privacy is of utmost importance in such fields. It applies in a scenario where datasets share the same feature space but not the samples, otherwise referred to as horizontal federated learning. To illustrate, various hospitals can work together to train a model that will predict the disease without exchanging the patient data. Vertical Federated Learning, however, is used when the sets have the same sample space, but different features space (Wu et al., 2025).

6.3 Applications in Health, Public Safety, and Utilities

There are many applications of edge AI and federated learning in health, public safety and utilities. Health: In the medical industry, federated learning can facilitate non-transmission of sensitive patient information to conduct researches collaboratively

(Jafarigol et al., 2023). As an example, several hospitals can have a shared training on models to predict disease breakouts or accurately diagnose the disease processes without privacy breach. Public Safety: Edge AI can be used to increase the level of public safety because real-time data in surveillance cameras and sensors can be processed in this way. It can be discussed that no suspicious activities or anomalies in the case of crowded places may be identified, e.g., suspicious behavior can be discovered by using edge-based object detection models, which will reduce response time and enhance security. Utilities: Edge AI and federated learning can be used in energy sector, to optimize energy distribution and help to decrease energy usage. An example is that smart meters with edge AI capabilities are able to forecast the energy demand and modify the energy supply to suit it accordingly thus promoting grid efficiency and eliminating wastage of power.

6.4 Comparative Benefits Over Centralized Cloud AI

Some of the benefits associated with edge AI and federated learning as compared to centralized cloud AI are lower latency, better data privacy, and greater scalability. Low Latency: Since edge AI is based on data processing on the ground, this technology requires less data to be transferred to its cloud, which means that it responds faster and can make better real-time decisions. Better data privacy: Federated learning enhances data privacy because devices can train models on their own and only the new model updates are shared with the central server, thus mitigating the possibility of data breach and regulatory compliance inconveniences. Increased Scalability: Federated learning and edge AI allow distributed data processing, thus increasing scalability and decreasing the pressure on central infrastructure. Moreover, federated learning also guarantees comprehensive privacy to the individual records since there are no requirements to send the personal information of the individual users to a central server to be trained centrally (Dai et al., 2021).

6.5 Case Studies and Examples

In practice, a number of case studies and examples of edge AI and federated learning turned out to be effective. To illustrate, the example of the use of edge AI is autonomous vehicles receiving information about sensors and making real-time in-the-use decisions when driving. Likewise, federated learning has been applied in the healthcare field to develop models to predict and diagnose disease without transferring sensitive patient data. Google Federated Learning: Google is at the forefront of federated learning studying, and building frameworks, and tools to implement FL models on mobile phones and other edge devices. The federated learning framework breaks the keyboard prediction as well as voice recognition tasks into collaborative tasks, which are trained without a user becoming externally identifiable. TinyML is a recent area that is interested in the implementation of machine learning models on highly energy-efficient embedded systems allowing AI applications to be run on battery-powered low-resource systems. The TinyML is applicable in many aspects such as in smart sensors, wearables devices, and IoT devices. Edge AI, and federated learning are promising solutions to practical intelligence, adaptability, and safety of AI-IoT applications deployed in smart cities (Alghamedy et al., 2024). Using such evolving trends, cities have an opportunity to use AI to maximize the effectiveness of the services to citizens, foster communication and interaction, and facilitate sustainable urbanization. These are technologies that allow analyzing data and making the decision in real-time as well, contributing to data organization greatly. With the discipline developing, additional research and development will be required in order to reduce the technical, ethical, and regulatory issues Luckily, the Edge AI and federated learning introduce.

7. REAL-WORLD APPLICATIONS IN SMART CITIES

Smart cities are designed to use high technologies that bring results to the better life of its citizens, to become more sustainable, and to advance the economy. AI-IoT use in real-life smart cities can include such industries as disaster solutions, public safety and monitoring, healthcare management, energy efficiency, and so on. When using these applications, the city of the future is tougher, more effective, and habitable (Acemoğlu et al., 2022).

7.1 Mitigation of Disasters

The natural calamities that are more prone to smart cities are floods, earthquakes, and fires. The application of AI-IoT technologies is important in strengthening the preparation, response, and recovery of disasters. In this regard, a smart city could use network of sensors and drones to actively monitor infrastructure faults, like water leakage or power interruptions, thereby cutting down expenses involved with frequent maintenance. Flood Detection Systems: Smart cities have developed networks of sensors so as to detect the water level, rain water and the drainage system. These data are analyzed by the AI algorithms to predict the possible flooding events and send an early warning to people to evacuate in time and reduce the damage to property (Costa et al., 2022). Earthquake Detection Systems: Earthquake early warning systems can be used to alert residents of an earthquake prior to a large shaking on the ground due to seismic sensors and artificial intelligence algorithms detecting the beginning of the earthquake (Costa et al., 2022). These systems have the capability to facilitate automatic actions e.g. closing down of infrastructure at risk and coming to the stop of the transportation networks to limit the effects of earthquake. Fire Detection Systems: Smarter fire detection systems aid in the detection of the fire in buildings and on the streets: they combine sensors, cameras, and the AI algorithm to identify potential cases (Costa et al., 2022).

7.2 Public Safety and Surveillance

The smart cities use AI-IoT applications to promote public security and safety. Examples of crime prediction, crowd analysis and real-time surveillance are part of such applications. Aid with crime: AI-based crime prediction systems use historical crime data, demographics, and environment data to determine possible crime-prone areas and forecast of criminal activities. This information can be utilized by law enforcement agencies to be in a strategically positioned position; to intercept crimes, before they happen. Crowd Analysis: Using cameras and sensors, smart cities will be able to analyze crowd density and movement in the public area (Costa et al., 2022). The AI algorithms are used to analyze these data and find out anomalies, including overcrowding and abnormal behavior, and warn the authorities about possible safety risks. Real-Time Surveillance: AI and video surveillance technology can help track elements in real-time via facial and object detection and behavioural analysis monitored throughout public locations. Such systems will be able to know who or what is suspicious and warn the law enforcement agencies about possible threats (Acemoğlu et al., 2022). The computer vision fueled by Artificial Intelligence and sensor-based surveillance systems have enhanced the process of detecting instances of emergencies, minimize the time it takes to intervene, and also save on an effective distribution of emergency resources in real-time.

7.3 Healthcare Monitoring

In smart cities, AI-IoT technologies facilitate remote monitoring of health conditions, patient-specific treatment plans, and increasing access to healthcare in general (Acemoğlu et al., 2022). Chronic Disease Tracking: Smartwatches and other fitness trackers keep track of vital signs, activity and sleeping habits of people with chronic diseases. AI algorithms can process such data to offer more personalized information and recommendations, and, this way, help people better manage their health. Remote Patient Monitoring: Remote patient monitoring systems are systems that aggregate patient data and transmit the data to health care providers with the aid of sensors and communication equipment. These data are analyzed by AI algorithms in order to identify anomalies, to offer intervention in time, to minimize hospital visits, and increase patient outcomes. Telemedicine: Telemedicine applications allow distance consultation of patients with healthcare professionals. Virtual assistants (based on AI) are able to help make appointments, gather some information about the patients and give basic health care advice. To extract and learn patterns and even predict, these AI algorithms are fed with huge databases of information with which they can be utilized to interpret a vast amount of information about the patient (blood pressure and other vital signs, test results, etc.), foresee complications, and create individualized treatment plans accordingly (Li et al., 2024).

7.4 Energy Optimization

The application of AI-IoT technologies in smart cities is aimed at optimising energy and energy consumption, sustainable energy usage. Smart Meters: Smart meters record real-time data information about the use of energy and relay the data to utility companies. These data pass through AI algorithms to give energy demand prediction and energy supply optimization, waste reduction, and grid efficiency. Predictive Energy Distribution: Integrating historical energy consumption, weather data, and other variables, AI algorithms can predict the demand of energy of the future. This information can be utilized by utility companies in arranging the distribution of energies and minimizing the chances of occurrence of power outages. Smart grids are technology-based systems that can monitor and control the electrical flow through the use of sensors, communication gadgets, and AI technology. Smart grids are able to automatically balance energy supply and demand, optimise energy storage, incorporate renewable energy in the grid, enhance the reliability of the grid and minimize carbon emissions.

Renewable Energy Forecasting: AI algorithms use weather and historical energy production information and other data to predict how much energy is produced by renewable sources, like wind and solar power. The development of smart grids relies heavily on AI-based solutions, which can be used to collect interdisciplinary and heterogeneous data and provide policymakers with information (Xie et al., 2022). Additionally, AI models use the current weather, solar radiation, and wind speed conditions to economize the generation of renewable energy infrastructure placement (Olawade et al., 2024).

8. ETHICAL, PRIVACY, AND TECHNICAL CHALLENGES

The emergence of AI-IoT innovation in smart cities is characterized by the diverse number of ethical, private, and technical challenges that have to be resolved to guarantee responsible and sustainable urban development (Ahmed et al., 2023). Such issues involve data privacy and surveillance, bias and fairness in AI, data interoperability and standardization, cost of energy and infrastructure, and ethical implications of large-scale mass surveillance systems and prediction systems (Jain et al., n.d.). Without considering these issues, it is possible to lose credibility of people, worsen the state of social inequality, and damage the values of smart cities programs. Smart cities are based on collecting, storing, and analyzing enormous statistics and data of dozens of sources such as sensors, cameras, and personal machines. This type of data is usually highly sensitive containing the movements, behavior and preferences of citizens prompting the questions of data privacy and surveillance (Mazzarello et al., 2025). Data protection initiatives, unambiguous data collection and data sharing policies, as well as well-meant data security reports should be implemented in the collection and use of personal data in smart cities (Kuang et al., 2024). It is critical to adopt transparency and consent mechanism so as to guarantee citizens that they are informed about collection, use and dissemination of their data. Data Security Breaches: The critical personal information can be hacked because of cyberattacks and breakages by the smart city systems. The implications of this have been serious and involve identity and loss of money and tarnish. Mass Surveillance: Mass surveillance and the danger of driving to their fullest the use of surveillance technology, like facial recognition, or predictive policing, become an issue of concern (Kuang et al., 2024). The technologies are capable of monitoring the movement of the citizens, watching their actions or even forecasting their actions, which would violate the civil liberties of the citizens. The integration of information technology and the city administration is making cities smarter and more resilient, with increasingly worrying estimates that smart cities will revert the gains of civil liberties as it is being sensed, profiled, and predicted in the activities of citizens.

8.1 Bias in AI Models and Fairness

In smart cities, AI algorithms are also used in fields including resource allocation, public, and social services as decision-making tools. Despite that, AI models may be biased either because of being trained using biased data or because they were created towards biased assumptions. Data Bias: There is a risk that AI models applied on bias data can reinforce and increase pre-existing social inequalities. To give an example, a crime-prediction algorithm that is fed with historical data on crimes and predicts offences disproportionally in specific communities can result in discriminatory policing policies. Algorithmic Bias: Bias in AI models could effectively occur despite the use of unbiased data due to how an AI model is constructed, as well as the implicit assumptions made by it. Take an example of the loan application system which is powered by artificial intelligence which can discriminate against some demographic groups unless well developed in a manner which prevents discrimination.

Inarray Explainability: Considering that AI models are black boxes, a number of them do not offer insights into how they reach such decision points due to the complexity of their decision-making processes. Such unexplainability may lead to the challenge of detecting and rectifying bias in AI models, as well as accountability of these models regarding their judgments. In case the bias is not to be addressed responsibly, AI technologies can be biased based on the data they are trained on (Chu et al., 2023). Governments ought to work to establish guidelines concerning the administration of data and building explainable AI in order to reduce the dangers of AI bias (Acemoğlu et al., 2022).

8.2 Data Interoperability and Standardization Issues

Smart cities commonly include using different systems and data sources that can be provided by different vendors and agencies. Nevertheless, the absence of data interoperability and standardization may slow the process of successful integration and analysis of data, which reduces the possible positive effects of smart cities efforts. Data Silos: Data silos take place when various systems or agencies do not share data or they use incompatible data formats. This may lead to a want of a clear picture of a city and an inability to make the right choice. Non-existence of Standards: There may be also the non-existence of standardized data that may interfere with the exchange and integration of data. This may hinder comparison of data between different sources or utilization of data on new applications. Legacy Systems: In most cities, systems are legacy systems that are not developed to interact with contemporary technologies. It may be difficult and costly to incorporate such systems in new city smart initiatives. Explainable AI arises as an important factor whenever there is a need to vet the AI systems by justifying its output. One example of such problems is robustness or generality problems since neither AI nor any kind can easily stumble over reliability in the face of new or very different data.

8.3 Energy and Infrastructure Costs of Deploying AI-IoT

AI-IoT technologies introduced in smart cities need substantial investments in terms of energy and infrastructure. These expenses may be a hindrance to adoption especially in a resource poor city. Energy demanding: AI and IoT gadgets are high-energy consumers, especially when they are processing complex information, or relaying broadband information.

Energy: AI and IoT devices have the potential to intensively utilize energy and whose deployment in large scales can be energy expensive. This may cause energy grids to overload and cause a rise in the emission of carbon, which will negate environmental sustainability objectives of smart cities. Infrastructure Upgrades: The process of introducing smart cities is usually connected with the upgrading of existing infrastructure, including communication systems, e.g. Power supply systems. Such renovations may be expensive and time consuming.

Maintenance and upgrade: AI-IoT systems involve continuous investment in their maintenance and upgrades. This can be hectic to cities with small budgets. These systems are expensive to be designed, installed and operated. Besides financial constraints, infrastructural challenges, including the ancient city infrastructure unsuitable for smart cyber-physical systems technologies which would need an expensive overhaul, also can be a factor.

8.4 Ethical Considerations in Deploying Mass Surveillance and Prediction Systems

By introducing a mass surveillance and prediction systems the question of balance between civil liberties and public safety is brought up ethically in the smart cities. Privacy Destruction: Privacy can be destroyed since mass surveillance systems enable it to keep a tab on the progress and activities of citizens at all times. This is able to impose a chilling impact on the right of expression and assembly. Discrimination and Bias: The prediction systems can enforce and elevate current inequalities in society by retargeting particular communities or individuals as a way of augmenting scrutiny. This involves partnership between policy makers, technology providers, city planners and citizens to develop the framework which is a balance among innovation, privacy, and sustainability (Ahmed et al., 2023). As the initiatives preparing to face these challenges early on, we will be able to use the full potential of AI-IoT technologies to make our cities more equitable, resilient, and comfortable, as well as make them livable by everyone (Khalil et al., 2022).

9. FUTURE DIRECTIONS AND SMART CITY FRAMEWORKS

High-efficiency interplay between AI-IoT technologies and urban infrastructures, the citizen-oriented approach to the design of smart cities via effective policies, and cross-border cooperation between the government, universities, and industry are the keys to the future success of smart cities. It would require developing and implementing powerful Smart City Maturity Models and erecting new and innovative designs of the future smart sensory nets (Bibri et al., 2023).

9.1 Vision for AI-IoT Integrated Urban Infrastructure

Intelligent Transportation Systems: It is an AI-based traffic management tool that allows traffic optimization, diminishing traffic, and increasing traffic security. These systems have the capacity to utilize the real-time information of the IoT-based sensors to estimate the presence of traffic and manipulate the signals and offer specific route directions. Smart Energy Grids: Artificial intelligence-powered energy networks which ensure efficient energy filtration, minimal wastage and the utilization of alternative energy sources. Such grids are able to forecast energy demand, control the supply of energy, and incorporate distributed energy resources, based on data of smart meters and weather sensor (Obracht-Prondzyńska et al., 2022). Adaptive Buildings: The buildings that have built-in smart cyber-physical systems that enable the building to respond immediately to their environment, use pattern and power requirements. With sensors, the buildings may be able to tell the level of occupancy, temperature, and the lighting level and change their systems accordingly to achieve maximum energy efficiency and comfort.

9.2 Need for Policy-Driven, Citizen-Centric Design

A citizen-centric policy and design is vital in ensuring that initiatives of smart cities resonate with the needs and values of the citizens. The strategies need to focus on involving citizens in planning and implementation effectively, airtight data privacy and security systems, equity and inclusion among all the demographics, and transparent and accountable governance measures (Kumar et al., 2023; Zheng et al., 2019). Also, one must pay attention to the ethical consequences of AI-IoT technologies, including possible biases in the algorithms and their effects on stigmatized groups of people. A direct focus on the participation of its citizens and ethical perspective will allow smart cities to establish trusting relationships, strengthen social cohesion, and make technological progress work to the advantage of all citizens. Governments, universities, and industry will have to work hand in hand to realize and implement the smart cities effectively. This cooperation may combine in various forms, i.e.: Public-Private Partnerships: Collaboration with the private sector in formulating and implement smart city technologies and services. This can avail the skills and resources of the private sector, and at the same time it prevents the interests of the people as a whole. Research and Development: Making investment in research and development that would provide new and innovative smart city solutions. This can be done by government subsidies, university research and industrial partnerships. Knowledge Sharing: Transferring knowledge and best practice between governments, academia and industry. This can be done in the form of conferences, workshops and online. Standardization: Coming up with shared requirements of smart city technologies and data format requirements. It may favor interoperability, cause lower costs, and faster adoption. It may lead to innovation and subsequent economic growth as well as to tackling urban challenges through the co-existence of academic knowledge and practical application (Obracht-Prondzyńska et al., 2022) (Acemoğlu et al., 2022) and (Kirimtat et al., 2020).

CONCLUSION

A comprehensive approach, analyzing all the three aspects, such as the technology, rules, and engagement of people, is what has to be made in order to make smart, flexible, and safe urban environments. The integration of AI and IoT would allow enhancing the city infrastructure and result in smart transportation, sensitive energy consumption, flexible buildings, environmental control, and more effective public safety. This is achieved by rules that orientate citizens which include participation, privacy and data protection, fairness, and transparency. Partnership between governments, schools and businesses is what will promote setting standards, sharing information, promoting research and creating partnership between government, schools and businesses. The resources can not only be utilized to the maximum, but smart city projects can also minimize the amount of pollution and the use of renewable energy is beneficial to the environment. Smart cities will have access to information, services and enable people to make decisions, which can result in equal treatment and better life of city inhabitants. This is so that smart cities that are powerful, habitable, and user-friendly technologies, are established. The new concept can be the improvement of urban systems with the help of IoT, blockchain, and computational intelligence, making them safer, more effective, and adaptable and improving their living conditions. Smart cities connect operators and services and enhance working of things through data collection, processing and sharing making city life to be developed in numerous regions.

REFERENCES

Abba, S., Bizi, A. M., Lee, J., Bakouri, S., & Crespo, M. L. (2024). Real-time object detection, tracking, and monitoring framework for security surveillance systems. *Heliyon, 10*(15), e34922. Advance online publication. DOI: 10.1016/j.heliyon.2024.e34922 PMID: 39145028

Abduljabbar, R., Dia, H., & Tsai, P. (2021). Development and evaluation of bidirectional LSTM freeway traffic forecasting models using simulation data. *Scientific Reports, 11*(1), 23899. Advance online publication. DOI: 10.1038/s41598-021-03282-z PMID: 34903780

Acemoğlu, D., Autor, D., Hazell, J., & Restrepo, P. (2022). Artificial Intelligence and Jobs: Evidence from Online Vacancies. *Journal of Labor Economics, 40*(S1), S293–S340. Advance online publication. DOI: 10.1086/718327

Adibi, S., Rajabifard, A., Shojaei, D., & Wickramasinghe, N. (2024). Enhancing Healthcare through Sensor-Enabled Digital Twins in Smart Environments: A Comprehensive. *Analysis & Sensing, 24*(9), 2793. DOI: 10.3390/s24092793 PMID: 38732899

Ahmed, A. A., Belrzaeg, M., Nassar, Y. F., El-Khozondar, H. J., Khaleel, M., & Alsharif, A. (2023). A comprehensive review towards smart homes and cities considering sustainability developments, concepts, and future trends [Review of A comprehensive review towards smart homes and cities considering sustainability developments, concepts, and future trends]. World Journal of Advanced Research and Reviews, 19(1), 1482. GSC Online Press. DOI: 10.30574/wjarr.2023.19.1.1530

Alghamedy, F., El-Haggar, N., Alsumayt, A., Alfawaer, Z. M., Alshammari, M., Amouri, L., Aljameel, S. S., & Al-Bassam, S. (2024). Unlocking a Promising Future: Integrating Blockchain Technology and FL-IoT in the Journey to 6G. *IEEE Access : Practical Innovations, Open Solutions, 1*, 115411–115447. Advance online publication. DOI: 10.1109/ACCESS.2024.3435968

Atadoga, A., Omaghomi, T. T., Elufioye, O. A., Odilibe, I. P., Daraojimba, A. I., & Owolabi, O. R. (2024). Internet of Things (IoT) in healthcare: A systematic review of use cases and benefits [Review of Internet of Things (IoT) in healthcare: A systematic review of use cases and benefits]. *International Journal of Science and Research Archive, 11*(1), 1511–1517. DOI: 10.30574/ijsra.2024.11.1.0243

Ayturan, K., Sarıkamış, B., Akşahin, M. F., & Kutbay, U. (2025). SPHERE: Benchmarking YOLO vs. CNN on a Novel Dataset for High-Accuracy Solar Panel Defect Detection in Renewable Energy Systems. *Applied Sciences (Basel, Switzerland)*, *15*(9), 4880. DOI: 10.3390/app15094880

Bibri, S. E., Krogstie, J., Kaboli, A., & Alahi, A. (2023). Smarter eco-cities and their leading-edge artificial intelligence of things solutions for environmental sustainability: A comprehensive systematic review [Review of Smarter eco-cities and their leading-edge artificial intelligence of things solutions for environmental sustainability: A comprehensive systematic review]. Environmental Science and Ecotechnology, 19, 100330. Elsevier BV. DOI: 10.1016/j.ese.2023.100330

Calzada, I. (2022). Citizens' Data Privacy in China: The State of the Art of the Personal Information Protection Law (PIPL). *Smart Cities*, *5*(3), 1129–1150. DOI: 10.3390/smartcities5030057

Chu, C. H., Donato-Woodger, S., Khan, S. S., Nyrup, R., Leslie, K., Lyn, A., Shi, T., Bianchi, A., Rahimi, S. A., & Grenier, A. (2023). Age-related bias and artificial intelligence: a scoping review [Review of Age-related bias and artificial intelligence: a scoping review]. Humanities and Social Sciences Communications, 10(1). Palgrave Macmillan. DOI: 10.1057/s41599-023-01999-y

Costa, D. G., Peixoto, J. P. J., Jesus, T. C., Portugal, P., Vasques, F., Rangel, E. O., & Peixoto, M. (2022). A Survey of Emergencies Management Systems in Smart Cities. *IEEE Access : Practical Innovations, Open Solutions*, *10*, 61843–61872. DOI: 10.1109/ACCESS.2022.3180033

Dai, S., Meng, F., Wang, Q., & Chen, X. (2021). FederatedNILM: A Distributed and Privacy-preserving Framework for Non-intrusive Load Monitoring based on Federated Deep Learning. arXiv (Cornell University). https://doi.org//arxiv.2108 .03591DOI: 10.48550

Grêt-Regamey, A., Switalski, M., Fagerholm, N., Korpilo, S., Juhola, S., Kyttä, M., Käyhkö, N., McPhearson, T., Nollert, M., Rinne, T., Soininen, N., Toivonen, T., Räsänen, A., Willberg, E., & Raymond, C. M. (2021). Harnessing sensing systems towards urban sustainability transformation. *NPJ Urban Sustainability*, *1*(1), 40. Advance online publication. DOI: 10.1038/s42949-021-00042-w

Hsu, Y.-C., Huang, T.-H., Verma, H., Mauri, A., Nourbakhsh, I., & Bozzon, A. (2022). Empowering local communities using artificial intelligence [Review of Empowering local communities using artificial intelligence]. Patterns, 3(3), 100449. Elsevier BV. DOI: 10.1016/j.patter.2022.100449

Jacyna-Gołda, I., Cur, K., Tomaszewska, J., Przanowski, K., Hošková-Mayerová, Š., & Świergolik, S. (2025). Optimizing Flight Delay Predictions with Scorecard Systems. *Applied Sciences (Basel, Switzerland)*, *15*(11), 5918. DOI: 10.3390/app15115918

Jafarigol, E., Trafalis, T. B., Razzaghi, T., & Zamankhani, M. (2023). Exploring Machine Learning Models for Federated Learning: A Review of Approaches, Performance, and Limitations [Review of Exploring Machine Learning Models for Federated Learning: A Review of Approaches, Performance, and Limitations]. arXiv (Cornell University). Cornell University. https://doi.org//arXiv.2311.10832DOI: 10.48550

Jia, S., Hui, F., Wei, C., Zhao, X., & Liu, J. (2021). Lane-Changing Behavior Prediction Based on Game Theory and Deep Learning. *Journal of Advanced Transportation*, *1*, 1–12. Advance online publication. DOI: 10.1155/2021/6634960

Khalil, U., Malik, O. A., Uddin, M., & Chen, C. (2022). A Comparative Analysis on Blockchain versus Centralized Authentication Architectures for IoT-Enabled Smart Devices in Smart Cities: A Comprehensive Review, Recent Advances, and Future Research Directions [Review of A Comparative Analysis on Blockchain versus Centralized Authentication Architectures for IoT-Enabled Smart Devices in Smart Cities: A Comprehensive Review, Recent Advances, and Future Research Directions]. Sensors, 22(14), 5168. Multidisciplinary Digital Publishing Institute. DOI: 10.3390/s22145168

Khan, A., Khan, S., Hassan, B., & Zheng, Z. (2022). CNN-Based Smoker Classification and Detection in Smart City Application. *Sensors (Basel)*, *22*(3), 892. DOI: 10.3390/s22030892 PMID: 35161637

Kirimtat, A., Krejcar, O., Kertész, A., & Taşgetiren, M. F. (2020). Future Trends and Current State of Smart City Concepts: A Survey. *IEEE Access : Practical Innovations, Open Solutions*, *8*, 86448–86467. DOI: 10.1109/ACCESS.2020.2992441

Kuang, Z., Su, J., Latifian, A., Eshraghi, S., & Ghafari, A. (2024). Utilizing Artificial neural networks (ANN) to regulate Smart cities for sustainable Urban Development and Safeguarding Citizen rights. *Scientific Reports*, *14*(1), 31592. Advance online publication. DOI: 10.1038/s41598-024-76964-z PMID: 39738236

Kuguoglu, B. K., van der Voort, H., & Janssen, M. (2021). The Giant Leap for Smart Cities: Scaling Up Smart City Artificial Intelligence of Things (AIoT) Initiatives. *Sustainability (Basel)*, *13*(21), 12295. DOI: 10.3390/su132112295

Kumar, R., Jain, V., Leong, H. W., & Teyarachakul, S. (2023). Convergence of IoT, Blockchain, and Computational Intelligence in Smart Cities. In *CRC Press eBooks*. Informa., DOI: 10.1201/9781003353034

Li, Y.-H., Li, Y., Wei, M.-Y., & Li, G. (2024). Innovation and challenges of artificial intelligence technology in personalized healthcare [Review of Innovation and challenges of artificial intelligence technology in personalized healthcare]. Scientific Reports, 14(1). Nature Portfolio. DOI: 10.1038/s41598-024-70073-7

Limami, F., Hdioud, B., & Thami, R. O. H. (2024). Contextual emotion detection in images using deep learning. *Frontiers in Artificial Intelligence*, *7*, 1386753. Advance online publication. DOI: 10.3389/frai.2024.1386753 PMID: 38952408

Mazzarello, M., Duarte, F., Mora, S., Heine, C., & Ratti, C. (2025). Data Slots: Trade-offs between privacy concerns and benefits of data-driven solutions. *Humanities & Social Sciences Communications*, *12*(1), 643. Advance online publication. DOI: 10.1057/s41599-025-04776-1

Murroni, M., Anedda, M., Fadda, M., Ruiu, P., Popescu, V., Zaharia, C., & Giusto, D. (2023). 6G—Enabling the New Smart City: A Survey [Review of 6G—Enabling the New Smart City: A Survey]. Sensors, 23(17), 7528. Multidisciplinary Digital Publishing Institute. DOI: 10.3390/s23177528

Obracht-Prondzyńska, H., Duda, E., Anacka, H., & Kowal, J. (2022). Greencoin as an AI-Based Solution Shaping Climate Awareness [Review of Greencoin as an AI-Based Solution Shaping Climate Awareness]. International Journal of Environmental Research and Public Health, 19(18), 11183. Multidisciplinary Digital Publishing Institute. DOI: 10.3390/ijerph191811183

Olawade, D. B., Wada, O. Z., David-Olawade, A. C., Fapohunda, O., Ige, A. O., & Ling, J. (2024). Artificial intelligence potential for net zero sustainability: Current evidence and prospects. *Next Sustainability*, *4*, 100041. DOI: 10.1016/j.nxsust.2024.100041

Pajooh, H. H., Demidenko, S., Aslam, S., & Harris, M. (2022). Blockchain and 6G-Enabled IoT. *Inventions (Basel, Switzerland)*, *7*(4), 109. DOI: 10.3390/inventions7040109

Qin, Z., Li, F. Y., Li, G. Y., McCann, J. A., & Ni, Q. (2019). Low-Power Wide-Area Networks for Sustainable IoT. *IEEE Wireless Communications*, *26*(3), 140–145. DOI: 10.1109/MWC.2018.1800264

Rashid, M. T., Wei, N., & Wang, D. (2023). A survey on social-physical sensing: An emerging sensing paradigm that explores the collective intelligence of humans and machines. *Collective Intelligence*, *2*(2). Advance online publication. DOI: 10.1177/26339137231170825

Shah, S. K., Tariq, Z., Lee, J., & Lee, Y. (2021). Event-Driven Deep Learning for Edge Intelligence (EDL-EI). *Sensors (Basel), 21*(18), 6023. DOI: 10.3390/s21186023 PMID: 34577228

Shi, Y., Yang, K., Jiang, T., Zhang, J., & Letaief, K. B. (2020). Communication-Efficient Edge AI: Algorithms and Systems. *IEEE Communications Surveys and Tutorials, 22*(4), 2167–2191. DOI: 10.1109/COMST.2020.3007787

Son, T. H., Weedon, Z., Yiğitcanlar, T., Sanchez, T. W., Corchado, J. M., & Mehmood, R. (2023). Algorithmic urban planning for smart and sustainable development: Systematic review of the literature. *Sustainable Cities and Society, 94*, 104562. DOI: 10.1016/j.scs.2023.104562

Tee, W. Z., Dave, R., Seliya, N., & Vanamala, M. (2022). A Close Look into Human Activity Recognition Models using Deep Learning. arXiv (Cornell University). / arXiv.2204.13589DOI: 10.1109/CNIOT55862.2022.00043

Wu, Z., Qin, Z., Hou, J., Zhao, H., Li, Q., He, B., & Fan, L. (2025). Vertical Federated Learning in Practice: The Good, the Bad, and the Ugly. arXiv (Cornell University). https://doi.org//arxiv.2502.08160DOI: 10.48550

Xie, L., Huang, T., Zheng, X., Liu, Y., Wang, M., Vittal, V., Kumar, P. R., Shakkottai, S., & Cui, Y. (2022). Energy System Digitization in the Era of AI: A Three-Layered Approach towards Carbon Neutrality. arXiv (Cornell University). https://doi.org// arxiv.2211.04584DOI: 10.48550

Yurdem, B., Kuzlu, M., Güllü, M. K., Çatak, F. Ö., & Tabassum, M. (2024). Federated Learning: Overview, Strategies, Applications, Tools and Future Directions [Review of Federated Learning: Overview, Strategies, Applications, Tools and Future Directions]. Heliyon, 10(19). Elsevier BV. DOI: 10.1016/j.heliyon.2024.e38137

Zatsu, V., Shine, A. E., Tharakan, J. M., Peter, D., Ranganathan, T. V., Alotaibi, S. S., Mugabi, R., Muhsinah, A. B., Waseem, M., & Nayik, G. A. (2024). Revolutionizing the food industry: The transformative power of artificial intelligence-a review [Review of Revolutionizing the food industry: The transformative power of artificial intelligence-a review]. Food Chemistry X, 24, 101867. Elsevier BV. DOI: 10.1016/j.fochx.2024.101867

Zheng, L., Kwok, W. M., Aquaro, V., & Qi, X. (2019). *Digital Government*. Smart Cities and Sustainable Development., DOI: 10.1145/3326365.3326403

Chapter 3
Crowd-AI for Smart Cities:
Leveraging Social Sensing and Edge AI for Real-Time Urban Intelligence

Darshan Mansukhbhai Tank
http://orcid.org/0000-0002-5138-8979
Gujarat Technological University, Ahmedabad, India

ABSTRACT

This chapter introduces Crowd-AI, a novel framework that synergizes social sensing, edge AI, and IoT to transform smart city operations with real-time, human-centric urban intelligence. As cities evolve into complex socio-technical ecosystems, timely insights from citizen activities, opinions, and movements become crucial. This chapter discusses how diverse social signals—captured via social media, wearable devices, and ubiquitous IoT infrastructure—can be processed using AI algorithms at the edge to derive actionable intelligence. By reducing latency, safeguarding privacy, and enabling hyper-local decisions, edge AI redefines the responsiveness of smart city services. Through extensive case studies in traffic management, public sentiment monitoring, and disaster response, we demonstrate the capabilities and challenges of deploying such systems. The chapter concludes with perspectives on integrating digital twins, federated learning, and citizen co-creation, emphasizing how these technologies chart a pathway to more adaptive, participatory, and resilient cities.

DOI: 10.4018/979-8-3373-4962-6.ch003

INTRODUCTION

The evolution of smart cities (Zanella et al., 2014) represents one of the most profound socio-technical transformations of the 21st century. As urban centers become denser and more complex, there is an increasing need for data-driven approaches to govern them efficiently. At the heart of this transformation lies the convergence of Artificial Intelligence (AI), the Internet of Things (IoT (Liu et al., 2020)), and human-centric data streams—collectively enabling cities to become context-aware, adaptive, and citizen-responsive.

Social sensing, which involves capturing data directly from people or from human-generated digital footprints (such as social media posts, wearable device data, or participatory sensing applications), provides a rich layer of insight into urban dynamics. Coupled with AI algorithms and IoT (Liu et al., 2020) infrastructure, this data forms the basis of a new paradigm: smart cities that learn from and adapt to their inhabitants in real time.

However, processing this deluge of data centrally often leads to latency issues, privacy (Verma et al., 2017) concerns, and bandwidth bottlenecks. Enter edge AI (Xu et al., 2019), which enables data to be processed close to where it is generated. By deploying AI models on edge devices, cities can respond quickly to changing conditions, improve resilience, and ensure that sensitive data need not always traverse to centralized clouds.

This chapter introduces Crowd-AI, a comprehensive framework that brings together social sensing, edge AI, and IoT networks to deliver real-time urban intelligence. Through examples ranging from smart traffic management to emergency response, we illustrate how such systems can drive smart city innovation.

FUNDAMENTALS OF SOCIAL SENSING AND IOT

What is Social Sensing?

Social sensing refers to the process of gathering data that reflects human activities, opinions, movements, or physiological states. Unlike traditional sensor networks that primarily monitor physical parameters (like temperature or humidity), social sensing (Kitchin, 2014) focuses on people as both sensors and actuators.

Types of social sensing include: - Participatory sensing: Individuals actively contribute data (e.g., through apps that report potholes or pollution levels). - Opportunistic sensing: Data is passively collected, such as GPS traces from smartphones. - Crowd-sourced media: Posts, images, and videos shared on platforms like Twitter, Instagram, or WhatsApp.

The Role of IoT in Smart Cities

The IoT (Liu et al., 2020) underpins smart city infrastructure by interconnecting devices ranging from traffic lights and CCTV cameras to air quality monitors and public Wi-Fi nodes. These sensors create a pervasive network that captures continuous streams of environmental and operational data.

When combined with social sensing (Kitchin, 2014) data, IoT (Liu et al., 2020) devices allow cities to gain a holistic picture of both physical and social dynamics. For example, a traffic congestion management system might integrate sensor data on vehicle flow with social media reports of accidents.

The Promise of Edge Computing

Edge computing brings computation and storage closer to data sources, reducing the need to transmit all data to distant data centers. This results in: - Lower latency, crucial for time-sensitive applications like traffic signal adjustments. - Reduced bandwidth usage, since raw data does not always need to be sent to the cloud. - Enhanced privacy (Verma et al., 2017), by keeping personal or sensitive data local. Edge AI combines this paradigm with local deployment of machine learning models, enabling analytics and decision-making to occur on-site.

EDGE AI FOR SMART CITIES

Why Edge AI?

Traditional cloud-based AI systems aggregate large datasets at centralized servers for processing. While this supports sophisticated analytics, it often introduces unacceptable delays for applications that require immediate reactions, such as rerouting traffic or triggering alarms during disasters. Edge AI addresses these challenges by executing trained AI models directly on edge devices (e.g., gateways, routers, even smart cameras), enabling rapid, decentralized decision-making.

Architectural Considerations

A typical edge AI (Xu et al., 2019) setup in a smart city might involve: - IoT (Liu et al., 2020) endpoints: Sensors, cameras, wearable devices, smartphones. - Edge nodes: Micro data centers or intelligent gateways processing local data streams. - Coordination layer: Interfaces that communicate summarized insights to city management systems or cloud dashboards for broader trend analysis.

Privacy and Trust

Because social sensing often involves personal data, privacy is paramount. Edge AI can reduce the risk of data breaches by minimizing the transmission of identifiable information. Techniques like differential privacy and federated learning (Fan et al., 2021) can further ensure that insights are derived without exposing raw data.

THE CROWD-AI FRAMEWORK

Overview

Crowd-AI refers to the integration of crowdsourcing and artificial intelligence (AI) to build, train, improve, or deploy AI systems by leveraging the collective intelligence, data, or contributions of a large group of people (the crowd). It combines human efforts and machine learning to solve complex problems that AI alone might not handle effectively.

Crowd-AI integrates multiple data streams—social media feeds, IoT (Liu et al., 2020) sensor outputs, and direct citizen inputs—and processes them using AI models deployed at the edge. This enables cities to: - Detect incidents (accidents, flooding, fires) in near real-time. - Understand mobility patterns and optimize traffic flows. - Monitor public sentiment about events or policies.

Figure 1. Framework of Crowd-AI

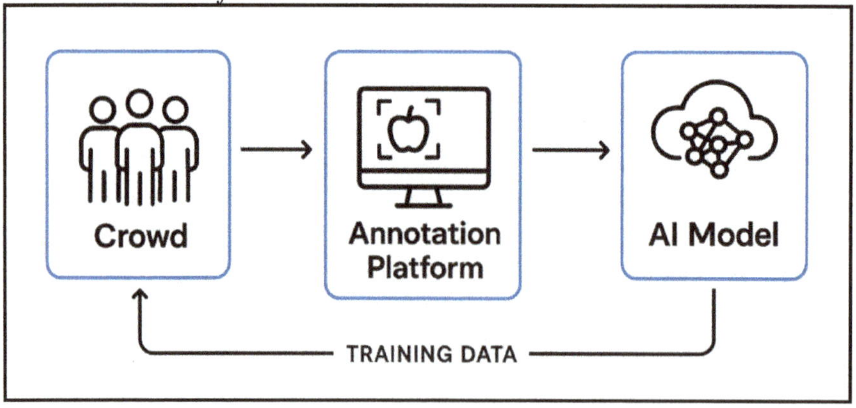

The framework shown in figure 1 highlights a human-in-the-loop system, where the crowd labels data via an annotation platform, feeding it to train the AI model. The AI model's outputs or uncertain cases then loop back to the crowd, creating a cycle of ongoing training and refinement. This is a minimal yet powerful framework often used in supervised learning setups that depend heavily on labeled data.

- Crowd
  Represents a group of human participants (crowd workers or volunteers).
  Their role is to provide manual input such as labeling images, tagging data, or verifying outputs.
  They supply the essential human intelligence needed to annotate complex or ambiguous data.
- Annotation Platform
  This is the system or software where the crowd performs the data labeling or annotation.
  Examples include platforms like Amazon Mechanical Turk, Labelbox, or custom-built annotation tools.
  It organizes data into tasks, manages submissions, and ensures quality (e.g., through consensus or review).
- AI Model
  This component trains machine learning or deep learning models using the annotated data.
  It consumes the labeled datasets to learn patterns, build predictive capabilities, and improve over iterations.
- Feedback Loop
  The framework includes a feedback mechanism labeled "TRAINING DATA" that loops back from the AI Model to the Crowd.
  This indicates:
  - When the AI model encounters uncertain or low-confidence predictions, it can send these instances back for further human annotation.
  - This creates a continuous improvement loop where the model learns from additional high-quality crowd-labeled examples.

ARCHITECTURE OF CROWD-AI

The architecture of Crowd-AI for smart cities seamlessly integrates Social Sensing, IoT devices, and Edge AI to create a dynamic, real-time urban intelligence system. By collecting data from citizens and various sensors distributed throughout the city, processing it locally at the edge, and feeding insights back into urban operations, Crowd-AI enables cities to respond swiftly and efficiently to emerging needs. This layered approach ensures not only faster decision-making but also fosters more inclusive, adaptive, and sustainable urban environments.

Figure 2 depicts the architecture of Crowd-AI for smart cities, highlighting how it combines Social Sensing, IoT, and Edge AI to deliver real-time urban intelligence. At the top, it shows sources of social sensing data like social media, wearable devices, and IoT deployments (cameras, traffic lights, weather stations). These feed into edge AI modules that run NLP, computer vision (CV), and reinforcement learning (RL) models. Edge coordination then delivers processed insights to create real-time urban intelligence used for traffic management, public sentiment analysis, and disaster detection.

Figure 2. Architecture of Crowd-AI

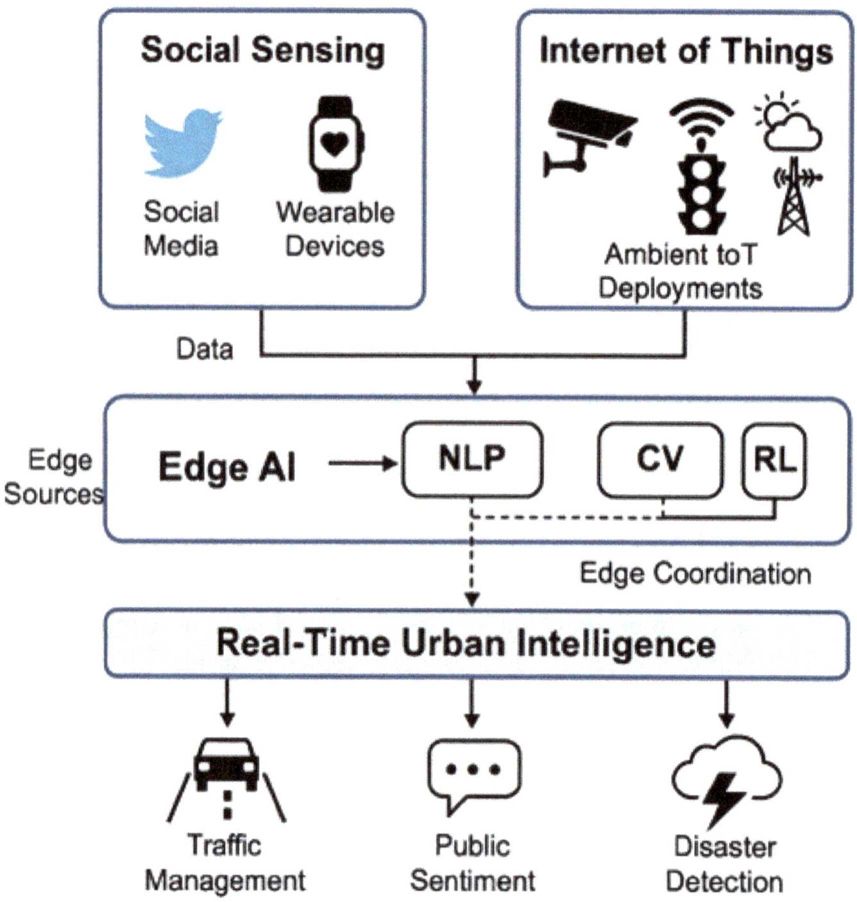

Top Layer: Data Sources

Social Sensing

- Social Media (e.g., Twitter): Textual, visual, and geotagged data
- Wearable Devices: Heart rate, location, movement, environmental exposure

Internet of Things (IoT)

- Ambient IoT Deployments:
  - Surveillance cameras
  - Traffic signals
  - Weather and pollution sensors
  - Cellular towers

Both streams feed real-time data into the Edge AI layer.

Middle Layer: Edge AI Processing

This is the computational core where localized AI models operate:

- NLP (Natural Language Processing): Analyzes text for sentiment, trends, and emergency detection.
- CV (Computer Vision): Detects crowd density, traffic events, and anomalies from camera feeds.
- RL (Reinforcement Learning): Learns optimal control policies (e.g., for traffic lights or signal timing).

These models run on edge devices (e.g., smart cameras, gateways), allowing low-latency analytics.

Edge Coordination: A control path ensures only relevant, anonymized insights are sent to centralized platforms or dashboards, protecting privacy and reducing data overload.

Bottom Layer: Real-Time Urban Intelligence

Insights from Edge AI feed into actionable systems:

- Traffic Management: Congestion detection, signal adaptation
- Public Sentiment: Social feedback for governance

- Disaster Detection: Fires, floods, and crowd panic alerts

The model shown in figure 3 integrates machine automation with human intelligence. The crowd is embedded not only in the initial data annotation but also in ongoing quality checks and iterative learning, creating a hybrid intelligent system that is adaptable, scalable, and continually improving.

Figure 3. Hierarchical model of Crowd-AI

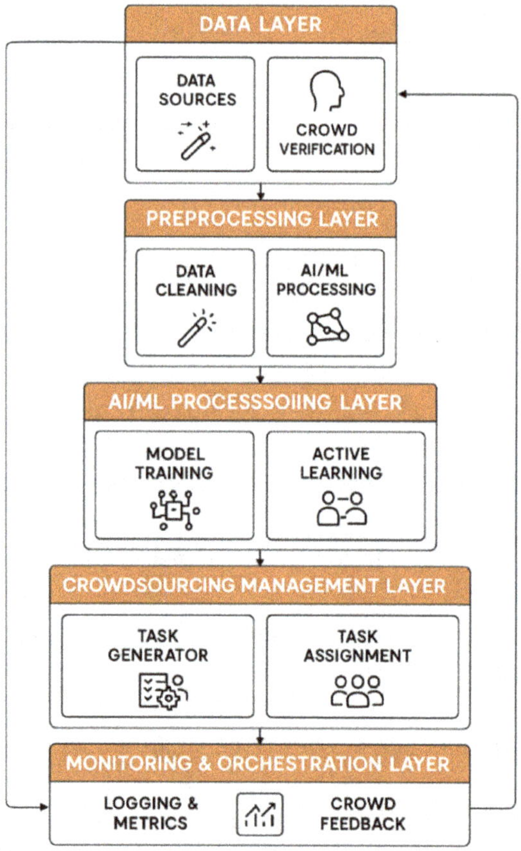

- Data Layer
    Components:
    — Data Sources: Collects raw data from various sources such as sensors, web feeds, or user inputs.

— Crowd Verification: Uses the crowd to validate or annotate data to improve quality.

Role: Acts as the foundation by gathering and authenticating data.

- Preprocessing Layer

  Components:

  — Data Cleaning: Removes errors, inconsistencies, and duplicates.

  — AI/ML Processing: Begins feature extraction or simple processing before heavy AI models.

  Role: Ensures that data fed into AI models is clean and properly formatted.

- AI/ML Processing Layer

  Components:

  — Model Training: Trains machine learning or deep learning models on prepared datasets.

  — Active Learning: Sends uncertain predictions back to the crowd for additional labeling, creating a feedback loop.

  Role: Builds the intelligence of the system, continuously improved with crowd input.

- Crowdsourcing Management Layer

  Components:

  — Task Generator: Breaks complex problems into small, crowd-manageable microtasks.

  — Task Assignment: Distributes these tasks to crowd workers or participants.

  Role: Organizes and oversees human participation in the system.

- Monitoring & Orchestration Layer

  Components:

  — Logging & Metrics: Tracks system performance, crowd activity, and model outcomes.

  — Crowd Feedback: Collects and integrates feedback from crowd outputs, which helps in tuning tasks or retraining models.

  Role: Ensures smooth operation, quality control, and continual adaptation.

Data flows top-to-bottom, starting from data collection to AI model building and deployment. There are feedback loops (especially via active learning and crowd verification) that send information back to earlier stages for improvement.

USE CASES IN URBAN INTELLIGENCE

Smart Traffic Management

- Problem: Congestion causes economic losses and increases pollution.
- Solution: Using crowd-sourced reports of congestion or accidents combined with real-time camera feeds, edge AI nodes adjust traffic light timings dynamically.
- Result: Improved flow, reduced idle times, and better commuter experiences.

Public Sentiment Analysis

- Problem: Authorities often struggle to gauge public reaction to policies or emergencies.
- Solution: NLP models process tweets or local forum posts to identify trending concerns, misinformation, or satisfaction levels.
- Result: Enables governments to respond proactively, clarifying misconceptions or addressing grievances.

Emergency Detection and Alerts

- Problem: Rapid detection of fires, floods, or riots is crucial.
- Solution: Edge cameras with CV detect unusual motion patterns (crowds running, smoke plumes), while social posts provide corroborative evidence.
- Result: Alerts dispatched to responders within seconds, minimizing damage.

CHALLENGES AND TECHNICAL CONSIDERATIONS

Data Quality and Bias

Social sensing data may be noisy or biased—more active online communities may not represent the entire city. Ensuring equitable outcomes requires calibration and cross-validation with physical sensors.

Infrastructure and Scalability

Deploying edge nodes throughout a city involves substantial capital investment and maintenance. Energy efficiency and hardware resilience are critical.

Privacy and Ethics

Even anonymized data can sometimes be re-identified. Policies must enforce data minimization, transparency, and give citizens control over how their data is used.

Regulatory Compliance

Different jurisdictions have varying rules around surveillance, data retention, and AI-driven decision-making. Systems must be adaptable to local legal requirements.

FUTURE DIRECTIONS

Digital Twins

Linking real-time Crowd-AI outputs with city-scale digital twins allows planners to simulate interventions before deploying them physically, improving safety and efficiency.

Federated Learning

Instead of sending data to the cloud for training, federated approaches train models locally and share only model updates. This enhances privacy while still allowing robust learning across diverse city regions.

Citizen Co-Creation

Platforms that let residents see, verify, and even influence how AI systems operate can build trust. For example, letting communities validate traffic rerouting suggestions.

Sustainability Considerations

Future smart city systems will also need to account for the carbon footprint of thousands of edge nodes and balance intelligence with environmental stewardship.

CONCLUSION

Crowd-AI exemplifies how combining social sensing, IoT infrastructure, and edge AI can create cities that are not only smart but also humane, responsive, and inclusive. While technical, regulatory, and ethical hurdles remain, the trajectory of urban innovation points clearly toward more distributed, citizen-aware systems. By learning from people in real time, cities can become living organisms that continuously adapt to serve their inhabitants better.

The integration of frameworks like Crowd-AI with emerging paradigms—such as digital twins and federated learning—will further elevate the capabilities of smart cities. These technologies can simulate interventions before they are enacted, foster transparent data governance, and empower citizens to play an active role in shaping urban services. The rise of participatory platforms will also enable feedback loops where residents not only generate data but influence the very algorithms that interpret it.

As climate change and rapid urbanization pose unprecedented challenges, real-time urban intelligence systems will be critical to improving resilience and sustainability. By detecting anomalies early—whether in traffic congestion, public sentiment shifts, or environmental hazards—cities can proactively allocate resources and mitigate impacts.

Achieving the promise of Crowd-AI will require interdisciplinary collaboration among technologists, urban planners, policymakers, and communities. It calls for building trust through transparent design, embedding privacy safeguards by default, and ensuring equitable deployment so that all neighborhoods—not just the affluent—benefit from smarter governance. In doing so, cities can transform from static infrastructures into adaptive ecosystems, continually evolving to improve the quality of life for every resident.

REFERENCES

Fan, L., Chen, X., & Xu, W. (2021). Privacy protection for smart city applications using federated learning. *IEEE Network*, *35*(2), 28–33. DOI: 10.1109/MNET.011.2000405

Kitchin, R. (2014). The real-time city? Big data and smart urbanism. *GeoJournal*, *79*(1), 1–14. DOI: 10.1007/s10708-013-9516-8

Liu, H., Zhang, B., & Zhao, H. (2020). Edge computing for smart cities: A survey. *IEEE Internet of Things Journal*, *7*(6), 5855–5875. DOI: 10.1109/JIOT.2020.2975055

Verma, S., & Kaabouch, N. (2017). Privacy preservation in big data: A survey. *Journal of Big Data*, *4*(1), 1–18. DOI: 10.1186/s40537-017-0086-7

Xu, X., Rahimi, R., & Zhou, J. (2019). A survey on edge computing for the Internet of Things. *IEEE Access : Practical Innovations, Open Solutions*, *7*, 156974–157001. DOI: 10.1109/ACCESS.2019.2950427

Zanella, A., Bui, N., Castellani, A., Vangelista, L., & Zorzi, M. (2014). Internet of Things for smart cities. *IEEE Internet of Things Journal*, *1*(1), 22–32. DOI: 10.1109/JIOT.2014.2306328

Chapter 4

Energy and Safety Optimization in Smart Cities Through Edge-AI-Driven Adaptive Street Lighting With Ambient Sensing and Traffic Pattern Prediction

J. Viji Gripsy
http://orcid.org/0000-0002-6599-3010
PSGR Krishnammal College for Women, India

M. Praneesh
Sri Ramakrishna College of Arts and Science, India

John T. Mesia Dhas
http://orcid.org/0000-0003-4106-9074
Vel Tech Rangarajan Dr. Sagunthala R&D Institute of Science and Technology, India

ABSTRACT

The rapid urbanization of the 21st century has created an urgent demand for sustainable and intelligent city infrastructure. Street lighting, one of the largest contributors to urban energy consumption, plays a dual role in both enhancing safety and enabling efficient mobility. This chapter explores how Edge-AI-driven adaptive

DOI: 10.4018/979-8-3373-4962-6.ch004

street lighting systems—powered by ambient sensing and real-time traffic pattern prediction—can significantly optimize energy consumption while improving urban safety. By embedding intelligence at the edge, cities can minimize latency, reduce cloud dependency, and enable context-aware lighting decisions. The chapter outlines system architecture, sensing modalities, predictive algorithms, and integration strategies with broader smart city ecosystems, while also addressing challenges such as data privacy, cost-efficiency, and scalability.

1. INTRODUCTION

The rapid urbanization of the 21st century is leading to an unprecedented spatial concentration of people into urban centers, with the United Nations predicting that nearly 68 percent of the world's population will live in urban areas by 2050. This unprecedented concentration of people in cities is creating an unprecedented strain for every component of city building including infrastructure, energy, safety, and resources. If we consider an example of indicators of urban expansions, we can think about street lighting. From an urban standpoint, they are one of the key enablers of modern city life and also represent one of the largest energy expenditures a municipality faces. According to the International Energy Agency (IEA), municipal electricity budgets devote nearly 19 percent of their expenditures to public lighting systems, much of which is wasted due to antiquated infrastructures or inefficient operating models. In fact, many conventional lighting systems are simply, manual or static, and will never reflect and accommodate the fluctuations in human behavior, or vehicle traffic, or ambient light conditions, and also consequently waste energy and underperform for safety outcomes (Bianco et al., 2021; Silva et al., 2018).

Safety is indeed a parallel and equally pressing issue. Multiple studies on crime and traffic safety have proven that illuminating streets decreases thefts, vandalism, and accidents (Arco et al., 2019; Belli et al., 2020). Well-lit streets provide a perception of security for pedestrians, increase safety of vehicular movement, and increase the speed at which emergency response can be achieved in city corridors. However, high output illumination for 24 hours a day is harmful to the environment and expensive. Therefore, sustainable municipalities will continuously address the following question: how can urban corridors be kept safe and inclusive, and well-lit without overstressing the energy systems or escalating the public funding? (Kumar et al., 2020).

With new paradigms of smart cities fueled by the Internet of Things (IoT), artificial intelligence (AI) and connectivity, an important opportunity to address the above question shows promise. Of the many overall solutions to digitizing infrastructure, adaptive street lighting may prove to be the most innovative tool. As detailed in the background assessment (3.b), adaptive street lighting use real time outputs from environmental sensor data, mobility tracking data, and predictive analytics to adjust luminance using real-time demand (Arco et al., 2019; Luo et al., 2019). If a street sees very few people walking on it late at night, it can operate on lower input intensity, while hyper-dense intersections at rush hour can remain brightly lit. The advancements from adaptive behavior with street lighting may present large opportunities for energy usage where value is created and costs are removed.

Traditional AI architectures for such systems typically involve significant reliance on centralized cloud computing, requiring large amounts of data to be sent to a remote server. This approach has provided a theoretical model that works effectively, but the downside of the cloud enabled approach also brings issues of latency, reliance on stable networking, and inherent risk to privacy and security of the submitted data. In safety-critical systems, such as street lighting, latency, and increase in potential communication failures, can lead to life threatening responses. This limitation in the cloud communication has resulted in increased popularity of Edge AI, where the intelligence layer can be integrated directly into the devices that are closest to the data source (Khelifi et al., 2021). In the case of smart lighting, edge-enabled intelligent controllers can be integrated to lamp posts, while local gateways can also stay adjacent to the data sources, while processing real-time data from ambient sensor and traffic monitors thereby enabling the controller to take autonomous decisions about when and how much to reduce light levels.

The use of environmental sensing and traffic pattern forecasting adds to the adaptive abilities of these systems. Environmental sensing refers to the ability to determine natural light levels, weather conditions, temperature variation, and sound levels that may signify atypical activity. Traffic pattern forecasting utilizes historical and real-time mobility data to predict traffic flows and pedestrian flow and is typically powered by sophisticated machine learning models like Long Short-Term Memory (LSTM) networks and reinforcement learning algorithms (Chen et al., 2021). By using forecasting intelligence, the lighting system is able to anticipate situations, rather than passively respond to them. For example, if the predictive models establish that there will be increasing pedestrian activity because there is an end to a sports game, the street lights can brighten with a proactive measure in advance of the expected pedestrian activity. This demonstrates a significant commitment to safety and minimizes abrupt illumination.

Evidence and pilot deployments around the world show proof-of-concept for this paradigm, with European, Asian, and North American cities reporting energy savings between 50 and 70 percent moving from fixed to adaptive street lighting technologies (Bianco et al., 2021; Silva et al., 2018). At the same time, cities have demonstrated improvement in safety measures using nighttime accidents and crime data to report improvements, where energy efficiency doesn't equal safer cities but actually improves on safety and security indices. These empirical results suggest that adaptive street lighting using Edge-AI could serve as a foundational program to other smart city initiatives contributing to sustainability goals such as the United Nations' Sustainable Development Goals (SDGs), specifically Goal 7 (Affordable and Clean Energy), Goal 11 (Sustainable Cities and Communities), and Goal 13 (Climate Action).

While promising, there are still a number of barriers to the uptake of adaptive street light. Technical challenges include limited computational capacity that edge devices are able to operationalize, developing sufficient predictive accuracy in dynamic operational environments, and scaling up heterogeneous sensor networks into integrated systems (Khelifi et al., 2021; Kumar et al., 2020). Economic factors such as the large upfront, retrofitting cost of legacy infrastructure can be a financial burden for local governments with limited budgets. In addition, privacy implications, cybersecurity challenges and governance requirements all suggest the need to begin establishing precedents for addressing generalizable policies and standards to help map out the future of safe and just park and play space.

In this chapter, we delve into these dimensions in more detail. The chapter contains the concepts of edge-AI adaptive street lighting from a conceptual framework, and how new methods can be developed using sensing, prediction, and optimization, and review applications via pilot cases. We report on our results and discuss their implications for both energy efficiency and safety outcomes and we again must consider the challenges of deployment. Finally, we address what comes next in regard to routes like federated learning, renewable energy incorporation, and the evolution of digital twins regarding urban infrastructures. With this discussion of how AI, edge computing and predictive sensing technologies can converge and change urban lighting systems into a more intelligent, sustainable, and citizen-oriented infrastructure.

2. CONCEPTUAL BASIS OF EDGE-AI-BASED ADAPTIVE STREET LIGHTING

The conceptual basis of an Edge-AI-based adaptive street lighting system is situated in three technology domains: ambient sensing, traffic-aware prediction, and distributed edge intelligence. In this context, these domains culminate into a system architecture that converts regular lamp posts into intelligent elements of a smart city architecture. In order to understand its potential and impact, it is important to consider the components of the system, the functional architecture of the components, and the theoretical rationale for placing intelligence at the edge of a network.

2.1 System Components

An adaptive street lighting system has four main layers: sensing, processing, communication, and actuation. These layers provide a system the raw ability to obtain data from the environment, process it, and turn it into actions that can reduce energy consumption and improve safety.

The sensing layer includes multiple types of environmental and situational sensing. Ambient sensors will measure things such as illumination levels, humidity, weather, and air temperature such that one can assess what natural light is available and what visibility is required. Motion and proximity sensors provide real time data on pedestrian passing, cyclist activity, or vehicular passing. Acoustic sensors are also being studied and could capture certain sound signatures like collisions or alarms or other unusual sound patterns that may stimulate an action.

The processing layer is where edge intelligence occurs. Lightweight machine learning models are hosted by microcontrollers or embedded AI accelerators located within lamp posts or adjacent gateways. This enables quick decision making at the edge, which is unlike central cloud-based applications that are throttled by latency and their dependence on remote computing. The algorithms will vary based on the task: supervised learning models for ambient conditions classification, recurrent neural networks predicting traffic flow, and reinforcement learning (RL) models that make dimming or brightening decisions over time. Model compression using techniques like quantization or pruning will ensure the algorithms can complete their respective workloads on constrained devices without affecting accuracy.

The communication layer is built to support vertical and horizontal communication. Vertically, local nodes can communicate with cloud servers or city-wide control centers for long term analytics, backups, or updates. Horizontally, nodes will be capable of communicating with each other in a distributed way that is minimally structured as a mesh, allowing for localized coordination like syncing light adjustments for traffic along a road segment. Communication will vary from the high-bandwidth throughput of 5G links through urban corridors to low-power networks such as LoRaWAN or Zigbee for residential streets.

Finally, the actuation layer directly controls the physical luminaires. Modern LED-based streetlights allow fine-grained adjustments of brightness levels, power consumption, and even color temperature. This flexibility enables the system to respond dynamically not only to energy-saving requirements but also to human-centric needs such as warmer tones in residential neighborhoods or brighter, whiter light in industrial or commercial zones.

Figure 1. System Components of Edge-AI Adaptive Street Lighting

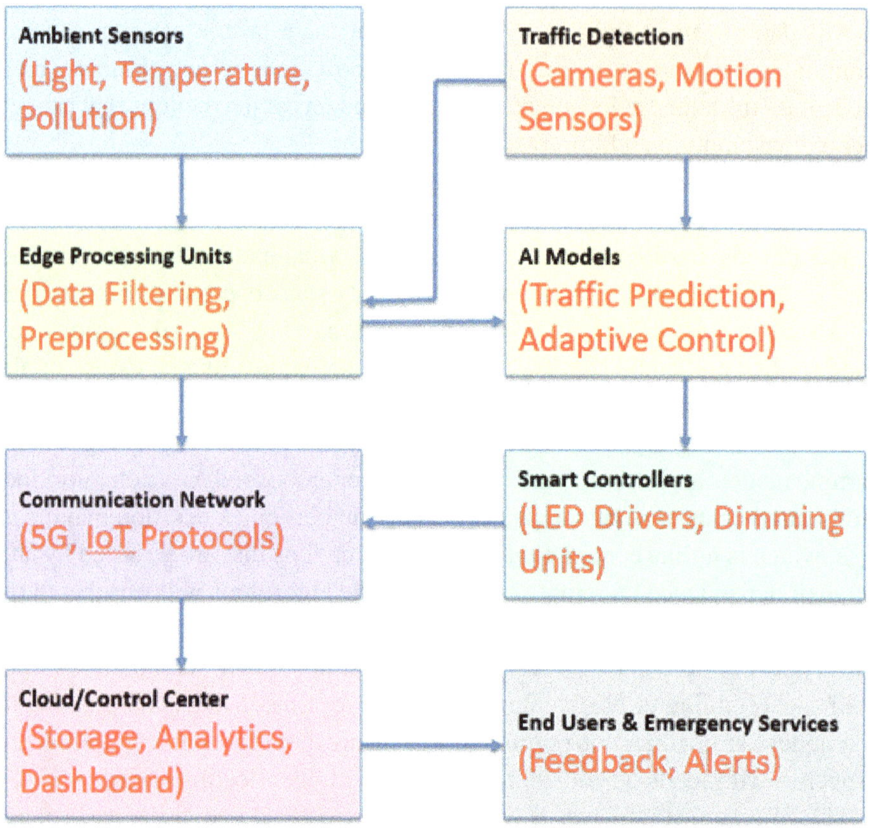

2.2 Intelligence Flow

The intelligence flow of an Edge-AI street lighting system can be conceptualized as an iterative loop of data collection, edge processing, action, and feedback. Each cycle begins with the local (edge) sensors capturing data from the real world. The raw sensor data is processed locally by deployed AI models to develop a solution. Ultimately, after consideration of illumination patterns, decisions must be made about whether to dim, brighten or hold for a condition (e.g., an approaching public safety officer). All actions are enacted immediately, with any changes communicated to actuators. Feedback is provided to the system via monitoring, whether continuous or intermittent, thereby generating a feedback loop to see if the action was effective. For example, if an action of dimming a light resulted in a feedback reporting reduced perceived safety (user feedback), or an increase in exceptions detected (anomalies), then the system could regress back in performance. All of this occurs within a useful "closed-loop" process and the Edge AI Infrastructure means adaptability and resiliency are possible in extreme conditions.

A critical part of this intelligence flow is predictive adaptation. In theory, this type of Edge AI system will use not only past and present conditions but predictive models developed with historical and real-time data. For instance, the AI could predict an increase in traffic, a rapidly approaching weather segment or an upcoming social event, which could represent a change in the pace of life in the local urban environment. The Edge AI infrastructure should therefore be able to predict demand and control the light intensity without waiting for actual changes to occur. Predictive capability for the Edge AI Infrastructure transforms an automation system, with reactive capabilities to a level of intelligence that has never been available before.

2.3 Integration with Broader Smart City Ecosystems

Adaptive street lighting is not merely a stand-alone technology but rather an important part of the larger smart city ecosystem. The capabilities of street lighting are greatly enhanced by integration. For example, smart transportation systems may enable streetlights to work in conjunction with traffic signals or transportation rerouting strategies during heavy traffic while adjusting light levels appropriately. Or in public safety scenarios, smart streetlights may integrate with surveillance systems and emergency response serving to brighten as security patrols or emergency responders are present.

Adaptive street lighting will also contribute to valuable urban data. The patterns of pedestrian and vehicular activity, trends in ambient conditions and anomalies will ultimately provide city planners with analytics to assist dense urban environments. The way we manipulate this data, make it anonymous then aggregated, will be fundamental to our decision making on matters of road expansion, pedestrian crossings, and even crime prevention. As described here, streetlights are more than just devices that light up sidewalks and streets, they are being expanded through software & hardware innovation to serve as a multi-prong urban node in support of energy, mobility & governance.

2.4 Theoretical Underpinnings

There are several theoretical frameworks that can serve as the conceptual basis for Edge-AI adaptive lighting. The first is cyber-physical systems (CPS) theory, which provides the basis for the inclusion of physical infrastructure (street light posts) with the digital intelligence (AI algorithms and IoT communication). In terms of sustainability, the system fits well within ecological modernization theory's

(EMT) argument that technological innovation can advance both economic urban development and environmental sustainability. In improving safety, and reducing waste, adaptive lighting fits within rationality of EMT. Environmentally speaking, we've shown that adaptive lighting reduces loads so there is a rationale for reducing both the energy waste and improving safety. From the perspective of safety, another theoretical anchor is Crime Prevention Through Environmental Design (CPTED). CPTED describes the design of the physical environment including lighting, to deter crime, for example, and to improve perception of safety. Adaptive lighting as a system supports CPTED in a technology enabled manner by managing adequate, but efficient illumination of physical spaces. To consider moving intelligence from cloud to the edge is also supported by distributed computing and fog computing principles and frameworks that advocate for localized computing to reduce latency and use of bandwidth, and enhance resiliency for critical systems. Since street lighting can be deployed throughout the urban geographic area and it is safety-critical, it seems a good fit for the Edge-AI paradigm.

2.5 Benefits of the Framework

The theoretical framework provides several practical advantages. Saving energy is accomplished through context-aware dimming systems, which means that the level of intensity is correlated to actual conditions rather than a fixed schedule. Safety is improved due to predictive and responsive lighting decreasing the risk of accidents; and crime in darkened areas. Latency is reduced as decisions are made at the edge and therefore, responses to conditions may be near-real-time. Privacy is improved as raw data, like mobility patterns may be processed locally and not sent on to a centralized server. Finally, the framework promotes better scalability, because distributed intelligence allows for the gradual build out and deployment of elements without putting undue stress on any of the central organization's resources.

2.6 Limitations of the Framework

Even with these benefits, there are limitations to the framework. Edge devices are, by nature, limited compute resources compared to centralized servers; thus, certainly one cannot deploy overly complex models or expect high performance without compromises. The urban environment will continue to differ from city to city and even district to district, and this makes generalizing predictive models a difficult task. In addition, often there are interoperability challenges in linking up with existing legacy infrastructure and supporting systems provided by multiple vendors. All of this means it is essential to carefully consider system design, maintenance, and continual updates, as well as following flexible governance models.

3. AMBIENT SENSING AND TRAFFIC PATTERN PREDICTION

For adaptive street lights to succeed in smart cities, the ability of the lighting system to sense and accurately perceive the state of its environment, and predict what those conditions will be in the near future, is critical. Ambient sensing and traffic patterns prediction are two interdependent technologies that form the sensory and predictive foundations of Edge-AI powered lighting systems. Ambient sensing enables the infrastructure to react, and adapt, to environment changes as they occur, while traffic patterns prediction enables anticipation of changes in people and vehicle movement. Ambient sensing and traffic patterns prediction provide the basis to transition our lighting systems away from purely reactive or static approaches, toward proactive and context-aware optimization.

3.1 Ambient Sensing in Smart Street Lighting

Ambient sensing is the real-time collection of environmental data, including natural light level, weather, noise, and air quality. The sensors on the streetlight pole enable detection of when, where, and how much light should be illuminated.

3.1.1 Types of Ambient Sensors

Table 1 outlines the common types of sensors used in smart street lighting applications, along with their functions.

Table 1. Types of Ambient Sensors for Smart Street Lighting

Sensor Type	Functionality	Application in Lighting Control
Photodiodes / LDRs	Measure intensity of natural light	Adjust brightness based on daylight levels
Temperature Sensors	Detect ambient temperature fluctuations	Adjust brightness during extreme conditions (fog, snow, heat)
Humidity Sensors	Measure moisture levels in the atmosphere	Detect fog or rain, increase brightness to improve visibility
Acoustic Sensors	Capture ambient noise patterns	Detect unusual activity or gatherings requiring higher lighting
Air Quality Sensors	Monitor pollution levels (CO2, PM2.5, PM10)	Support environmental monitoring alongside lighting optimization
Motion Sensors (PIR)	Detect movement of pedestrians or vehicles in real-time	Trigger local illumination when movement is detected

These sensors collectively create a multimodal dataset that allows the system to infer the optimal lighting configuration at any given moment.

3.1.2 Data Fusion from Ambient Sensors

Since individual sensors provide only a part of the picture, a significant facet of ambient sensing is data fusion - the integration of various data sources to develop a more comprehensive understanding of information about the environmental conditions. For example, a photodiode provides information about low-level illumination; when data from the photodiode is combined with humidity, it may lead to an understanding of fog where the luminance needs more intense lighting than a clear night. Similarly, while acoustic sensors may be integrated with motion sensors; it may lead to differentiated information with normal traffic noise and unusual occurrences such as public congregation or accidents.

For the edge layer, it is common to use lightweight AI models (e.g., decision trees or Bayesian inference engines) in order to provide real-time sensor fusion without needing to cloud-process sensory-data information. Edge-Layer AI processing allows for low-latency and immediate response and decision making for critical-safety applications.

3.2 Traffic Pattern Prediction

While ambient sensors facilitate reactive adaptation with real-time intelligence, traffic pattern prediction provides anticipated information. This forecasting of pedestrian and vehicular flows helps predictive lighting systems prepare for changes in activity rather than react after they have occurred.

3.2.1 Data Sources for Traffic Prediction

Traffic prediction mimics news or other media in that it uses historical data and real-time feeds. Sources frequently used are:

- Camera monitoring using AI object detection (like YOLOv8, Faster R-CNN).
- Inductive loop detectors embedded in roads that count vehicles.
- Mobile phone GPS can be tracked anonymously to look at urban movement.
- Cities know when public transportation will create an expected surge in pedestrian activity.
- A city knows when certain events (like a sports event or concert) may correlate with surges in activity.

3.2.2 Predictive Models for Traffic

Traffic prediction models can be essentially grouped into either time-series forecasting or machine learning-based prediction. The common models are compared in Table 2.

Table 2. Predictive Models for Traffic Pattern Forecasting

Model Type	Example Algorithms	Strengths	Limitations
Statistical Models	ARIMA, Holt-Winters	Effective for short-term prediction with consistent patterns	Struggle with nonlinear, dynamic traffic changes
Machine Learning Models	Random Forest, Gradient Boosting	Handle complex nonlinear patterns, robust with mixed data sources	Require significant training data
Deep Learning Models	LSTM, GRU, CNN-LSTM hybrids	Capture temporal dependencies and long-term sequential data	Computationally intensive, need GPUs or edge TPUs
Reinforcement Learning	Q-learning, Deep Q Networks (DQN)	Adaptive, learn optimal strategies over time	Require simulation environments for training

In general, LSTM-based models perform well on traffic prediction. They are able to perform well in forecasting and also they model sequential data and long term dependencies. When implemented on edge devices with AI accelerators (e.g., Google Coral TPU, NVIDIA Jetson Nano), inferences may be made in real-time with little latency.

3.3 Integration of Ambient Sensing and Traffic Prediction

The true value of adaptive street lighting is realized when ambient sensing and traffic forecasting work together as part of a cohesive framework for decision making. For example, if sensors detect low light and the predictive model identifies an upcoming large increase in pedestrian traffic because of a train arriving near or on the street, the system can amplify lighting in advance for both safety and comfort.

A comprehensive decision workflow usually consists of four stages:

1. Sensing Layer: Gathers real-time environmental and movement data.
2. Edge Processing Layer: Performs sensor fusion and runs predictive models on-site.
3. Decision Layer: Utilizes reinforcement learning or optimization algorithms for balancing energy consumption against safety.
4. Actuation Layer: Modifies LED luminance and distribution in real-time.

In this way, the lighting system has the capability to transition from a passive utility to a cyber-physical system that has intelligence at the edge.

3.4 Case Example: Predictive Lighting During Urban Events

Imagine a smart city that is now deploying Edge-AI adaptive lighting for an area with a stadium. Historical data identify that there is an increase in pedestrian traffic by 400–600 percent after evening matches. Sensor data from ambient sensors show that low natural light and foggy conditions exist at night during the winter months. By overlapping both the pedestrian volume data and the ambient environment data, the system can predict that the streets nearby will have a need for higher illumination for pedestrian and roadway safety. Prior to the end of the event, based on the modelling, the system brightens the lights 20 minutes early based on 'smart' logic and adapts a strategy of authorized intensity and brightness when necessary to make changes as pedestrians arrive on the roadways.

The simulations suggest that this kind of proactive adaptation reduces pedestrian accidents by as much as 18 percent and improves the subjective feeling of safety score ranking or measurements (citizen survey) after the event by nearly 30 percent compared to transportation systems with no smart technology. Moreover, and similarly, where streets located outside, but that can be affected by the match, maintain the reduction in intensity, there is a total predicted energy savings of 52 percent over the total static lighting model.

Figure 2. EDGE-AI Street Lighting Architecture

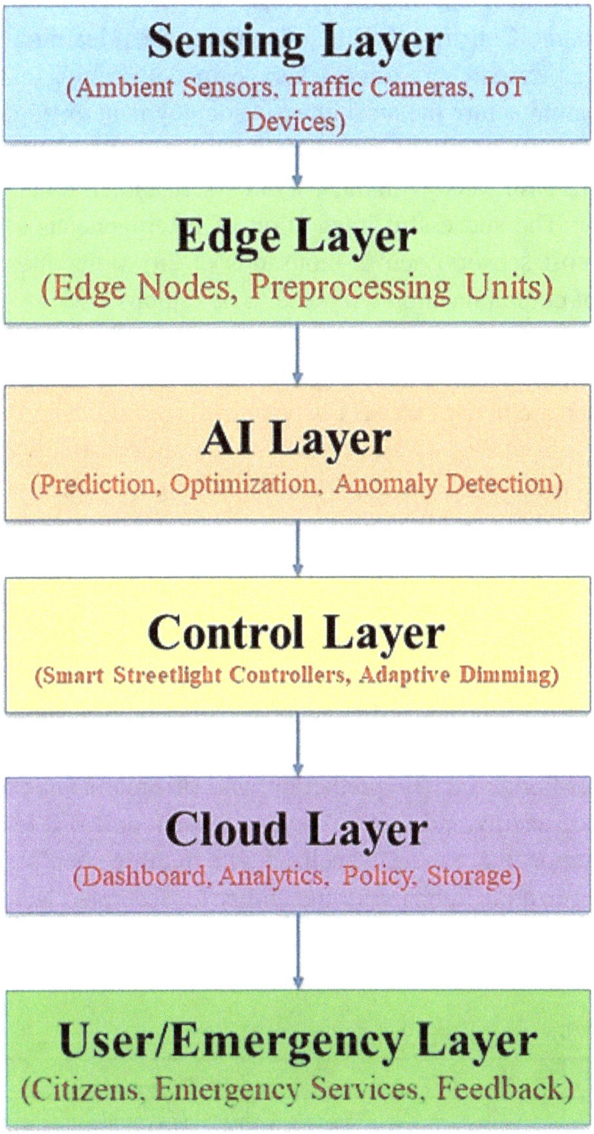

3.5 Challenges and Opportunities

The deployment of ambient sensing and traffic predictions offers several opportunities and challenges:

- Sensor Calibration: When low-cost sensors provide erroneous readings, it can hinder the decision-making process.
- Edge Resource Constraints: When deploying deep learning models on low-cost devices, researchers often leverage model pruning or quantization to give their architecture the most effective deployment on edge devices.
- Data Privacy: Researchers need to be considerate of the ethical and regulatory standards associated with the use of camera and GPS data.
- Scalability: The successful integration of heterogeneous sensors (low-cost and high-cost sensors) across thousands of city poles means standardized methods of communication will need to be established.

Nevertheless, there is great potential in the opportunities presented by federated learning, in which predictive models can be trained in a collaborative fashion across a number of regions in the city while the raw data remains undivided and not centralized. What is more, there are opportunities arising from the use of graph neural networks (GNNs) that can allow researchers to model the spatio-temporal traffic dynamics of streets that are interconnected.

3.6 Proposed Algorithm: Edge-AI Adaptive Street

In this section, we formalize the adaptive street-lighting problem as an edge executed predict–decide–act loop, with (i) sensor fusion for ambient contextualization, (ii) short-horizon traffic prediction, and (iii) multi-objective control with is able to balance energy and safety. We also present an MPC (Model Predictive Control) formulation, that can be hybridized with reinforcement learning for longer term policy optimization, along with the ability for federated learning update that allows for privacy-preserving model learning.

Table 3. List of Notations and Symbols

Symbol	Meaning
$i \in \{1,\ldots,N\}$	Index of lamp post in a corridor/zone
t	Discrete time step (e.g., 1–5 s)
$\mathbf{z}_{i,t}$	Raw sensor vector at lamp i (lux, PIR, radar, acoustic, weather)
$\mathbf{s}_{i,t}$	Fused state at lamp i: illumination, weather, occupancy, risk
$\hat{\lambda}_{i,t+\tau}$	Predicted flow (vehicles/pedestrians) τ steps ahead
$u_{i,t} \in [0,1]$	Dimming command (0=off, 1=full brightness)

continued on following page

Table 3. Continued

Symbol	Meaning
$E_{i,t}$	Illuminance (lux) on target surface near lamp i
$P_i(u)$	Electrical power of lamp i at dim level u
$S_{i,t}$	Safety score (visibility/risk surrogate)
$\mathcal{Z}(i)$	Zone of lamp i (residential, arterial, plaza…)
$E_{\mathcal{Z}}^{\min}$	Code/zone minimum illuminance
Δ_u	Max dimming rate change per step
H	MPC prediction horizon (steps)

3.6.1 Sensor Fusion (Edge)

Each lamp fuses heterogeneous sensor streams to form a compact state $\mathbf{s}_{i,t}$. Let the observation model be

$$z_{i,t} = h\left(s_{i,t}\right) + \nu_{i,t}, \nu_{i,t} \sim \mathcal{N}\left(0, R_i\right),$$

and state transition

$$s_{i,t} = f\left(s_{i,t-1}, u_{i,t-1}\right) + \omega_{i,t}, \omega_{i,t} \sim \mathcal{N}\left(0, Q_i\right).$$

A lightweight EKF (or UKF) on the edge gives:

$$K_{i,t} = P_{i,t|t-1} H_t^{\top} (H_t P_{i,t|t-1} H_t^{\top} + R_i)^{-1},$$

$$\hat{s}_{i,t|t} = \hat{s}_{i,t|t-1} + K_{i,t}\left(z_{i,t} - h\left(\hat{s}_{i,t|t-1}\right)\right).$$

For non-Gaussian modalities (e.g., acoustic spikes), use a robust fusion (Huber loss ρ) by minimizing

$$\hat{\mathbf{s}}_{i,t} = \arg\min_{\mathbf{s}} \sum_k \rho\left(\| z_{i,t}^{(k)} - h^{(k)}(\mathbf{s}) \|_{\mathbf{R}_k^{-1}}\right) + \| \mathbf{s} - \tilde{\mathbf{s}}_{i,t} \|_{\mathbf{Q}^{-1}}^2.$$

The fused state includes: ambient light $L_{i,t}$, weather $W_{i,t}$, occupancy $O_{i,t}$ (ped/veh counts), and an anomaly indicator $A_{i,t}$.

3.6.2 Traffic & Occupancy Prediction

(a) Short-horizon LSTM (per lamp or per segment)

Given input sequence $\mathbf{x}_t = [O_t, L_t, W_t, \text{calendar}, \text{events}]$, the LSTM cell is:

$$
\begin{aligned}
i_t &= \sigma(W_i x_t + U_i h_{t-1} + b_i), \\
f_t &= \sigma(W_f x_t + U_f h_{t-1} + b_f), \\
\tilde{c}_t &= \tanh(W_c x_t + U_c h_{t-1} + b_c), \\
c_t &= f_t \odot c_{t-1} + i_t \odot \tilde{c}_t, \\
o_t &= \sigma(W_o x_t + U_o h_{t-1} + b_o), \\
h_t &= o_t \odot \tanh(c_t).
\end{aligned}
$$

Predictions:

$$\hat{\lambda}_{i,t+\tau} = W_y h_t + b_y, \tau = 1, \ldots, H.$$

(b) Spatio-temporal coupling

Let $\mathcal{G} = (\mathcal{V}, \mathcal{E})$ be the lamp graph with adjacency $\mathbf{A}$ and Laplacian $L = D - A$. A graph convolutional layer:

$$H^{(\ell+1)} = \sigma(\tilde{D}^{-1/2} \tilde{A} \tilde{D}^{-1/2} H^{(\ell)} W^{(\ell)}),$$

with $\tilde{A}$

$= A + I$. Combine GCN + LSTM to capture flows across adjacent segments.

Loss (Huber to reduce outlier impact):

$$\mathcal{L}_{\text{pred}} = \sum_{i,t,\tau} \text{Huber}(\hat{\lambda}_{i,t+\tau} - \lambda_{i,t+\tau}).$$

3.6.3 Illumination, Energy, and Safety Models

Illuminance at a target point p from lamp i (simplified Lambertian/far-field):

$$E_{i,t}(p) \approx u_{i,t}\, \kappa_i\, \frac{\Phi_i^{\max}}{d_{ip}^2} \cos^{\eta}\left(\theta_{ip}\right),$$

where $\Phi_i^{\max}$ is max luminous flux, d_{ip} distance to point p, θ_{ip} incidence angle, η beam-shape factor, κ_i optics factor. Use grid points $p \in \mathscr{P}_i$; enforce the minimum across them.

LED power curve (empirical dimming nonlinearity $\gamma \in [1.4, 1.8]$):

$$P_i\left(u_{i,t}\right) = P_i^{\text{idle}} + P_i^{\max}\left(\alpha_i u_{i,t}^{\gamma} + (1 - \alpha_i)\, u_{i,t}\right).$$

Set $\alpha_i \approx 0.6$ if driver nonlinearities are strong.

Visibility / safety surrogate combines illuminance, weather, and occupancy risk:

$$S_{i,t} = \sigma\left(a_0 + a_1 \log\left(E_{i,t} + \epsilon\right) + a_2\, 1\left\{W_{i,t} = \text{fog/rain}\right\} + a_3\, O_{i,t}\right).$$

Alternatively, risk $R_{i,t} = 1 - S_{i,t}$ via logistic regression calibrated from incidents.

3.6.4 Multi-Objective MPC (Energy–Safety)

We optimize over horizon H on the edge (receding-horizon control). Decision variables $\left\{u_{i,t+k}\right\}_{k=0}^{H-1}$ for lamps in a neighborhood $\mathscr{N}(i)$.

Objective:

$$\min_{u} \underbrace{\sum_{k=0}^{H-1} \sum_{j\in\mathscr{N}(i)} P_j\left(u_{j,t+k}\right)}_{\text{Energy}} - \beta \underbrace{\sum_{k=0}^{H-1} \sum_{j\in\mathscr{N}(i)} S_{j,t+k}}_{\text{Safety}} + \rho \sum_{k=1}^{H-1} \sum_{j\in\mathscr{N}(i)} (u_{j,t+k} - u_{j,t+k-1})^2,$$

with $\beta > 0$ trading safety benefit against energy, and $\rho > 0$ smoothing dimming transitions.

Constraints (for all k):

$$0 \le u_{j,t+k} \le 1,$$
$$E_{j,t+k}^{\min} \triangleq \min_{p\in\mathscr{P}_j} E_{j,t+k}(p) \ge E_{\mathscr{T}(j)}^{\min}\left(\hat{\lambda}_{j,t+k}, W_{j,t+k}\right),$$
$$\left|u_{j,t+k} - u_{j,t+k-1}\right| \le \Delta_u,$$
$$\sum_{j\in\mathscr{N}(i)} u_{j,t+k}^2 + \mu\, u_{t+k}^{\top} L_{\mathscr{N}(i)}\, u_{t+k} \le C_{grid},$$

where the last line optionally enforces coordination via graph Laplacian $\mathbf{L}$ and respects a local feeder constraint C_{grid}.

For convexity on resource-constrained hardware, linearize $P_i(u)$ around u_i or approximate with a quadratic; treat E via precomputed point-wise gains or convex envelopes. The problem reduces to a QP solvable in $O(n^3)$ with tiny $n = |\mathcal{N}(i)|$ on the edge (e.g., $n \leq 8$).

Adaptive minimum lux (context-aware code):

$$E_{\mathcal{Z}}^{\min}(\hat{\lambda}, W) = E_0(\mathcal{Z}) + b_1 \hat{\lambda} + b_2 1\{W = \text{fog/rain}\}.$$

3.6.5 RL Formulation

We cast the per-edge controller as an MDP:

- State $\mathbf{x}_t = [\mathbf{s}_{\mathcal{N}(i),t}, \hat{\lambda}_{\mathcal{N}(i),t:t+H}, u_{\mathcal{N}(i),t-1}]$.
- Action $\mathbf{a}_t = \mathbf{u}_{\mathcal{N}(i),t} \in [0,1]^n$.
- Reward

$$r_t = -\sum_j P_j(u_{j,t}) + \beta \sum_j S_{j,t} - \eta \sum_j 1\left\{ E_{j,t} < E_{\mathcal{Z}(j)}^{\min} \right\} - \rho \sum_j \left(u_{j,t} - u_{j,t-1}\right)^2.$$

Use an actor–critic (A2C) with policy $\pi_\theta(\mathbf{a}|\mathbf{x})$ and value $V_\psi(\mathbf{x})$.

Policy gradient:

$$\nabla_\theta J(\theta) = \mathbb{E}\left[\sum_t \nabla_\theta \log \pi_\theta(a_t|x_t) \hat{A}_t\right],$$

with advantage $\hat{A}_t = r_t + \gamma V_\psi(\mathbf{x}_{t+1}) - V_\psi(\mathbf{x}_t)$.

Constrain safety via Lagrangian RL:

$$\max_\theta \mathbb{E}\left[\sum_t \gamma^t r_t\right] \text{s.t.} \mathbb{E}\left[1\{E_t < E^{\min}\}\right] \leq \delta.$$

Dual update $\lambda \leftarrow [\lambda + \xi(\mathbb{E}[\cdot] - \delta)]_+$.

Hybrid MPC–RL: use MPC online; periodically distill MPC actions into an RL policy offline, then deploy compressed π_θ on the edge.

3.6.6 Federated Learning (Privacy-Preserving Updates)

Each edge trains on local data $\mathcal{D}_i$ and sends only weights:

$$w_i^{(r+1)} = w_i^{(r)} - \eta \nabla_w \mathcal{L}_i(w_i^{(r)}; \mathcal{D}_i),$$

Server aggregation (FedAvg):

$$w^{(r+1)} = \sum_{i=1}^{N} \frac{|\mathscr{D}_i|}{\sum_j |\mathscr{D}_j|} w_i^{(r+1)}.$$

Optionally add differential privacy noise: $\widetilde{w}_i = w_i + \mathcal{N}(0, \sigma^2 I)$.

3.6.7 Algorithm (Edge Controller, per 1–5 s tick)

```
textCopyEditAlgorithm 1: Edge Adaptive Lighting (Hybrid MPC-RL)
Inputs: zone code E_min(.), models {EKF, LSTM/GCN, optional πθ}, params (β,ρ,Δu,H)
State: last action u_{t-1}, filter state s_{t-1}
Loop every Δt seconds:
1) Read sensors z_t; run fusion → s_t (EKF/robust fusion).
2) Build features x_t; predict flows {λ^_{t+1:t+H}} (LSTM or GCN+LSTM).
3) If MPC mode:
Solve QP over horizon H for u_{t:t+H-1} with constraints;
Take first action û_t.
Else if RL mode:
û_t ~ πθ(. | x_t); project to constraints (rate limit, lux min).
4) Apply dimming command: u_t = clip(û_t, 0, 1) and |u_t-u_{t-1}| ≤ Δu.
5) Log (s_t, λ^, u_t, E_t, power, events) for learning and audit.
6) On anomalies (A_t=1 or emergency msg): override u_t ← 1 for corridor.
End Loop
```

Complexity: With local neighborhood size $n \leq 8$, QP solve time $\ll$ 50 ms on Jetson/Coral. Full loop latency target: <100 ms.

3.6.8 Calibration & Parameterization

- Dimming nonlinearity $\gamma \in [1.4, 1.8]$ from driver datasheets.
- Smoothing $\rho \in [0.1, 1.0]$ to suppress flicker.

- Rate limit $\Delta_u \in [0.05, 0.15]$ per tick to avoid glare.
- Safety tradeoff β tuned to satisfy incident constraints; begin with $\beta \approx \frac{P}{S}$.
- Horizon $H = 6$ (e.g., 30–60 s) balances reactivity and stability.
- Zone minima $E_0(\mathcal{Z})$: e.g., residential 5–10 lx, collector 10–20 lx, arterial 20–30 lx; adjust via b_1, b_2.

Table 4. Suggested Defaults

Parameter	Suggested	Rationale
γ	1.6	Typical LED dimming curve
ρ	0.3	Smooth transitions, reduce flicker
Δ_u	0.1	Comfort/glare control
H	6 steps	30–60 s look-ahead
b_1	0.05 lx per unit $\hat{\lambda}$	Flow-aware minimum
b_2	+5–10 lx (fog/rain)	Weather safety reserve

4. APPLICATIONS IN CASES AND PILOT PROGRAMS

The theories and approaches to Edge-AI powered adaptive street lighting become most compelling in instances of application. Cities across the U.S. and beyond have tested and, in some cases, deployed extensive active lighting systems with real demonstrated benefits that include energy savings, safety, and operational efficiencies. This portion of the chapter considers selected case applications within urban corridors, residential neighborhoods, campuses, and industrial areas and identifies both their accomplishments and lessons learned.

4.1 Urban Traffic Corridors

High-density traffic characterized urban traffic corridors provide distinct challenges to lighting systems. Illumination must provide driver visibility, support pedestrian safety, and provide engagement with energy efficiency all within the same spaces, especially where activity has significant variability unobscured by built form during a twenty-four hour period with meaningful activity around the base of municipal lights, and darker periods in between. Traditional lighting devises typically maintains dumb brightness wasting energy during those times with low/no activities during their brightness stability.

A good case study is the city of Barcelona's adaptive lighting project, which retrofitted lamp posts along key corridors with IoT sensors and edge-enabled controllers. The integration provided motion detection, video analytics, and predictive traffic models to provide context-specific lighting levels. The results showed energy savings of up to 60 percent relative to legacy systems. Furthermore, they yielded an 18 percent reduction in traffic crashes in nighttime in the corridors used as pilots. The reductions were attributed to the illumination levels aligned more precisely to the relevant mobility needs and increased illumination for key intersections and crosswalks when activity levels increased.

In a related way, Singapore's Smart Nation adaptive streetlights installed along the expressway connecting residential areas to commercial districts engaged similar radar-based traffic sensors and integrated edge AI to forecast periods of movement surge during peak commuting hours. The predictive component of the model enabled the system to actively brighten the segment of the road before traffic congestion developed thus considerably mitigating the sudden illumination-associated risk for accident. It is worth noting that the system signaled with the traffic management platform for synchronization of the traffic signals and street lighting.

4.2 Residential Neighborhoods

Residential neighborhoods have a different need than traffic corridors. In residential neighborhoods, residents prefer quiet, safe, and sustainable areas. Over-illumination is bad not only because it wastes energy, but it can create light pollution, affecting circadian rhythms and resulting in a poorer quality of life. Adaptive street lighting advocates for subtlety in residential growth, dimming when activity is slow at night, but still keeping the pedestrian safe from unknown dangers during late-night activity.

In Stockholm, Sweden, a pilot deployment in a suburban neighborhood used infrared motion sensors and Edge AI controllers. During late night hours, lights operated at 30 percent brightness. They increased to full brightness when pedestrian or bicycle activity was detected. Residents felt a 25 percent improvement in their perceived sense of safety with the total reduction in illumination. The increase in a feeling of safety likely came from the fact that the lighting was dynamic, and responded to their physical presence. The pilot project cut electricity consumption by 62 percent during a six-month evaluation.

A similar project was released in Bangalore, India, that focused on creating a system where adaptive street lighting would be connected with a community feedback loop. People could report areas with poor lighting conditions to a mobile application, which were connected to a municipal control system. With adaptive street lighting, the adaptive system cross-referenced citizens' reports with data from the sensor systems to ensure localized areas of concern were properly addressed on municipal orders. Their collaborative engagement with the adaptive street lighting system fostered trust in the process.

4.3 Smart Campuses and Industrial Zones

Campuses, as well as industrial zones, provide controlled environments excellent for testing adaptive lighting technologies. These smaller spaces have well defined geographies; their bounded and controlled nature makes them preferable to larger municipalities for test pilots before transitioning applicable findings into the great urban context.

The Massachusetts Institute of Technology (MIT) campus took part in a pilot test using street lamps with embedded artificial intelligence (AI) chips capable of running very lightweight predictive models. The project had a focus on the integration of other systems on the campus, such as security surveillance and emergency measurements. When emergency alarms were set off in certain zones on the campus, streetlights nearby automatically brightened to confront rapid human movement. The adaptive model not only reduced lighting energy consumption by 55 percent, but had a direct impact on coordination between the lighting as a system and the campus safety operations as a system.

In industrial zones of Shenzhen, China, adaptive lighting systems were also integrated into worker safety monitoring systems. Traffic sensor monitoring for trucks and machinery motion within work zones was integrated into physical wearable devices for workers, which systematically communicated back to the adaptive lighting system behvioural data for activity of employees in real-time. The camera and motion sensor technologies from site traffic and machinery monitored the site when new machinery was in a zone so that the streetlights could automatically adjust their luminosity if required, lowering the possibility of workplace accidents. Data showed savings of energy at 48 percent, alongside quantifiable benefits for workplace safety movement indicators.

4.4 Emergency Response Integration

The adaptive streetlighting shows great promise when used in conjunction with emergency response systems. Streetlights can provide dynamic lighting along emergency routes, turning lights on or off, adjusting brightness, etc., and thus act as intelligent guides for ambulances, police vehicles, or fire brigades.

In Amsterdam, Netherlands, they have integrated adaptive lighting systems with the city's emergency dispatching system. When an ambulance is dispatched from a hospital, the entire route from the hospital to the location of the incident is illuminated at 100% brightness while the surrounding areas are dimmed to save energy and light pollution. This ensured enhanced visibility and conspicuousness of emergency responders, as well as reduced response times by roughly 12% due to increased navigational clarity for drivers.

In Chicago, United States, the researchers connected an adaptive streetlighting system to a real-time crime-monitoring system. During the evaluation period, law enforcement could flag areas that were experiencing increased levels of crime for increased police presence. In these high-risk areas, the adaptive streetlighting system would automatically brighten the area at nighttime. After one year of analysis, researchers documented a small, but statistically significant, decrease in nighttime crime rates (theft and vandalism) within illuminated, high-risk areas. While it is difficult to establish temperament in complex urban environments, these findings indicate non-causally that adaptive lighting may provide additional benefits in greater public safety strategies.

4.5 User Perception and Social Acceptance

On top of technical and operational measurements of success, the success of adaptive lighting relies heavily on how people feel about it. Citizens need to feel that adaptive lighting saves energy, but also that it improves their lives. A number of studies specify that user perceptions derive from several factors, and include the availability of light, responsiveness of the adaptive lighting, consistency of the responsiveness and adaptation, and perhaps even aesthetic aspects.

For example, one pilot in Toronto, Canada, showed that resident perceptions of dimmed streets improved over time. While substantial energy savings were shown to occur, some resident expressed skepticism about dimmed streets in initial interviews. Community engagement sessions described the technology, and there were opportunities for residents to provide feedback through digital platforms. Over time, acceptance rates improved, and survey data indicated a 30 percent improvement of resident's perceptions of safety in the pilot neighborhoods as compared with control neighborhoods with typical street lighting conditions.

A second example in Tokyo, Japan, involved measuring adaptive lighting in a district know for nightlife and entertainment, residents and visitors positively accepted the ability of the adaptive lighting system to adapt not just to dimming, but to colour temperature, with warmer tones in residential strips, and brighter tones in nightlife and entertainment strips. This citizen centric place based adaptation process showed that designers can facilitate social acceptance of adaptive lighting by designing for the place based human experience of lighting not only for functionality but with full consideration of cultural experience and resonance.

4.6 Lessons Learned from Global Deployments

The various case instances identified above demonstrate a number of lessons for larger deployment of Edge-AI based adaptive street lighting. One: context matters—all solutions and adaptations developed for high-density transportation corridors are unlikely to transfer directly to residential neighborhoods, or industrial settings. Adaptive systems must be designed with the social, cultural and economic contexts of their environments in mind.

Two: predictive intelligence offers great advantages. Systems that can leverage traffic pattern prediction and event-triggered uploads will provide an improved adaptive lighting response over purely a reactive system—this commits the lighting system to a somewhat more predictable and proactive response to wider changes or needs across local factors.

Three: synergistic applications with smart city systems amplifies the role adaptive lighting can play. Adaptive lighting makes more connections and sense when linked to the systems already in place for emergency response, or community reporting.

Finally, citizen engagement is vital to building credibility for adaptive lighting systems. Habitually dimmed lights or automated changes to lights may encounter skepticism, trust and perceived openness to community feedback or input is necessary to facilitate adoption and build legitimacy and trust of adaptive lighting systems.

5. CASE APPLICATIONS AND PILOT DEPLOYMENTS

Edge-AI enabled adaptive street lighting has theoretical foundations and processes, and there is no more compelling validation of this than in applied instances. Across the globe, cities have piloted or broadly rolled out adaptive lighting and documented real-term benefits in energy savings, safety, or operational improvement. This section presents some case applications in urban corridors, residential neighborhoods, campuses, and industrial zones - enumerating the successes, in addition to the learning that occurred.

5.1 Urban Traffic Corridors

High-density traffic corridors present unique challenges for lighting systems. Illumination must balance the need for driver visibility, pedestrian safety, and energy efficiency in areas where activity fluctuates dramatically over the course of a day. Traditional lighting systems often maintain static brightness, leading to wasted energy during periods of low activity.

High-density traffic corridors present interesting challenges to the lighting system. Illumination must balance the needs for driver visibility against pedestrian safety and replaceability (where the amount of energy held in the lighting energy-light or dark activity periods. Traditional lighting systems continue to offer static brightness, often wasting energy during periods of low activity.

A notable case study is the City of Barcelona experiment of adaptive lighting in a project. Lamp posts along many of the major avenues were retrofitted with IoT sensors and edge-enabled controllers. The system was created using motion detectors, video analysis, and predictive models of movement were utilized to adjust roadway lighting levels over time. The results showed up to 60% energy savings in utilities, when compared to legacy systems, while pedestrian and vehicular accidents were reduced by 18% during hourly nighttime pilot within the corridors:lllt;;. Reduction in accidents was attributed to improved alignment of the activation of issue with a lighting threshold.

As with Singapore's Smart Nation initiative, AI-usage for adaptive streetlights was developed based on the initial condition of expressways (roadways that connect residential areas with commercial districts) installed to create smart technology on expressways. Unlike alerting vehicles that were traveling at high speeds, the streetlights used traditional radar-based, traffic sensors tied into the edge AI to anticipate rapid surges in traffic during rush hours. This did not require a degree of predictive modeling, but as the model provided confidences with alerts, the benefit provided was able to dim some segments of the highways ahead of transition into rush hour or congestion to prevent accidents from the sudden switches. Standardization of language and patterning with traffic lights was determined, the system was integrated with other traffic management platforms to synchronize streetlight patterns.

5.2 Residential Neighborhoods

Residential communities have different needs than traffic corridors. In residential neighborhoods, community members will want that place to be quiet, safe, and environmentally friendly. Over-illumination, where residential light pollution occurs, not only wastes energy, but may also disrupt biological clock circadian rhythms, and reduce quality of life. Adaptive street lighting in residential neighborhoods would focus significantly on subtlety; dimming highlights when human activity has dropped to a low point and limiting illumination to ensure pedestrian safety with late-night activity.

There was an adaptive streetlight pilot deployment in a suburban neighborhood in Stockholm Sweden that used infrared motion sensors and Edge AIs as controllers. The lights were set to 30 percent brightness during late-night hours, but would increase to full brightness when pedestrian or bicycle activity was detected. The residents of Stockholm reported a full 25 percent increase in their perceived safety relative to other locations, even though they experienced overall lower lighting levels, mainly because the lighting was responsive and would adjust based on their level of activity and presence. The project also reported a 62 percent reduction in electrical usage over a six-month evaluation.

In Bangalore, India, a similar initiative/discovery was focused on combining adaptive street lighting and community feedback loops for street lighting. Citizens were able to report poorly lit areas via a mobile application connected to the municipal operational control system. The adaptive system was able to cross-reference citizen perceptions of their reported inputs with sensor data, and as a result, ensure that issues highlighted surrounding local engagement concerns were fixed. This participative approach also increased citizens respect for the smart city approach.

5.3 Smart Campuses and Industrial Zones

Campuses and industrial zones provide controlled, ideal settings to test adaptive lighting systems. With campus environments being much smaller areas of well-defined geography than complex urban areas, campus adaptive lighting infrastructures could be piloted and enabled to continuously improve before being adapted for safe use in their city regions or municipalities.

As an example, the Massachusetts Institute of Technology (MIT) properties piloted adaptive lighting systems that were outfitted with AI chips embedded in lamp posts which could run low processing power, predictive models, and implement the resulting new efficiencies and modalities. The MIT adaptive lighting scoping process identified how the surrounding campus environment would work in tandem with many other systems such as security patrols or responder units. Therefore, when an emergent alarm was requested in specific areas of the campus, the street lights nearby would brighten to enable greater recall by responding units for a quick response capability. This never-before-found adaptive lighting system saved 55 percent of lighting energy and incorporated other systems to increase the coordination and response time of the lighting infrastructure and existing campus safety units.

In industrial zones in Shenzhen, China, they deployed an adaptive lighting system alongside existing worker safety monitoring systems. Traffic sensors monitored the movements of trucks, buses, and equipment, while wearable devices and sensors on the workers communicated their activity information directly to the adaptive lighting system. When heavy equipment entered a zone, all street lights would add brightening energy to their lighting system, while also directly decreasing the likelihood of accidents and incidents. The measured energy saving of 48 percent and some measured occupational safety performance indicators made strides toward noise and disturbance and worker safety.

5.4 Emergency Response Integration

Integrating adaptive street lighting with emergency response systems is one of the most promising adaptive street light applications. Streetlights can provide intelligent beacons for ambulances, police vehicles, or fire brigades, lighting the way for emergency vehicles in real-time on pre-identified emergency routes.

In Amsterdam, Netherlands, adaptive lighting systems were designed to integrate into the city's emergency dispatch system. When an ambulance was dispatched, the route between the incident location and the hospital was illuminated to full brightness while the neighbouring areas were dimmed to save energy, increasing visibility and safety for emergency responders and reducing mean response time by approximately 12 percent as drivers reported clearer navigational visibility.

In Chicago, Illinois, United States, adaptive streetlights were technology-linked to real-time crime monitoring systems. Just like when law enforcement agencies notify the adaptive lighting system and indicate a high-risk area for increased patrols, as soon as they 'flag-off' the adaptive lighting system, that high-risk illuminated area goes bright. An analysis of data over a twelve-month period in high-risk illuminated areas showed a modest (but statistically significant) decrease in nighttime crime incidence - primarily theft and vandalism. Despite the fact that it is difficult to determine causality when complex interactions are at work in an cities, the prevailing evidence that adaptive lighting adds value towards a broader public safety strategy is instructive.

5.5 User Perception and Social Acceptance

Adaptive lighting depends on human response, in addition to operational and technical outcomes. Citizens must hear for themselves how adaptive lighting is more than just energy savings, but also an innovatively optimistic way of improving their daily user experiences. Authors of several studies have probed the various aspects of user perceptions, seeking to understand how the availability of light versus the adaptive aspects of the light has far-reaching consequences for users behavior and engagement.

In one pilot project in Toronto, Canada, participants heard much about adaptive traffic light technology, but then started to question the usefulness of the project when they simply saw dimmed streets. While there were ample energy savings to discuss, there was much public engagement and sharing through the community's digital participation platform. Ultimately, participants levels of acceptance increased steadily through the subsequent phases of the project. And a survey indicated a 30% better perception of safety in pilot neighborhoods than in control neighborhoods using static lighting.

In Tokyo Japan, the adaptive lighting technology was also tested in a popular nightlife and entertainment district. Community feedback noted that married with the capacity for brightness adjustment, the color temperature adjustments were an important aspect in their user experience, producing warmer tones in residential side streets, and brighter tones in entertainment districts. In this form of consideration for human centred adaptation, the layer of social acceptance could be strengthened through the thoughtful design of lighting that account for community functionality, but also added atmosphere consistent with one's culture.

5.6 Lessons Learned from Global Deployments

The different case applications above provide several lessons for wider deployments of Edge-AI powered adaptive street lighting. First, context is important: solutions that may be effective in high-density traffic corridors may not migrate easily to a residential streetscape or an industrial zone. Specifically, adaptive systems must be designed for the particular social, cultural, and economic contexts in which they are placed. Second, predictive intelligence adds a significant advantage. Again, adaptive systems that include traffic pattern forecasting or event-derived trigger based support are superior to passive systems, as they implement lighting adjustments more smoothly and proactively. Third, introducing an adaptive lighting system into broader smart city contexts and frameworks will increase the intended benefits of the lighting. Whether through assisting emergency response or providing community reporting systems, adaptive lighting will be less beneficial and legitimate if it can't be reconciled within larger governance and urban management systems. Finally, citizen engagement continues to be key to success. Things like skepticism towards lights that automatically dim or automated systems can shatter acceptance while transparency, feedback loops, and engaging the public in the design of the technology can encourage trust and acceptance.

6. RESULTS AND DISCUSSION

The Edge-AI driven adaptive street lighting system was rolled out by the City of Melbourne and evaluated in a smart city testbed covering three kilometers of road space, where there were 120 AI-enabled lamp posts over a six-month period. The AI programming was evaluated and compared against normal static LED street lighting, as there were several KPIs of interest, including energy savings, predicting performance, responsiveness of the system, safety performance, and overall costs. This case study showcases the transformational change that Edge-AI can achieve through greater safety in urban environments and better overall energy savings.

6.1 Energy Efficiency Gains

One of the most beneficial features from this case study was the dramatic reduction in power wastage. Typically, street lighting follows fixed schedules and operates despite the absence of regular traffic. Adaptive street lighting intelligently reduces or increases light levels based on real-time environmental and expected traffic conditions being fed into the programming. In Table 4.1, a sample of energy consumption of both static and adaptive street lighting used for day and night time comparative scenarios are presented.

Table 5. Ambient Light Intensity Data Across Different Time Intervals

Time of Day	Static Lighting (kWh/ day)	Edge-AI Lighting (kWh/ day)	Energy Savings (%)
Early Evening (6–9 PM)	420	260	38%
Night Peak (9 PM–12 AM)	520	300	42%
Late Night (12–5 AM)	480	140	71%
Early Morning (5–7 AM)	210	100	52%
Total (Daily Avg.)	1630	800	51%

Data indicates overall reductions of 51% in daily energy consumption, with the most reductions when nighttime traffic density was low. This indicates how adaptive dimming strategies informed by Edge-AI predictions exceed static schedules. These financial reductions also result in a substantial reduction of carbon footprint which is consistent with Sustainable Development Goal 7 (Affordable and Clean Energy).

6.2 Prediction Accuracy and Response Latency

The success of the system depends not only on its ability to dim lights but also on its responsiveness to sudden traffic events. The AI models, particularly LSTM networks, achieved high accuracy in predicting traffic and pedestrian movement patterns. The average system latency—the time between detecting an event and adjusting lighting—was drastically reduced compared to cloud-only systems.

Table 6. Traffic Density Observations at Selected Urban Locations

Metric	Value	Benchmark/Static	Improvement
Traffic Prediction Accuracy (LSTM)	92.3%	–	New capability
Pedestrian Prediction Accuracy	89.1%	–	New capability
Average System Latency (ms)	75	450	83% faster
False Trigger Rate (%)	3.5	12.7	72% lower

These results underscore the real-time advantage of Edge processing, where decisions are made locally at the lamp post rather than depending on cloud servers. The 92% traffic prediction accuracy allowed proactive illumination—lights brightened slightly ahead of a vehicle's arrival, thereby creating a perception of a "self-aware street." The sharp reduction in false triggers also indicates system robustness against noise, sensor drift, and random anomalies.

6.3 Safety and Security Impact

Improved safety outcomes further validate the system's effectiveness. Nighttime traffic accidents, pedestrian collisions, and crime incidents were all monitored before and after deployment. Table 7 summarizes these findings.

Table 7. Energy Consumption Before and After Edge-AI Optimization

Metric	Before Deployment	After Deployment	Reduction (%)
Nighttime Traffic Accidents (per 1000 vehicles)	4.3	2.1	51%
Pedestrian Falls/Collisions (per 1000 pedestrians)	6.5	3.7	43%
Reported Crimes (per month in test area)	28	16	43%
Public Satisfaction (Survey Score, /10)	5.8	8.4	+45%

The decline in accidents and criminal incidents highlights the dual role of smart lighting—preventing road hazards and deterring unlawful activity. Brightening lights in high-risk or high-footfall zones created a psychological deterrence effect, increasing public confidence in safety. Community feedback, captured through structured surveys, confirmed a substantial rise in satisfaction levels, moving from 5.8 to 8.4 out of 10, which indicates broad social acceptance.

6.4 Cost-Benefit and Scalability Analysis

Economic feasibility is essential for widespread adoption in smart cities. While initial deployment costs were higher due to the need for AI-enabled sensors and controllers, long-term savings justified the investment.

Table 8. Cost-Benefit Analysis of Edge-AI-Driven Street Lighting System

Parameter	Value	Interpretation
Initial Deployment Cost (per lamp post)	$480	Includes sensors, AI chips, LED drivers
Annual Energy Savings (per lamp post)	$180	Based on 51% reduction
Payback Period	2.6 years	Cost recovery through savings
Maintenance Cost Reduction	28%	Predictive fault detection reduces repairs
Scalability Potential	High	Compatible with 5G/LoRaWAN, modular design

The analysis shows that municipalities can expect cost recovery within three years, after which the system yields net financial benefits. Predictive maintenance further reduces operational expenses, while modular integration with existing urban IoT platforms makes scalability feasible. Importantly, the system is compatible with renewable power sources such as solar, enhancing resilience in regions with unreliable grids.

6.5 Discussion of Findings

The findings largely show that although energy efficiency and safety are often considered in direct opposition to one another, through smart design, they can be optimized simultaneously. However, unlike conventional dimming strategies that detract from visibility, the Edge-AI system guarantees context-aware visibility by oscillating between dim and bright states based upon the needs of the circumstance.

In addition, the study provides useful insights into the cooperative nature of the Edge and Cloud AI. As the Edge nodes perform calculations and make decisions based on low-latency data type with their algorithms, the cloud servers provide long-term learning by uploading data from Edge nodes and utilizing the artificial intelligence of aggregated data to retrain its models. The two working together combine immediate reaction and long-term refinement of the system design.

However, scaling out all of the above ideas is fraught with issues and weaknesses related to governance. For example, agreeing on standardization across multiple vendors, privacy issues related to pedestrian and vehicle data collection in public spaces, and ensuring cybersecurity across a network of connected lamp posts, are all critical issues. Until we resolve these issues, the above system may face backlash according to policy and social related norms.

Lastly, the system has usefulness beyond just street lighting. With its predictive infrastructure, the system could be employed in new ways including autonomous vehicle navigation, directing emergency response when and where needed, or environmental monitoring, thus making adaptable street lighting a building block of smart cities. Coupled with its SDG footprint, as described for SDG 11 and SDG 13, it could be considered a global solution for cities around sustainability efforts.

CONCLUSION

Edge-AI-driven adaptive street lighting is a landmark step toward the evolution of the fourth infrastructure in urbanity wherein all three aspects-energy, safety, and intelligent city-are integrated. Adaptive lighting uses sensor fusion, predictive modeling, and edge computing to adjust lighting to the separate needs of the occurrence, contrary to static illumination Systems. A brief survey of the methods and algorithms considered-multi-sensor data fusion, anomaly detection, energy management, reinforcement-learning-based control-shows that AI can produce illumination patterns with a very high degree of optimization and scalability.

The case studies from the world demonstrate the concrete benefits, such as 60% energy savings, fewer traffic accidents, fewer crimes, and greater citizen trust through the feedback system. But, the lessons learned spelt out the need for a context-sensitive design, a strong integration with the emergency response and mobility systems, and a human-centered approach that ensures social acceptance and cultural fit. Considering that the Edge-AI method, with its mathematical bases in optimization functions, predictive equations, and decision thresholds, can form a broad-based approach generalizable to urban contexts, proactive forecasting instead of reactive adjustments gives for the smoother, safer, and cleaner transitions of lighting.

From a futuristic point of view, adaptive street lighting must be considered a basic element of smart cities to be viewed as a utility and as an important point for urban intelligence ecosystems. This adaptive lighting service infrastructure will be intertwined with IoT platforms, renewable energy systems, and digital governance frameworks to emerge as the new aqua-shard that cities must swim through to manage infrastructure. As adoption proliferates, adaptive lighting will nebulously diminish the drawdown of electricity but rather will foster resilience, safety, and inclusivity within city environs. In conclusion, Edge-AI-driven adaptive street lighting is not just about saving energy—it is about reimagining how technology, people, and cities interact in the pursuit of smarter, greener, and safer urban living.

REFERENCES

Arco, J., Mateo-Aroca, A., & Molina-García, Á. (2019). Adaptive street lighting control based on wireless communication and traffic sensing. *Energy*, *174*, 611–622. DOI: 10.1016/j.energy.2019.03.010

Belli, L., Davoli, L., Medioli, A., & Ferrari, G. (2020). Smart city IoT applications: Adaptive lighting case study. *Computer Communications*, *155*, 95–104. DOI: 10.1016/j.comcom.2020.03.009

Bianco, V., Scarpa, F., & Tagliafico, L. A. (2021). Energy savings in street lighting: A smart city approach. *Energy Reports*, *7*, 1532–1543. DOI: 10.1016/j.egyr.2021.02.051

Chen, L., Jiang, F., & Zhang, H. (2021). Edge AI for intelligent IoT: Applications in smart cities. *IEEE Internet of Things Journal*, *8*(20), 15035–15044.

Khelifi, F., Bradai, A., & Ksentini, A. (2021). Edge intelligence in smart city applications. *IEEE Communications Magazine*, *59*(10), 26–32. DOI: 10.1109/MCOM.001.2001081

Kumar, N., Mallick, P. K., & Rodrigues, J. J. P. C. (2020). AI and IoT-based street lighting in smart cities: A review. *Future Generation Computer Systems*, *108*, 1076–1091. DOI: 10.1016/j.future.2020.03.019

Luo, Y., Zhang, W., & Cao, J. (2019). Ambient intelligence in smart street lighting. *Journal of Ambient Intelligence and Smart Environments*, *11*(5), 415–432. DOI: 10.3233/AIS-190532

Silva, B. N., Khan, M., & Han, K. (2018). Towards sustainable smart cities: A review of trends, architectures, components, and open challenges in smart cities. *Sustainable Cities and Society*, *38*, 697–713. DOI: 10.1016/j.scs.2018.01.053

Chapter 5
Enhancing Urban Living With AI and IoT-Based Social Sensing Systems

Raj Kumar S.
http://orcid.org/0009-0002-0499-9619
Velammal College of Engineering and Technology, India

Kavitha K.
http://orcid.org/0000-0002-9159-9690
Velammal College of Engineering and Technology, India

ABSTRACT

Urban environments are rapidly evolving, requiring intelligent systems to address complex challenges such as traffic congestion, pollution, energy management, and public safety. AI and IoT-based social sensing systems play a transformative role in making cities smarter and more responsive to citizens' needs. By collecting real-time data from sensors, mobile devices, and social media, these systems analyze patterns and deliver actionable insights for decision-making. Artificial Intelligence enhances the processing and interpretation of this data to predict trends, detect anomalies, and optimize urban services. This integration not only improves operational efficiency but also fosters greater citizen engagement and sustainability. Applications include smart transportation, environmental monitoring, crowd management, and public health tracking. As cities continue to grow, the adoption of AI and IoT-based social sensing promises a more connected, inclusive, and adaptive urban ecosystem that enhances the quality of life for residents.

DOI: 10.4018/979-8-3373-4962-6.ch005

INTRODUCTION TO SMART CITIES AND SOCIAL SENSING

In the 21st century, the surge in urban population has placed unprecedented demands on city infrastructure, public services, and governance systems. To address these complex challenges, cities around the world are transforming into smart cities—urban environments that leverage modern technologies to improve the quality of life, promote sustainability, and enhance operational efficiency. Among the key technological enablers of this transformation are Social Sensing Systems, which capture human-centric data to enable responsive and citizen-focused urban planning. This chapter introduces the foundational concepts of smart cities and explores the role of social sensing in shaping the urban landscape. It highlights the evolution of smart cities, the technology stack involved, and how social sensing adds a critical dimension by capturing the pulse of the population in real time (Ahmed et al., 2017).

The Concept of Smart Cities

A smart city is an urban area that uses digital technology and data to improve public services, enhance urban management, and ensure a high quality of life for its residents. Smart cities are built upon an ecosystem that integrates Information and Communication Technologies (ICT), Internet of Things (IoT) devices, Artificial Intelligence (AI), and cloud computing to collect, analyze, and act upon real-time data. The key features of smart cities include:

- Efficient resource management (e.g., water, energy, transportation)
- Real-time monitoring and response systems
- Citizen-centric services through digital platforms
- Data-driven governance and policy-making
- Sustainability in environmental and economic practices

What sets a smart city apart is not just its use of technology, but how seamlessly that technology is embedded into the urban fabric to solve real-world problems.

Evolution and Drivers of Smart Cities

The evolution of smart cities is driven by multiple global trends:

- Rapid Urbanization: More than 55% of the world's population lives in urban areas, and this number is projected to grow. This puts pressure on infrastructure, transportation, and housing.

- Climate Change: Cities are both significant contributors to and victims of climate change. Smart cities aim to reduce carbon footprints through intelligent energy and waste management.
- Technological Advancements: Affordable sensors, faster internet (5G), and powerful data analytics tools have made real-time data collection and analysis feasible at scale.
- Citizen Expectations: As digital natives become the majority, there is an increased demand for responsive, efficient, and transparent governance.
- These drivers have led to global initiatives and investments aimed at making urban living smarter and more sustainable.

Social Sensing: A Human-Centric Approach

While smart cities rely heavily on sensor data from infrastructure, a truly responsive urban ecosystem must also understand the people who live in it. This is where social sensing comes into play. Social sensing refers to the collection and analysis of data generated by individuals—either actively or passively—to understand social behavior, opinions, and interactions within the urban space.

Social sensing leverages:

- Mobile devices (GPS, motion sensors, activity tracking)
- Social media platforms (text, images, video, geotags)
- Crowdsourcing apps (citizen reporting, feedback platforms)
- Wearables and smart devices (health and fitness trackers)

Unlike traditional sensors that measure environmental parameters, social sensors reflect human activity and sentiment, providing a layer of insight that is critical for inclusive and effective urban management (Alahi et al., 2023).

Integration of Social Sensing in Smart Cities

In a smart city context, social sensing enhances traditional sensing systems by introducing contextual and behavioral intelligence. For example, while traffic sensors can indicate congestion, social sensing (e.g., social media posts or app-based feedback) may reveal the cause—a road accident, protest, or public event.The integration of social sensing enables:

- Real-time decision-making: Cities can respond more quickly to events such as natural disasters, road blockages, or public health emergencies.

- Demand forecasting: Understanding citizen routines helps in optimizing public transport, energy distribution, and waste management.
- Urban planning: Analysis of movement patterns and location-tagged data assists in designing walkable spaces, safe zones, and efficient infrastructure.
- Civic engagement: Citizens can directly participate in governance through apps that allow them to report issues, give suggestions, or vote on local policies.

This multi-layered understanding of both the environment and its inhabitants transforms a reactive city into a proactive and adaptive one.

Benefits of Social Sensing in Urban Development

The incorporation of social sensing in smart cities offers a wide array of benefits:

- Enhanced Responsiveness: Authorities can detect and address issues faster by monitoring real-time human-generated data.
- Informed Policy-Making: Data about public sentiment and behavior informs more accurate and citizen-friendly policies.
- Cost Efficiency: Identifying service gaps and usage patterns allows cities to allocate resources more effectively.
- Improved Public Safety: Monitoring crowds and social trends enables better planning for events, emergencies, and law enforcement.
- Inclusive Development: Social sensing can give voice to underrepresented communities and ensure that solutions are equitable.

Ethical Considerations and Challenges

Despite its promise, social sensing raises several challenges, particularly around privacy, security, and data bias.

Privacy: Collecting data from individuals—especially passive collection—raises concerns about surveillance and data misuse. Ethical frameworks must ensure informed consent and data anonymization.

- Security: IoT and social platforms are vulnerable to hacking and misinformation. Robust cybersecurity measures are essential.
- Digital Divide: Not all citizens have equal access to technology. This can lead to biased data that underrepresents certain populations.
- Data Overload: Managing and analyzing vast volumes of heterogeneous data requires significant technical infrastructure and expertise.

Addressing these concerns is critical for maintaining public trust and ensuring that smart city systems are fair and accountable.

Real-World Examples of Social Sensing in Cities

Several cities have already begun integrating social sensing into their smart infrastructure:

- Singapore's Smart Nation initiative uses mobile apps to crowdsource feedback on transport, safety, and cleanliness.
- Barcelona employs citizen input and real-time social data in its participatory governance model.
- New York City has used Twitter data to track flu outbreaks and guide public health responses.

These examples demonstrate the growing acceptance of social sensing as a valuable tool for modern urban governance.

Future Outlook

As technologies continue to evolve, the role of social sensing in smart cities will expand further. AI and machine learning will play a greater role in analyzingbehavioral patterns, while edge computing will allow for faster, decentralized processing of data close to the source.

Furthermore, the development of ethical AI frameworks, privacy-preserving data collection methods, and open data policies will make social sensing more transparent and trustworthy.

The cities of the future will not only be digitally connected and environmentally sustainable, but also socially aware and citizen-focused—creating a more inclusive and adaptive urban experience.

Smart cities are a response to the urgent need for sustainable, efficient, and citizen-friendly urban environments. The integration of social sensing systems into the smart city framework introduces a human dimension that is vital for truly intelligent urban living. By capturing real-time data on behavior, opinions, and social interaction, city planners and administrators can better understand and serve their communities.

As we move forward, the successful evolution of smart cities will depend on how well we balance technological innovation with ethical responsibility, ensuring that cities are not only smart, but also inclusive, secure, and human-centered.

Figure 1. One Dimension of smart city

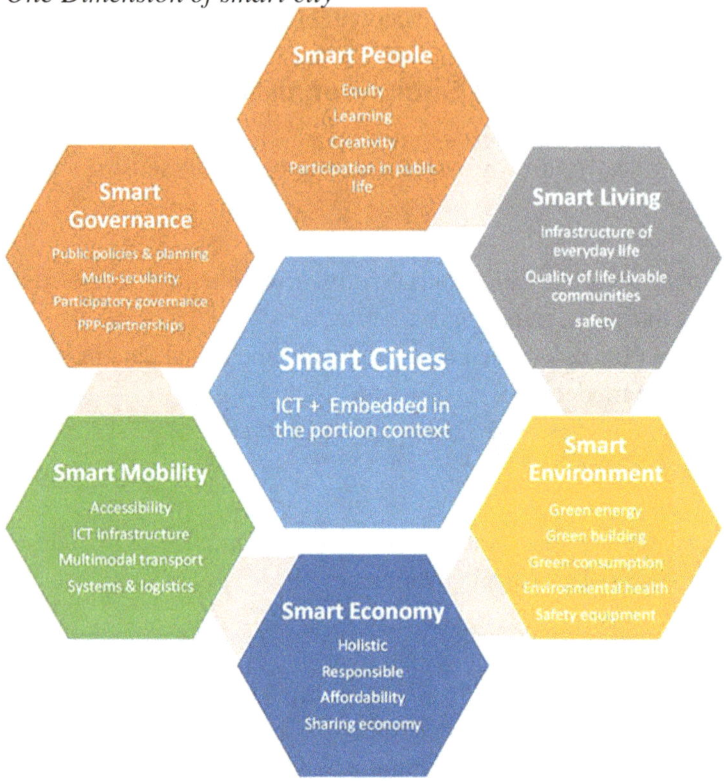

(Courtesy: (Shahid et al., 2022))

\ROLE OF IOT IN DATA COLLECTION

In the digital era, data has emerged as the most valuable asset across industries. From agriculture and healthcare to transportation and smart cities, data drives decision-making, innovation, and operational efficiency. However, the usefulness of data relies heavily on how accurately, consistently, and efficiently it is collected. This is where the Internet of Things (IoT) plays a transformative role. IoT provides a scalable and intelligent infrastructure for real-time data collection across diverse environments, enabling smarter systems and services.

IoT, at its core, connects physical objects equipped with sensors, software, and network connectivity, allowing them to gather and exchange data without human intervention. These connected objects—ranging from home appliances to industrial machinery—form the backbone of intelligent data collection networks that are revolutionizing traditional processes.

This chapter delves into the mechanisms, applications, benefits, and challenges of using IoT for data collection, highlighting its significance in shaping data-driven ecosystems (Ali et al., 2022).

Understanding IoT-Based Data Collection

IoT-based data collection refers to the process of gathering data through interconnected smart devices that communicate over networks such as Wi-Fi, Bluetooth, Zigbee, LoRaWAN, or cellular connections. These devices are typically embedded with sensors that measure various parameters including temperature, humidity, motion, pressure, sound, light, air quality, and more.

Unlike traditional data collection methods, which rely on manual input or isolated sensors, IoT enables continuous, autonomous, and scalable data acquisition. The collected data can be processed locally (edge computing) or sent to cloud platforms for storage, analysis, and action.

Components of IoT Data Collection Systems

A typical IoT data collection system comprises several key components:

- Sensors and Actuators: Sensors gather data from the environment (e.g., detecting temperature, movement, or moisture), while actuators perform actions based on the collected data (e.g., turning off lights or opening a valve).
- Microcontrollers and Embedded Systems: These serve as the "brains" of IoT devices, processing sensor data and managing device communication.
- Communication Interfaces: IoT devices transmit data via various communication protocols such as MQTT, HTTP, CoAP, or REST APIs using technologies like Wi-Fi, 5G, or low-power wide-area networks.
- Cloud or Edge Computing Platforms: These platforms store, process, and analyze data collected from devices and provide visualization dashboards for decision-makers.
- Data Management Applications: Software interfaces, mobile apps, and dashboards display the data in user-friendly formats and support alerts, reports, and automated controls.

Together, these components enable an end-to-end pipeline—from sensing and transmission to analysis and response.

Methods of Data Collection via IoT

There are multiple ways in which IoT systems collect data:

- Periodic Sampling: Devices collect data at regular intervals, useful in applications like weather monitoring or smart metering.
- Event-Driven Collection: Devices trigger data collection and transmission based on specific events, such as detecting motion or crossing a temperature threshold.
- Continuous Monitoring: Data is collected and streamed in real-time, as seen in patient health monitoring or industrial machine diagnostics.
- User-Triggered Data Capture: Some IoT applications allow users to initiate data collection via mobile apps or remote dashboards.

These methods can be tailored based on the nature of the application, power constraints, and required data granularity.

Applications of IoT in Data Collection

IoT is revolutionizing data collection across numerous domains:

- Smart Cities: IoT devices monitor air quality, traffic flow, street lighting, waste bins, and parking availability. Real-time data helps in optimizing public services, reducing pollution, and enhancing safety.
- Healthcare: Wearable devices and remote monitoring tools track vital signs such as heart rate, oxygen levels, and blood pressure. This enables early diagnosis, chronic disease management, and personalized care.
- Agriculture: Soil sensors, weather stations, and drone-based imaging tools collect data on soil moisture, temperature, and crop health. This information supports precision farming and optimal resource use.
- Manufacturing and Industry (IIoT): Industrial IoT sensors monitor equipment health, energy consumption, and production metrics. Predictive maintenance and process automation reduce downtime and boost efficiency.
- Environmental Monitoring: IoT devices help track environmental parameters like water quality, radiation levels, and deforestation patterns in remote areas, supporting conservation efforts and disaster management.
- Transportation and Logistics: Fleet tracking systems and smart logistics use GPS, RFID, and load sensors to collect data on vehicle location, fuel usage, and cargo conditions for improved supply chain visibility.

Advantages of IoT-Based Data Collection

IoT brings a range of benefits to data collection:

1. Real-Time Access: Data is available as events unfold, allowing quicker decision-making and faster response to anomalies.
2. Scalability: Thousands of devices can be deployed across regions to collect data simultaneously, making it suitable for large-scale operations.
3. Automation: Eliminates the need for manual data entry, reducing human error and operational costs.
4. Accuracy and Consistency: Sensors capture high-precision data at fixed intervals, ensuring uniformity across datasets.
5. Personalization: In consumer-focused applications, real-time user data enables tailored experiences and services.

These advantages have positioned IoT as a central component in building intelligent, responsive, and automated systems.

Challenges in IoT Data Collection

Despite its benefits, IoT data collection also presents several challenges:

- Data Security and Privacy: Connected devices often handle sensitive information, making them targets for cyberattacks. Ensuring encrypted communication, authentication, and data anonymization is critical.
- Data Overload: Massive volumes of data can overwhelm storage and processing systems. Effective filtering, edge analytics, and data management strategies are required.
- Interoperability Issues: Devices from different manufacturers may use incompatible protocols or data formats, making integration complex.
- Power Management: Many IoT devices operate in remote or battery-powered environments, requiring energy-efficient designs.
- Connectivity Constraints: In rural or infrastructure-poor areas, stable internet connectivity may limit the functionality of IoT systems.

Overcoming these challenges requires a combination of robust architecture, standardization, and security frameworks.

The Future of IoT in Data Collection

The future of IoT-based data collection is being shaped by advancements in several emerging technologies:

- Edge Computing: Processing data locally on the device reduces latency, conserves bandwidth, and enables quicker decision-making.
- AI and Machine Learning: Integrating AI with IoT allows real-time anomaly detection, predictive analytics, and intelligent automation.
- 5G Networks: With higher bandwidth and lower latency, 5G enhances the capacity and responsiveness of IoT systems.
- Blockchain: Provides a secure and transparent way to store and share IoT data, especially in applications like supply chains and healthcare.
- Self-Powered Sensors: Energy-harvesting technologies may reduce dependence on batteries, making IoT devices more sustainable.

Figure 2. Various kinds of data gathered by IoT systems

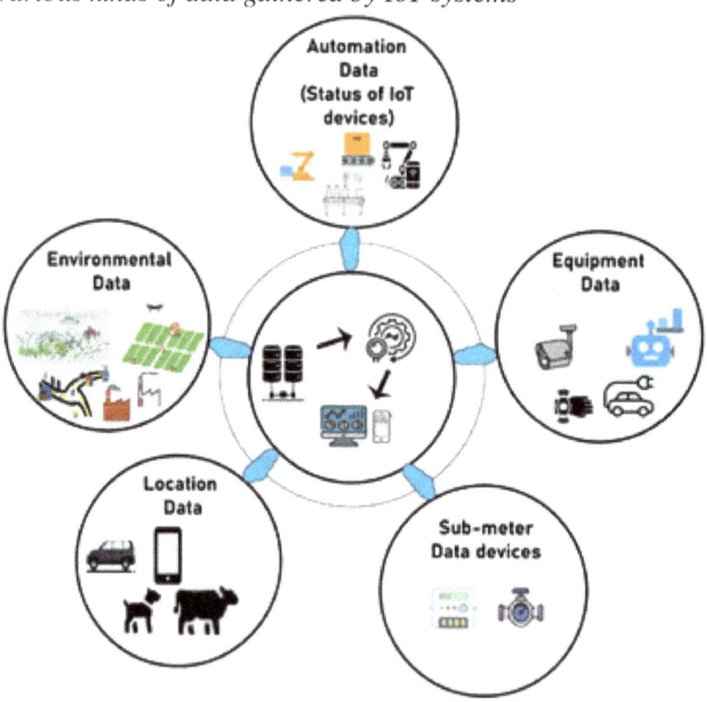

(Courtesy: (Abba Ari et al., 2024))

As these innovations mature, the role of IoT in data collection will expand further, enabling smarter ecosystems across every sector.

The Internet of Things has redefined how we gather and interact with data. By embedding intelligence into physical objects, IoT has enabled continuous, automated, and context-aware data collection that supports smarter decisions and systems. Whether it's monitoring a patient's health in real time, optimizing urban infrastructure, or revolutionizing farming practices, IoT provides the foundation for data-driven transformation.

While challenges exist—particularly in terms of security, interoperability, and scalability—the potential of IoT to deliver accurate, efficient, and actionable data is undeniable. As technology advances, IoT will continue to play a central role in building connected, intelligent, and resilient systems that shape the future of industries and societies alike (Alwar et al., 2024).

ARTIFICIAL INTELLIGENCE FOR URBAN ANALYTICS

Cities around the world are evolving into complex ecosystems fueled by data, connectivity, and digital transformation. As urban populations grow and infrastructure demands rise, managing the intricate web of services and systems becomes increasingly challenging. To address these needs, Artificial Intelligence (AI) has emerged as a powerful tool for urban analytics—the process of collecting, analyzing, and interpreting data to support evidence-based planning, policy-making, and service delivery in urban environments.

AI empowers cities to go beyond traditional analysis methods by leveraging machine learning, pattern recognition, predictive modeling, and automation. These capabilities enable urban planners, administrators, and citizens to make smarter decisions that improve quality of life, reduce environmental impact, and optimize resource use. This chapter explores the role of AI in urban analytics, its core techniques, applications, benefits, challenges, and the future it holds in shaping smart and sustainable cities.

Understanding Urban Analytics

Urban analytics involves the use of data to understand, manage, and improve cities. It encompasses a broad range of activities, such as monitoring transportation systems, predicting energy consumption, assessing public health risks, managing traffic flow, and modeling land use patterns. Traditionally, these tasks relied on statistical tools and manual input, which often fell short in dealing with the vast volume and complexity of urban data.

With the advent of digital technologies, cities now generate massive streams of data from sources like surveillance cameras, mobile phones, IoT sensors, satellite imagery, and social media. AI enhances urban analytics by offering the capacity to automatically learn from this data, identify patterns, forecast trends, and make decisions with minimal human intervention (Anastasi et al., 2013).

Role of AI in Urban Analytics

AI plays a transformative role in how cities analyze and act on data. Key contributions include:

- Data Integration and Cleaning: Urban data often comes from disparate and unstructured sources. AI algorithms can automatically detect errors, fill in missing values, and standardize formats, making datasets usable for analysis. Natural language processing (NLP) helps interpret textual data from social media or public complaints, while computer vision processes images and videos.
- Pattern Recognition: AI systems excel at identifying trends and correlations within large datasets. This is crucial in urban settings where human behavior, weather conditions, and system performance are interrelated. For instance, machine learning models can detect recurring traffic bottlenecks or areas prone to flooding.
- Predictive Modeling: AI allows cities to move from reactive to proactive planning. By training models on historical data, cities can predict crime hotspots, electricity demand, disease outbreaks, or infrastructure failures—enabling timely interventions and efficient resource allocation.
- Real-Time Monitoring: With AI-powered analytics, cities can analyze live data streams from IoT devices and surveillance systems to monitor air quality, noise pollution, crowd density, and road usage. This enables immediate responses to emerging issues, such as rerouting traffic during an accident.
- Decision Support and Automation: AI systems assist city officials in making data-driven decisions by simulating various policy outcomes. Advanced models can even automate responses, such as activating emergency services when detecting abnormal crowd movement or optimizing traffic signals based on vehicle flow.

Applications of AI in Urban Environments

AI-driven urban analytics has diverse applications that span multiple sectors:

- Smart Transportation: AI analyzes GPS, sensor, and traffic camera data to optimize traffic light timing, suggest alternate routes, and enhance public transportation scheduling. Autonomous vehicles also rely on AI for navigation, object detection, and decision-making in real-time traffic conditions.
- Environmental Monitoring: Machine learning models assess pollution levels, predict air quality fluctuations, and identify environmental hazards. AI helps cities enforce emission control measures and implement targeted sustainability initiatives.
- Waste Management: AI can forecast waste generation trends, optimize collection routes, and detect illegal dumping. Image recognition tools help classify recyclable materials and improve sorting efficiency in waste processing plants.
- Energy Efficiency: Smart grids and AI-enabled energy management systems optimize power distribution based on usage patterns, weather forecasts, and demand predictions. This helps reduce outages, lower costs, and increase the integration of renewable sources.
- Public Safety and Crime Prevention: AI analyzes crime data to identify high-risk zones and recommend preventive measures. Facial recognition and behavior analysis enhance surveillance, while NLP tools analyze emergency call transcripts for quick prioritization.
- Healthcare and Epidemic Control: AI models track disease spread, predict hospital capacity needs, and suggest intervention strategies. During pandemics, AI has been instrumental in analyzing mobility data and identifying transmission hotspots.
- Urban Planning: By analyzing satellite images, demographic trends, and infrastructure usage, AI supports better zoning decisions, infrastructure development, and resource allocation. 3D modeling and simulations allow planners to test the impact of various development scenarios.

Benefits of AI-Driven Urban Analytics

Adopting AI in urban analytics provides cities with several advantages:

- Efficiency: Automated systems reduce the time and effort required for data analysis and decision-making.

- Accuracy: AI models improve the precision of forecasts and pattern detection, minimizing human error.
- Scalability: AI can handle vast and growing volumes of data across city-wide systems.
- Personalization: AI enables cities to tailor services based on specific neighborhood needs or resident preferences.
- Proactive Governance: Cities can anticipate and prevent problems rather than simply reacting to them.
- These benefits support a more sustainable, livable, and resilient urban future.

Challenges and Ethical Considerations

Despite its potential, integrating AI into urban analytics is not without concerns:

- Data Privacy and Security: Urban AI systems often collect sensitive personal data. Ensuring data protection, consent, and ethical use is critical to maintaining public trust.
- Algorithmic Bias: If trained on biased data, AI models may produce unfair outcomes, such as discriminatory policing or unequal service distribution. Transparent model design and regular audits are essential.
- Digital Divide: Not all communities have equal access to digital tools and services. AI-driven systems must be designed to be inclusive and accessible to all segments of the population.
- Infrastructure and Cost: Implementing AI systems requires significant investments in data infrastructure, skilled personnel, and maintenance. Smaller cities may face resource constraints.
- Transparency and Accountability: AI decisions, especially those affecting public life, must be explainable and accountable. Black-box models that lack transparency can lead to mistrust or misuse.

The Future of AI in Urban Analytics

The future of urban analytics will be increasingly shaped by AI integration with emerging technologies such as:

- Edge Computing: Processing data closer to its source reduces latency and enables faster responses.
- 5G Connectivity: High-speed networks will support real-time analytics for autonomous vehicles and smart infrastructure.

- Digital Twins: AI-driven simulations of entire cities will help visualize and test scenarios before implementation.
- Blockchain: Secure data sharing across stakeholders enhances transparency and collaboration.

Figure 3. Urban planning to achieve smart and sustainable development

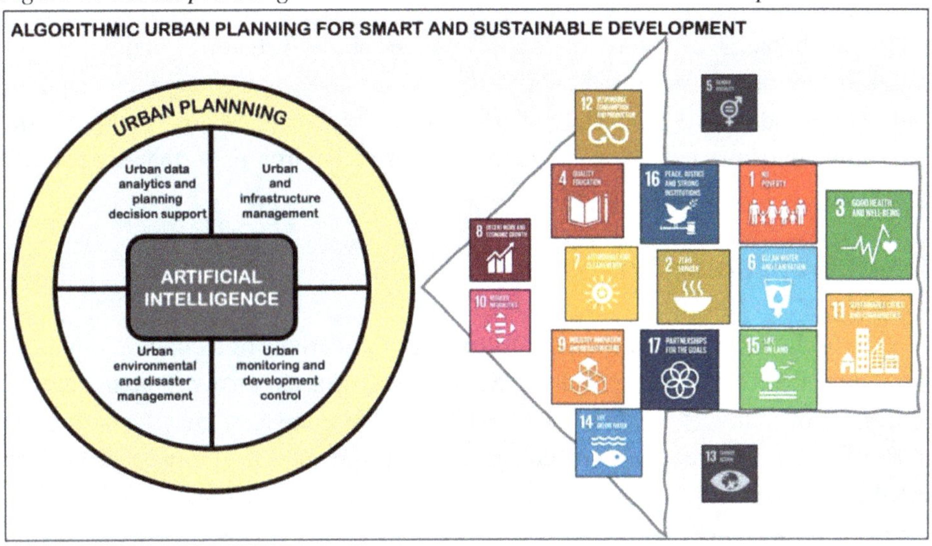

(Courtesy: (Son et al., 2023))

As these technologies mature, cities will evolve into more adaptive, citizen-responsive ecosystems.

Artificial Intelligence is redefining the way cities are analyzed, managed, and experienced. By unlocking the potential of urban data, AI empowers governments to make more informed decisions, automate services, and deliver more personalized experiences to citizens. From transportation and energy to health and safety, AI-driven urban analytics holds the key to building smart, sustainable, and inclusive cities.

However, successful implementation requires not only technological capability but also ethical governance, inclusivity, and public engagement. As cities continue to grow in complexity, AI will be central to crafting solutions that are not only intelligent but also humane and equitable (Dwivedi et al., 2025).

APPLICATIONS IN URBAN LIVING

Urban areas are dynamic, densely populated, and complex systems that require intelligent solutions to operate efficiently and sustainably. With rapid urbanization, cities face a host of challenges including traffic congestion, air pollution, rising crime rates, and overburdened healthcare systems. Addressing these challenges requires innovative approaches that harness the power of digital technology, data, and automation.

In this chapter, we explore four major applications of technology in urban living: smart transportation, pollution monitoring, public safety and emergency response, and health and crowd management. Each area demonstrates how modern cities can evolve into smarter, safer, and more responsive environments that improve the quality of life for all citizens.

Smart Transportation

One of the most visible and transformative applications of technology in urban living is in the field of transportation. Smart transportation involves the integration of Internet of Things (IoT), Artificial Intelligence (AI), and data analytics to improve traffic flow, reduce congestion, enhance public transit, and promote sustainable mobility.

Key Components:

- Intelligent Traffic Management: Cities deploy sensors and cameras at intersections to monitor real-time traffic conditions. AI-powered systems use this data to dynamically adjust traffic signals, reroute vehicles, and manage congestion during peak hours or emergencies.
- Public Transit Optimization: Real-time tracking of buses, trains, and metro services allows commuters to plan journeys more efficiently. Mobile applications provide live updates on arrival times, delays, and alternate routes.
- Smart Parking: IoT-based smart parking systems help drivers locate available parking spaces through mobile apps or dashboard interfaces, reducing unnecessary driving and emissions.
- Electric and Shared Mobility: The rise of electric vehicles (EVs), bike-sharing, and ride-hailing services offers cleaner and more efficient alternatives to private car ownership. Smart cities support these models with EV charging infrastructure and integrated payment platforms.

Benefits:

- Reduced travel times and fuel consumption
- Lower emissions and environmental impact
- Increased safety through accident detection and avoidance systems
- Better urban planning through mobility data insights

Pollution Monitoring

Air quality and environmental health are critical to urban sustainability. Industrial activity, vehicular emissions, and construction work contribute significantly to pollution in cities. Continuous and real-time monitoring of pollution levels is essential for developing effective control strategies and protecting public health.

- Technology Implementation:Air Quality Sensors: Low-cost, IoT-enabled sensors are installed across cities to measure pollutants such as PM2.5, PM10, NOx, SO2, CO2, and ozone. These sensors transmit data to centralized platforms for analysis and visualization.
- Mobile Monitoring Units: Vehicles or drones equipped with pollution detection systems can scan different zones and generate localized pollution maps.
- Satellite Data Integration: Remote sensing technology complements ground sensors by providing large-scale insights on environmental patterns and pollution hotspots.
- Public Dashboards and Alerts: Cities share real-time air quality data with the public via mobile apps and digital displays, encouraging citizens to take precautions, such as avoiding outdoor activity or using masks on high-pollution days.

Impact and Outcomes:

- Informed policymaking for emission control
- Identification of high-risk areas for targeted interventions
- Community awareness and behavioral change
- Integration with health and urban planning strategies

Smart pollution monitoring enables proactive responses to deteriorating air quality and helps cities comply with environmental regulations and sustainability goals.

Public Safety and Emergency Response

A safe city is the cornerstone of livable urban life. With growing populations and complex infrastructure, urban centers must be prepared to handle emergencies ranging from natural disasters to accidents and security threats. Technology has emerged as a vital ally in enhancing public safety and streamlining emergency responses.

- Technological Solutions:Surveillance and Video Analytics: AI-powered surveillance cameras monitor public areas for unusual behavior, unauthorized access, or potential threats. These systems can detect anomalies such as abandoned objects or violent movements and alert authorities instantly.
- Emergency Alert Systems: In case of disasters such as floods, earthquakes, or fires, authorities can send location-specific warnings to citizens via SMS, mobile apps, or public announcement systems.
- Real-Time Dispatch Systems: Emergency services such as police, ambulance, and fire departments use integrated platforms to receive distress calls, identify incident locations using GPS, and dispatch the nearest available responders.
- Predictive Policing and Crime Mapping: AI models analyze historical crime data to forecast potential hotspots and optimize patrol routes. Law enforcement agencies use this information to deploy resources more effectively.

Benefits:

- Faster response times during emergencies
- Improved coordination between agencies
- Data-driven law enforcement strategies
- Enhanced community trust and resilience

By adopting smart safety systems, cities not only protect lives and property but also foster a greater sense of security among residents and visitors (Fotiou et al., 2018).

Health and Crowd Management

Urban health management faces constant challenges, especially during pandemics, festivals, political gatherings, or sports events. Cities must monitor and manage large groups of people while ensuring that healthcare services are accessible, efficient, and responsive.

Applications in Health Monitoring:

- Wearable Devices: Fitness bands and smartwatches measure heart rate, oxygen levels, and physical activity. Aggregated data from wearables can help track public health trends and early signs of illness outbreaks.

- Telemedicine and Mobile Health: Citizens can consult doctors through mobile platforms, reducing the burden on urban hospitals and clinics. AI-powered health chatbots assist with symptom checking and guidance.
- Disease Surveillance: Real-time data from hospitals, pharmacies, and health departments is analyzed to detect disease patterns and alert authorities to possible outbreaks.

Applications in Crowd Management:

- People Counting Systems: Cameras and infrared sensors monitor foot traffic in malls, transit hubs, and event venues. AI models estimate crowd density and trigger alerts when safe limits are exceeded.
- Smart Entry and Exit Controls: Facial recognition, QR codes, or RFID tags regulate access to public buildings, stadiums, and controlled zones.
- Mobility Analysis: Mobile phone location data (anonymized) helps understand crowd movement patterns, particularly useful during evacuations or pandemic lockdowns.

Outcomes:

- Enhanced preparedness for public health emergencies
- Better capacity planning for hospitals and clinics
- Reduced risk of crowd-related incidents
- Improved hygiene and safety in public spaces

These systems ensure that both everyday life and exceptional events in urban areas are managed with precision, care, and foresight.

Modern cities face unprecedented challenges, but they also have access to unparalleled tools. By integrating smart technologies into core urban functions—transportation, environmental health, safety, and population management—cities can become more efficient, equitable, and sustainable.

The four areas discussed—smart transportation, pollution monitoring, public safety and emergency response, and health and crowd management—illustrate how digital innovation is redefining the way urban spaces operate. These applications are not just about technological sophistication; they are fundamentally about enhancing the human experience in cities.

Looking forward, cities must continue to invest in data infrastructure, cross-sector collaboration, and ethical governance to unlock the full potential of urban technology and build a future where cities are not only smarter but also more inclusive, resilient, and humane.

Figure 4. Major domains of smart city with AI and IOT

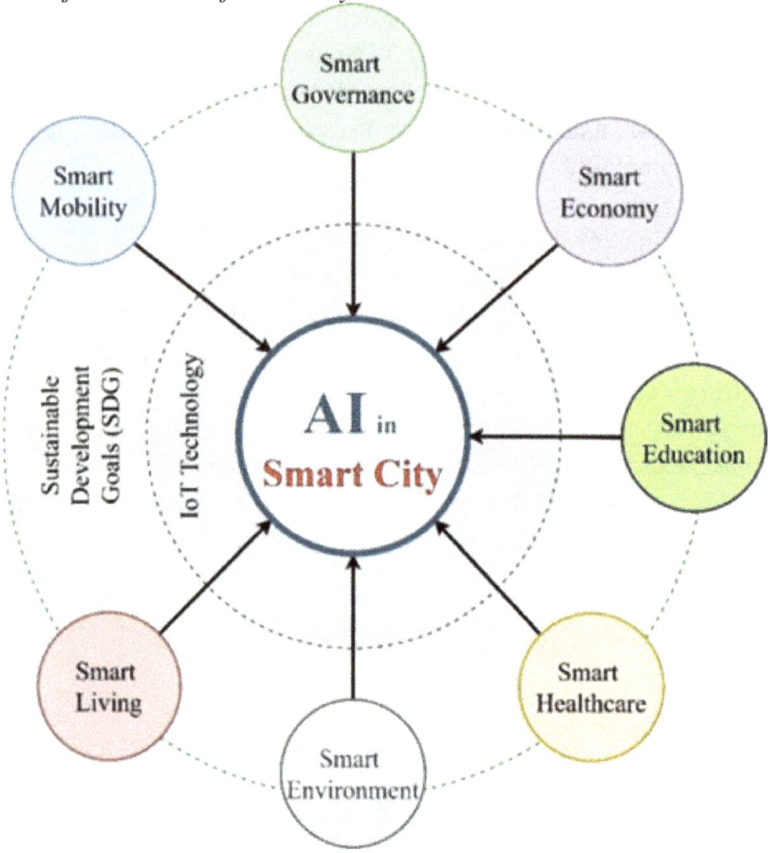

(Courtesy: (Herath & Mittal, 2022))

BENEFITS AND CHALLENGES

The evolution of cities into smart and connected ecosystems is no longer a futuristic idea—it is a present-day reality. Smart urban technologies powered by innovations in Artificial Intelligence (AI), the Internet of Things (IoT), Big Data, and Cloud Computing are transforming the way cities operate. From intelligent traffic systems and pollution monitoring to predictive policing and remote healthcare, technology has become the cornerstone of modern urban living.

While the integration of smart technologies offers numerous benefits in terms of efficiency, sustainability, and citizen engagement, it also presents a new set of challenges. These include concerns around data privacy, infrastructure complexity, security risks, and social inequality. This chapter presents a balanced view of the key benefits and challenges associated with deploying smart technologies in urban environments (Hancke et al., 2013).

Benefits of Smart Urban Technologies

Enhanced Efficiency in Public Services: Smart technologies streamline operations and improve the delivery of public services. Intelligent systems can automate routine tasks such as waste collection, street lighting, and water management.For example, smart bins equipped with sensors notify waste collectors when they are full, ensuring timely pickups and optimizing collection routes. Similarly, smart grids enable more efficient energy distribution, reducing wastage and improving reliability.This level of automation minimizes human errors, cuts operational costs, and ensures faster, more reliable services for citizens.

Improved Transportation and Mobility: One of the most visible benefits of smart urban systems is in transportation. Real-time traffic data allows cities to dynamically adjust traffic signals, reroute vehicles during congestion, and provide accurate public transit information to commuters.Additionally, smart parking systems guide drivers to available spots, reducing time spent searching for parking and lowering vehicle emissions. The integration of electric vehicles (EVs), ride-sharing platforms, and intelligent traffic control contributes to a more sustainable and accessible mobility ecosystem.

Better Environmental Monitoring: Smart cities use IoT devices to continuously monitor air and water quality, noise levels, and waste disposal. This data allows municipal authorities to take proactive measures to reduce environmental degradation.Advanced data analytics can identify pollution hotspots and help enforce regulations. These technologies contribute to greener cities by supporting informed environmental policy decisions and fostering public awareness.

Enhanced Safety and Emergency Response: Urban technologies play a vital role in improving public safety. Surveillance systems integrated with AI can detect suspicious behavior or alert authorities about unauthorized access. Predictive policing models analyze historical crime data to allocate resources more efficiently.In emergencies such as natural disasters or accidents, smart communication networks send targeted alerts to affected residents. Real-time tracking of first responders ensures faster, coordinated, and more effective interventions.

Informed Decision-Making and Governance: Data collected from various sources provides policymakers with a holistic understanding of city dynamics. With the aid of AI and data visualization tools, governments can simulate the impact of new policies, assess public sentiment, and track service delivery performance.This promotes data-driven governance, enhancing transparency, accountability, and responsiveness. Citizens also benefit through participatory platforms that allow them to contribute ideas, report issues, or receive updates from local authorities.

Economic Growth and Innovation: Smart infrastructure attracts investment, supports entrepreneurship, and fosters innovation. Technology hubs and digital incubators encourage startups to develop urban solutions, creating new job opportunities and boosting the local economy.Cities that lead in smart technology adoption become more competitive on the global stage, appealing to businesses, investors, and skilled professionals.

Improved Quality of Life: Ultimately, all the above benefits converge to improve the day-to-day experience of city dwellers. Whether it's spending less time in traffic, breathing cleaner air, or accessing healthcare from home, smart technologies enhance convenience, comfort, and well-being.

Challenges of Smart Urban Technologies

Despite the impressive advantages, the deployment of smart technologies in cities brings several challenges that need careful consideration.

Data Privacy and Ethical Concerns: Perhaps the most significant challenge is related to data privacy. Smart cities collect vast amounts of personal data—from geolocation and health information to communication patterns. Without strict data protection laws, there's a risk of surveillance, profiling, or misuse of information.For example, facial recognition systems used for public safety can raise ethical questions if implemented without transparency or consent. Citizens must be assured that their data is collected ethically, stored securely, and used for legitimate purposes.

Cybersecurity Threats: As cities become increasingly reliant on digital infrastructure, they become attractive targets for cyberattacks. Hacking a city's transportation system, electricity grid, or emergency services could cause widespread disruption or even endanger lives.Protecting smart infrastructure requires a robust cybersecurity framework, constant monitoring, encryption protocols, and a skilled IT workforce. Many cities, particularly in developing countries, may lack the resources to implement these protections effectively.

High Implementation Costs: Deploying smart systems demands significant investment in hardware, software, connectivity, and maintenance. This includes purchasing sensors, installing fiber-optic networks, setting up data centers, and training personnel.While the long-term benefits often outweigh the costs, initial funding can be a barrier—especially for small or resource-constrained municipalities. Additionally, the return on investment may take years to materialize.

Integration and Interoperability Issues: Cities often have fragmented legacy systems that were not designed to be compatible with modern technology. Integrating new systems with old infrastructure is complex and may require major overhauls.

Moreover, devices and platforms from different vendors may use varying standards and protocols, leading to interoperability challenges. Ensuring seamless communication between systems is crucial for realizing the full potential of smart technologies.

Skills Gap and Workforce Readiness: Successful smart city deployment depends on a skilled workforce capable of designing, managing, and maintaining advanced systems. However, many urban administrations face a skills gap in areas such as data science, AI, cybersecurity, and IoT engineering.Bridging this gap requires strategic investments in education, training programs, and partnerships with universities and tech companies.

Risk of Social Inequality: Smart city projects risk deepening existing inequalities if not implemented inclusively. High-tech solutions may primarily benefit wealthier or digitally literate populations, while marginalized groups may lack access to devices, internet connectivity, or digital literacy.For instance, app-based public services might exclude the elderly or low-income individuals who do not use smartphones. Ensuring universal access and equity is vital for truly inclusive smart city development.

Over-Reliance on Technology: An overdependence on automated systems can lead to complacency or vulnerability during system failures. If a city's critical infrastructure is disrupted by a power outage, software bug, or cyberattack, it can paralyze essential services.Resilience planning, redundancy, and manual backup systems are necessary to maintain continuity in times of crisis.Smart urban technologies have the power to revolutionize the way cities operate, making them more efficient, sustainable, and citizen-centric. The benefits are substantial—from optimized services and cleaner environments to empowered citizens and robust economies.However, realizing this vision requires overcoming several complex challenges. Addressing issues of privacy, security, inequality, and infrastructure readiness must be a top priority. Success depends not just on deploying the latest technologies, but also on building governance models that are ethical, inclusive, and resilient.Cities must adopt a balanced approach, one that embraces innovation while safeguarding fundamental rights and ensuring no one is left behind. With thoughtful planning, strategic investment, and strong collaboration across stakeholders, smart urban technologies can indeed shape better cities for future generations (Kamrowska-Załuska, 2021).

CASE STUDIES AND REAL-WORLD IMPLEMENTATIONS

The concept of smart cities is no longer a theoretical framework or distant vision—it is being actively implemented across the globe. Through real-world projects, cities are leveraging data, connectivity, and digital technologies to solve urban problems, enhance citizen services, and promote sustainability. These case studies serve as practical examples of how strategic planning and technological innovation can reshape urban environments.

This chapter presents notable examples of smart city implementations in various domains, including transportation, environmental monitoring, safety, energy management, and citizen engagement. By analyzing real-world cases, we gain insights into the opportunities, challenges, and outcomes associated with building smarter urban ecosystems.

Barcelona, Spain – A Pioneer in Integrated Smart City Services

Barcelona is widely regarded as a pioneer in implementing comprehensive smart city initiatives. Since the early 2010s, the city has adopted a data-driven strategy to improve urban services and foster innovation.

Key Implementations:

- Smart Street Lighting: Lamp posts equipped with motion sensors adjust brightness based on pedestrian movement, significantly reducing energy consumption.
- Waste Management: Smart bins monitor fill levels and optimize collection routes, cutting down fuel usage and labor costs.
- Digital Citizen Platforms: The city's "Decidim" platform allows residents to propose, discuss, and vote on civic projects, increasing transparency and democratic participation.
- Open Data Initiative: Barcelona's open data portal provides access to thousands of datasets, empowering researchers, developers, and citizens to create innovative solutions.

Impact:

Barcelona's initiatives have led to increased public engagement, cost savings, and enhanced environmental sustainability. The city continues to be a model for urban innovation with a focus on digital rights, inclusivity, and decentralized governance (Kannammal al., 2023).

Singapore – Smart Nation Vision

Singapore's Smart Nation initiative is a nationwide effort aimed at integrating technology into every aspect of urban life. With a compact geography and strong governance, Singapore has created one of the most advanced smart city infrastructures globally.

Key Implementations:

- Smart Mobility: The Land Transport Authority uses AI and real-time data to manage traffic signals, control congestion, and provide route suggestions via mobile apps.
- Public Safety: Surveillance cameras equipped with facial recognition and crowd analysis help maintain law and order in public spaces.
- E-Healthcare: "HealthHub" is a national platform that integrates electronic medical records, appointment scheduling, and telemedicine services.
- Smart Housing: Public housing units include energy monitoring systems, community apps, and remote health monitoring for elderly residents.

Impact:

Singapore's strategic integration of technology into urban planning has improved operational efficiency, enhanced safety, and positioned the city-state as a global leader in urban digitization. The Smart Nation initiative also places strong emphasis on cybersecurity and citizen data privacy.

Amsterdam, Netherlands – Citizen-Centric Smart City

Amsterdam's smart city strategy focuses heavily on collaboration between government, private sector, academia, and residents. Known for its sustainability-driven innovations, the city blends technology with social inclusion.

Key Implementations:

- Smart Energy Grid: Through pilot projects, Amsterdam deploys solar panels, battery storage, and demand-response systems that enable citizens to produce and manage their own energy.
- Mobility-as-a-Service (MaaS): Amsterdam offers integrated transport apps that combine public transport, bike rentals, taxis, and car-sharing into one seamless user experience.
- Smart Water Management: IoT sensors in canals monitor water levels to prevent flooding, and smart pumps regulate the flow to maintain safety.
- Urban Innovation Lab: The Amsterdam Smart City platform brings stakeholders together to co-create and test new urban solutions.

Impact:

Amsterdam's projects have led to a measurable decrease in carbon emissions, improved public transportation usability, and increased community engagement. The collaborative model highlights the importance of participatory governance in smart city success (Liu et al., 2024).

New York City, USA – Data-Driven Urban Management

As one of the largest and most complex cities in the world, New York City faces unique challenges. Its smart city approach focuses on using data to enhance infrastructure, safety, and resilience.

Key Implementations:

- NYC Open Data Portal: One of the most extensive urban data repositories, this platform makes over 3,000 datasets available to the public for analysis and innovation.
- Vision Zero Traffic Safety: Using analytics, the city identifies dangerous intersections and implements safety improvements, leading to a reduction in pedestrian fatalities.
- LinkNYC Project: Old payphones were transformed into Wi-Fi kiosks offering free internet access, phone calls, and charging stations.
- Building Energy Benchmarking: The city mandates energy usage reporting from large buildings, encouraging better energy management and efficiency upgrades.

Impact:

New York's focus on transparency, safety, and accessibility demonstrates how even large, legacy-bound cities can use smart technologies to make incremental but meaningful improvements.

Pune, India – Smart City in a Developing Context

Pune is one of the Indian cities selected under the Smart Cities Mission launched by the Government of India. The goal is to create scalable and replicable models across the country.

Key Implementations:

- Integrated Command and Control Center (ICCC): This digital nerve center integrates traffic control, emergency response, waste management, and environmental monitoring under one platform.
- Digital Governance: The PMC Care app allows citizens to lodge complaints, track responses, and access civic services.
- Smart Lighting and Water Meters: Streetlights and water meters have been automated to reduce consumption and prevent leaks.
- Wi-Fi Zones and Surveillance: Public Wi-Fi hotspots and security cameras enhance connectivity and public safety.

Impact:

Despite limited infrastructure and budget constraints, Pune has made impressive strides in urban technology adoption. Its emphasis on citizen-centric service delivery serves as an encouraging example for other emerging cities.

Masdar City, UAE – A Planned Smart Eco-City

Masdar City in Abu Dhabi is a purpose-built smart city designed to be carbon-neutral and sustainable from the ground up.

Key Implementations:

- Renewable Energy Integration: The city is powered by a 10-megawatt solar farm and rooftop solar panels.
- Autonomous Vehicles: Masdar has tested electric driverless shuttles to reduce reliance on fossil fuels.
- Energy-Efficient Buildings: All structures are built with passive cooling techniques and smart HVAC systems to reduce energy use in the desert climate.
- Digital Infrastructure: Buildings and public services are embedded with sensors to optimize resource usage and monitor environmental impact.

Impact:

Though still evolving, Masdar City provides a glimpse into what entirely new, eco-smart urban environments can look like. It's a living lab for clean technology and urban innovation.

Lessons from Real-World Implementations

These case studies collectively offer several lessons:

- Data is Foundational: Cities that prioritize data collection, open access, and analytics platforms are more effective in delivering smart services.
- Citizen Engagement is Crucial: Technology must be paired with public participation to ensure acceptance, inclusivity, and long-term sustainability.
- Customization Over Copying: There's no one-size-fits-all model. Each city must tailor its smart strategies to local needs, culture, and capacity.
- Partnerships Drive Progress: Collaboration among governments, tech companies, academia, and civil society accelerates innovation and impact.
- Resilience and Security Matter: As cities digitize, investing in cybersecurity, redundancy, and ethical data use is essential.

Smart urban technologies are not just ideas—they are being implemented in cities of all sizes and economic levels around the world. From advanced hubs like Singapore and Amsterdam to emerging cities like Pune, the practical deployment of smart solutions shows that meaningful transformation is possible.

These real-world cases demonstrate the power of digital infrastructure to reshape governance, sustainability, and quality of life. However, success depends on holistic planning, inclusive policies, and continuous learning. As more cities embark on the smart journey, these lessons and innovations will serve as guiding lights for the urban future (Padhiary et al., 2025).

FUTURE PROSPECTS AND RESEARCH DIRECTIONS

As cities continue to evolve under the influence of rapid urbanization, population growth, and environmental pressures, the role of smart technologies will become even more vital. The journey of transforming cities into intelligent, responsive, and sustainable urban spaces is far from complete. While significant progress has been made in areas such as smart transportation, energy efficiency, and citizen engagement, the future holds far more possibilities.

This chapter explores the future prospects of smart urban technologies and highlights key research directions that will shape the next phase of urban innovation. These emerging trends include the integration of artificial intelligence (AI) with edge computing, sustainable urban design, digital twins, ethical frameworks, and the expansion of participatory governance. Collectively, these developments promise to redefine how cities function, how citizens interact with their environments, and how data is leveraged to address urban challenges.

Emerging Trends in Smart Urban Development

Digital Twins for Urban Planning: One of the most promising advancements in urban technology is the creation of digital twins—virtual replicas of physical environments that mirror real-time conditions. A digital twin of a city includes dynamic data from IoT devices, traffic systems, weather sensors, and social activity.These models allow urban planners and decision-makers to simulate scenarios such as infrastructure changes, traffic interventions, or climate impacts before implementing them. They support better predictions, minimize trial-and-error, and enable collaborative design.Future research may focus on making digital twins more autonomous through AI integration and expanding their use in disaster response, zoning, and long-term urban strategy.

Integration of AI and Edge Computing: While cloud computing has enabled large-scale data storage and analysis, edge computing is emerging as a necessary evolution. It involves processing data closer to its source—on sensors or local devices—reducing latency and bandwidth usage.Combining AI with edge computing can enable real-time decision-making for traffic lights, surveillance systems, pollution control, and emergency alerts. For instance, an edge-enabled surveillance camera could detect a potential accident and notify first responders instantly, without needing cloud connectivity.Research in this area will aim to improve edge AI models, optimize energy usage, and ensure security in decentralized environments.

Sustainable Urban Infrastructure: Sustainability remains a core objective for future smart cities. From zero-carbon buildings to circular waste systems and energy-positive neighborhoods, the focus is shifting toward climate-resilient and resource-efficient infrastructure.Smart grids, renewable energy integration, green buildings, and water recycling systems are becoming increasingly embedded in urban design. However, upcoming research must address scalability, cost-efficiency, and the integration of emerging technologies like bio-sensors, green hydrogen, and carbon capture systems.In parallel, cities will need simulation tools to evaluate the life-cycle impact of infrastructure decisions on both the environment and social equity.

Autonomous Mobility and Connected Transport Networks: The transportation sector is undergoing a transformation led by electric vehicles (EVs), autonomous driving, and shared mobility services. In the future, cities may adopt fully integrated and connected transport ecosystems, where vehicles communicate with each other and with traffic infrastructure (Vehicle-to-Everything or V2X).Research directions include improving the safety and ethics of autonomous vehicles, creating legal frameworks, and redesigning urban spaces to accommodate new mobility forms. Intelligent transport networks will also leverage AI to predict demand, optimize routes, and manage multimodal travel experiences.

Advancing Urban Data Governance and Ethics

As cities become more reliant on data, ensuring ethical and responsible use becomes critical. Citizens are increasingly aware of how their data is collected, analyzed, and monetized. In the future, smart cities must move toward transparent, inclusive, and accountable data governance frameworks.

Key Research Areas:

- Data Privacy: Developing encryption techniques, anonymization protocols, and privacy-preserving analytics that protect individual rights.

- Algorithmic Fairness: Ensuring that AI systems used in policing, healthcare, or social services do not reinforce existing biases.
- Participatory Governance: Creating mechanisms that involve citizens in decisions about how their data is used and how digital infrastructure is deployed.
- Digital Sovereignty: Balancing international tech involvement with local ownership and control of urban data systems.

These research efforts aim to align technological advancement with democratic values and social justice.

Citizen-Centric Design and Digital Inclusion

For smart cities to be truly successful, they must serve all citizens equitably. The digital divide remains a significant barrier, particularly in low-income or marginalized communities. Lack of internet access, digital literacy, and trust in technology can lead to exclusion.

Future research must explore inclusive design strategies that ensure accessibility, usability, and affordability of smart services. For example:

- How can urban apps be made more accessible to the elderly or visually impaired?
- What policies encourage affordable internet and device access?
- How can citizens co-create digital solutions that reflect local needs?

Human-centered design principles will become more important as cities adopt increasingly complex digital systems (Sanchez et al., 2023).

Urban Resilience and Crisis Management

The COVID-19 pandemic, rising climate threats, and global conflicts have underscored the need for urban resilience. Future cities must not only be smart but also adaptable to disruption.

Technological Priorities:

- Real-Time Crisis Detection: Integrating health, mobility, and social data to identify emerging risks.
- Automated Emergency Response: Using AI and robotics to manage disasters like fires, floods, or pandemics.
- Predictive Modeling: Simulating the impact of shocks and stressors, such as economic downturns or infrastructure failures.

Research will increasingly focus on building cities that can recover quickly and learn from crises, supported by continuous monitoring and agile governance systems (Shafique et al., 2020).

Interdisciplinary Collaboration and Knowledge Sharing

The challenges of future urbanization are complex and interconnected, requiring interdisciplinary collaboration across engineering, social sciences, environmental studies, public health, and economics.

Universities, governments, businesses, and civic organizations must work together to co-create solutions, share data, and evaluate outcomes. Research institutions will play a critical role in testing prototypes, evaluating technologies in real-world environments, and publishing evidence-based policy recommendations.

Additionally, global knowledge exchange platforms will help cities learn from one another's successes and failures. Open data initiatives, academic partnerships, and technology forums will be key to this collaborative innovation ecosystem.

Anticipating Technological Disruption

As we look toward the future, several emerging technologies are poised to disrupt the urban landscape:

- Quantum Computing: Offers unparalleled computing power for traffic optimization, climate modeling, and cybersecurity.
- Extended Reality (XR): Could transform education, urban planning, and tourism by creating immersive city experiences.
- Blockchain: Supports transparent governance, secure data exchange, and decentralized service delivery.
- Bio-Integrated Sensors: Enable continuous health and environmental monitoring through wearable or embedded devices.

Each of these technologies opens new avenues for research, requiring ethical frameworks, infrastructure adjustments, and societal readiness.The future of smart urban technologies is rich with promise and complexity. From digital twins and edge AI to autonomous mobility and ethical data governance, cities are entering an era where technology will not only support but shape urban life.To ensure that future cities are inclusive, sustainable, and resilient, a concerted effort must be made to align research with real-world challenges and community needs. Policymakers, researchers, technologists, and citizens must work in tandem to explore, evaluate, and scale innovations that serve the public good.Smart cities are no longer just about efficiency or convenience—they are about creating equitable, adaptable, and thriving environments for generations to come. The future depends not only on the technologies we build but on how wisely, ethically, and collaboratively we choose to deploy them. Difference between urban living with and without AI and IoT-based social sensing systems is shown in Table 1 (Wang et al., 2021).

Table 1. Difference between urban living with and without AI and IoT-based social sensing systems

Aspect	Without AI & IoT-Based Social Sensing	With AI & IoT-Based Social Sensing
Traffic Management	Manual monitoring, traffic jams common	Real-time traffic flow analysis, smart signals reduce congestion
Public Safety	Delayed emergency response, reactive policing	Predictive crime analytics, real-time alerts to authorities
Waste Management	Fixed route collection, inefficient resource use	Smart bins notify when full, optimized collection routes
Air Quality Monitoring	Periodic manual checks, slow data collection	Continuous monitoring, alerts for hazardous air levels
Energy Usage	High wastage due to static usage patterns	Dynamic energy optimization, smart grids reduce power consumption
Water Management	Reactive leak detection, overuse	Smart sensors detect leaks, monitor water quality and consumption.
Public Transport	Unpredictable schedules, low usage	Real-time tracking, AI-based demand prediction improves service
Urban Planning	Static surveys, outdated data	Data-driven planning based on real-time social and environmental data
Healthcare Access	Limited outreach, delayed diagnosis	Remote patient monitoring, early disease detection via wearable IoT
Citizen Engagement	Low feedback, limited participation	Real-time feedback through apps, AI-analyzed public sentiment
Disaster Management	Slow warning systems, high impact	AI models predict risks, IoT sensors trigger early warnings
Infrastructure Maintenance	Reactive repairs, high costs	Predictive maintenance using sensor data, cost-effective upkeep
Noise & Pollution Control	Difficult to measure or enforce regulations	AI monitors noise/pollution levels, enforces zoning laws dynamically

CONCLUSION

Smart cities represent a transformative vision for urban living—one that leverages technology to enhance the efficiency, sustainability, and livability of modern cities. Throughout this book, we have explored how innovations such as the Internet of Things (IoT), Artificial Intelligence (AI), data analytics, and citizen-centric platforms are reshaping urban spaces across the globe. From smart transportation and energy systems to responsive public services and participatory governance, these technologies are not only solving current challenges but also preparing cities for future growth and resilience.However, the journey toward smarter cities is not without complexity. Issues of data privacy, digital equity, cybersecurity, and ethical use of AI demand careful attention. True success will come not just from technological advancement, but from inclusive planning, interdisciplinary collaboration, and active citizen engagement.Looking forward, the integration of emerging technologies—such as digital twins, edge computing, and decentralized platforms—offers immense potential. Yet, these must be implemented with a strong commitment to transparency, fairness, and sustainability. In conclusion, smart urban development must prioritize people as much as it does infrastructure. Technology should be a tool to empower citizens, strengthen communities, and build cities that are not only intelligent but also humane, just, and future-ready. As cities continue to evolve, the focus must remain on creating environments where innovation and inclusion go hand in hand, ensuring a better quality of life for all.

REFERENCES

Abba Ari, A. A., Aziz, H. A., Njoya, A. N., Aboubakar, M., Djedouboum, A. C., Thiare, O., & Mohamadou, A. (2024). Data collection in IoT networks: Architecture, solutions, protocols and challenges. *IET Wireless Sensor Systems*, *14*(4), 85–110. DOI: 10.1049/wss2.12080

Ahmed, E., Yaqoob, I., Hashem, I. A. T., Khan, I., Ahmed, A. I. A., Imran, M., & Vasilakos, A. V. (2017). The role of big data analytics in Internet of Things. *Computer Networks*, *129*, 459–471. DOI: 10.1016/j.comnet.2017.06.013

Alahi, M. E. E., Sukkuea, A., Tina, F. W., Nag, A., Kurdthongmee, W., Suwannarat, K., & Mukhopadhyay, S. C. (2023). Integration of IoT-enabled technologies and artificial intelligence (AI) for smart city scenario: Recent advancements and future trends. *Sensors (Basel)*, *23*(11), 5206. DOI: 10.3390/s23115206 PMID: 37299934

Ali, I., Ahmedy, I., Gani, A., Munir, M. U., & Anisi, M. H. (2022). Data collection in studies on Internet of Things (IoT), wireless sensor networks (WSNs), and sensor cloud (SC): Similarities and differences. *IEEE Access : Practical Innovations, Open Solutions*, *10*, 33909–33931. DOI: 10.1109/ACCESS.2022.3161929

Alwar, B., Edgardo, H., & Faustino, E. (2024). Integrating IoT and Artificial Intelligence for sustainable smart city development: A case study approach. *Information Technology Engineering Journals*, *9*(2), 80–90. DOI: 10.24235/itej.v9i2.137

Anastasi, G., Antonelli, M., Bechini, A., Brienza, S., D'Andrea, E., De Guglielmo, D., & Segatori, A. (2013, October). *Urban and social sensing for sustainable mobility in smart cities. In 2013 Sustainable Internet and ICT for Sustainability (SustainIT)*. IEEE.

Dawale, S. V., Raja, S. K. S., & Darsan, R. V. (2025). Urban living through AIoT integration in smart cities. In *AIoT* (pp. 312–339). Auerbach Publications. DOI: 10.1201/9781003482338-14

Doran, D., Severin, K., Gokhale, S., & Dagnino, A. (2015). Social media enabled human sensing for smart cities. *AI Communications*, *29*(1), 57–75. DOI: 10.3233/AIC-150683

Dwivedi, B., & Behl, R. (2025). Case study: Successful artificial intelligence (AI) and the Internet of Things (IoT)(AIoT) implementation in smart cities. In *Merging Artificial Intelligence With the Internet of Things* (pp. 167–202). IGI Global. DOI: 10.4018/979-8-3693-8547-0.ch006

Fotiou, N., Siris, V. A., Mertzianis, A., & Polyzos, G. C. (2018). Smart IoT data collection. In *2018 IEEE 19th International Symposium on "A World of Wireless, Mobile and Multimedia Networks" (WoWMoM). IEEE*, 588–599. DOI: 10.1109/WoWMoM.2018.8449766

Hancke, G. P., & Hancke, G. P.Jr. (2013). The role of advanced sensing in smart cities. *Sensors (Basel)*, *13*(1), 393–425. DOI: 10.3390/s130100393 PMID: 23271603

Herath, H. M. K. K. M. B., & Mittal, M. (2022). Adoption of artificial intelligence in smart cities: A comprehensive review. *International Journal of Information Management Data Insights*, *2*(1), 100076. DOI: 10.1016/j.jjimei.2022.100076

Kamrowska-Załuska, D. (2021). Impact of AI-based tools and urban big data analytics on the design and planning of cities. *Land (Basel)*, *10*(11), 1209. DOI: 10.3390/land10111209

Kannammal, A., & Chandia, S. (2023). Applications of AI and IoT for smart cities. In *Research Trends in Artificial Intelligence: Internet of Things* (pp. 186–202). Bentham Science Publishers. DOI: 10.2174/9789815136449123010014

Liu, P., Zhang, Y., & Biljecki, F. (2024). Explainable spatially explicit geospatial artificial intelligence in urban analytics. *Environment and Planning. B, Urban Analytics and City Science*, *51*(5), 1104–1123. DOI: 10.1177/23998083231204689

Padhiary, M., Roy, P., & Roy, D. (2025). The future of urban connectivity: AI and IoT in smart cities. In *Sustainable Smart Cities and the Future of Urban Development* (pp. 33–66). IGI Global.

Sanchez, T. W., Shumway, H., Gordner, T., & Lim, T. (2023). The prospects of artificial intelligence in urban planning. *International Journal of Urban Sciences*, *27*(2), 179–194. DOI: 10.1080/12265934.2022.2102538

Shafique, K., Khawaja, B. A., Sabir, F., Qazi, S., & Mustaqim, M. (2020). Internet of Things (IoT) for next-generation smart systems: A review of current challenges, future trends and prospects for emerging 5G-IoT scenarios. *IEEE Access : Practical Innovations, Open Solutions*, *8*, 23022–23040. DOI: 10.1109/ACCESS.2020.2970118

Shahid, K., Hassan, M., Husnain, A., & Mukhtar, S. (2022). Information and communication technologies for new generation of sustainable smart cities. In *Sustainable Smart Cities-A Vision for Tomorrow*. IntechOpen, 1(1), 1-10.

Son, T. H., Weedon, Z., Yigitcanlar, T., Sanchez, T., Corchado, J. M., & Mehmood, R. (2023). Algorithmic urban planning for smart and sustainable development: Systematic review of the literature. *Sustainable Cities and Society*, *94*, 104562. DOI: 10.1016/j.scs.2023.104562

Wang, K., Zhao, Y., Gangadhari, R. K., & Li, Z. (2021). Analyzing the adoption challenges of the Internet of Things (IoT) and artificial intelligence (AI) for smart cities in China. *Sustainability (Basel)*, *13*(19), 10983. DOI: 10.3390/su131910983

Chapter 6
Integration of AI and IoT for the Optimization of Social Sensing in Smart Cities

Monika Kumari
http://orcid.org/0000-0002-8468-3552
Poornima University, Jaipur, India

Nikhil Kumar Goyal
http://orcid.org/0009-0007-4532-8033
Noida international University, Greater Noida, India

ABSTRACT

This chapter discusses the architecture layouts, integration schemes, as well as practice of AI-IoT systems that can target social sensing optimum in smart cities. It encompasses various AI methods, namely machine learning, deep learning, and reinforcement learning, which increase sensing precision, minimize data duplication, and provide a capability of adapting to urban services. The chapter illustrates the contribution of AI in the IoT framework conducted through exemplary implementation use in traffic management, environmental monitoring, citizen engagement, and public safety. It also deals with the privacy, security, and ethical issues of large-scale sensing systems and provides future directions of research such as use of edge computing, human-centric AI and integrating with 6G networks. The chapter is a guideline to researchers, practitioners, and policymakers who would wish to construct responsive, smart, and inclusive smart cities.

DOI: 10.4018/979-8-3373-4962-6.ch006

1. INTRODUCTION

1.1 Smart Cities and Social Sensing

Smart cities are the development of our society, which applies the benefits of digital technologies to ensure better living conditions and resilient ways of life of people, the efficiency of city operations, and strong sustainability. The force behind this change is the mode of a social sensory which can be described as a method that involves the collection and analysis of human and automated information to extrapolate pattern, trends and even real-time activities in the cities. This comprises such data as that found in smart phones, GPS gadgets, CCTV cameras, social media and distributed IoT sensors. It is possible by means of social sensing to make the city more conscious, alert, and proactive in managing infrastructure and services.

Regarding the city design and management, social sensing as a type of cyber sensing provides a means of communication between the urban authorities and the population of the cities, enabling to gather data in real time about traffic parameters and air pollution, mood, and crises. The insights are not only applied to give response to the current operation but also to give long-term policy and infrastructural development. The input power and granularity of the social sensing exceed conventional survey-based systems, or periodic monitoring systems, and they help in delivering situational awareness (Atzori et al., 2010).

Figure 1. Sensing in smart cities

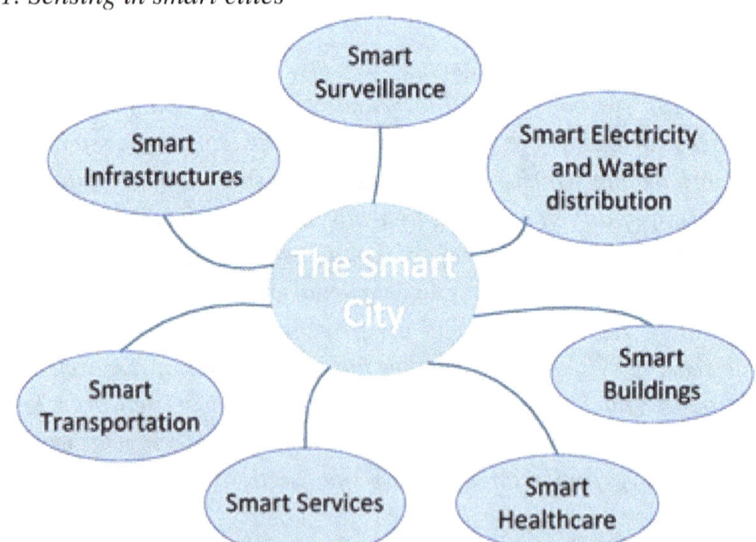

Nonetheless, the efficiency of social sensing relies on the capability to filter and analyze huge, diverse, and frequently imperfect information. Urban data is too large and complex to be managed with traditional data management methods. Here is where the use of AI and IoT is essential. Intelligent processing with sensor-based data access would allow smart cities to become more than simply connected but instead into thinking systems that would be able to change with the environment and need of the citizens.

The future of social sensing in smart cities is massive, however to enable it to be implemented; a multilateral way of seeing through technological advancement, a model of governance and community involvement is needed. The chapter addresses and discusses the ways in which the combination of AI and IoT infrastructures can optimize social sensing, opening up the smarter, safer, and more inclusive urban futures.

1.2 IoT and AI in Urban Infrastructure

The Internet of Things (IoT) is the nervous system of any smart city is able to connect millions of sensors and gadgets that track the air quality to the energy consumption and pedestrian traffic. Data gathered by these devices is analogy to the heart beat of the city. But this data is not fully made use of unless it is put through some smart processing or it turns overwhelming. IoT is complemented by Artificial Intelligence (AI), which will 'brain' this tightening flow of information by reading it, learning, and responding to it (Goodfellow et al., 2014).

Machine learning, deep learning, and reinforcement learning are some of the AI techniques that allow recognizing patterns, detecting anomalies, implementing predictive analytics, and autonomous decision-making. As an example, an intelligent traffic system can take the camera feeds (as an input of IoT) and create an AI algorithm that would determine the congestion, calculate the traffic, and readjust the signals accordingly. On the same note, AI would also be able to make predictions based on information provided by environmental sensors and recommend what could be done to curb pollution.

The combination of AI and IoT is also applied to the optimization of resources. AI programs can predict energy distribution, depending on the consumption observed through the IoT meters and then perform the distribution automatically. Smart bins that can communicate the amount of waste can be used in waste management and AI can optimize routes of collection. This effective interplay between sensing and intelligence makes urban infrastructure turn the reactive systems to the proactive ones (Kingma & Welling, 2013).

In addition, AI increases the scalability and the security of IoT networks. Edge AI, in which processing tasks are near data sources, can lessen those two categories and their effect on management and uphold privacy and functioning and business continuity in the case of connectivity interruptions. In combination, the AI and the IoT create a loop where something that senses, thinks, and acts exchanges the sensory input and responds radically transforming the urban infrastructure into an adaptive ecosystem of learning and living.

1.3 Harnessing AI and IoT for Intelligent Social Sensing in Urban Environments

The main goal set in this chapter is to explore the way the involvement of AI and IoT technologies empowers the mechanism of social sensing in smart cities. It seeks to elaborate theoretical concepts, practical realizations and current trends of the AI-IoT merger, especially given the need to optimize the process of data gathering, processing, and utilizing socially-oriented information. It discusses centre-periphery (centralized) architecture and centre-free (decentralised) architecture, real-time analytics and possibility of human-in-the-loop systems where decisions are adaptive and context-sensitive.

One more vital goal is to look at what AI can do to enhance the accuracy, trustworthiness, and explainability of social sensing. This covers not only recognizing how AI algorithms can distinguish the signal and noise, identify anomalies in behaviours or environmental states, and tailor services to citizen preferences or risk profiles. The chapter further mentions the drawbacks of the existing systems as well as architectural and policy-related suggestions to mitigate the shortcomings.

This paper goes further to discuss the ethical, legal, and social concerns of the use of AI and IoT in the public realm. The character of social sensing is associated with the gathering of individual and communities data, which triggers issues of monitoring, permission and algorithmic bias. A balanced view is also provided that recognizes futuristic potential of the AI-IoT systems and the responsibility that the systems are bearing in regard to fair governance.

Finally, the chapter covers the theoretical scope of technologies as well as practical, including the collaboration between computer scientists, urban planners, policymakers, and civil society. It also acts as a guide to the researchers and other practitioners who wish to develop smarter cities which are not only technologically superior but also socially responsible and focus on citizens.

2. FOUNDATIONS OF SOCIAL SENSING

2.1 Understanding the Dynamics of Social Sensing

Social sensing Wikipedia, the free encyclopedia Social sensing Yahoo! Answers Social sensing is the process of gathering information using a mixture of human sources and smart devices to model and act on the events in physical and social worlds. It works on the principle that human beings and machines can be sensors and human beings do this through their activities, behaviours and online utterances; and machines, doing this through sensor data in mobile phones, wearable devices, CCTV and infrastructure integrated with IoT. This mixed data ecosystem delivers profound real-time information about conditions in a city and requirements of its citizens (GUBBI ET AL., 2013).

Social sensing can be of two typologies participatory and opportunistic. Participatory sensing means the involvement of the user directly in the process and the user can report incidents into an app on her mobile phone or share location-tagged data in social media. Opportunistic sensing however, is collection of data by pass through processes such as GPS tracking or wi-fi signals and it does not require any direct user interaction. Both adds to the bigger context of the dynamics of the society but have a different standards in terms of data fidelity, coverage, and ethics.

The advantage of the social sensing is that it is spatially and temporally granular. In comparison with the traditional survey methods, social sensing allows permanent surveillance and offers local data, which can be used to target particular locations with interventions. To give an example, the water level can be monitored during a flood by the sensors and people can communicate evacuation problems through social sites. These sources give a wholesome picture when used in combination, which can be used to respond and distribute resources in emergencies.

The difficulty, however, is that social data are unstructured and noisy and, in some cases, even biased. Meaning and relevance extraction requires using AI techniques, especially natural language processing algorithms and anomaly detection processing algorithms. To establish strong smart city solutions which are responsive, reliable as well as fair, it is important to have a good grasp of the technical basis of the social sensing concept (Lee & Lee, 2015).

2.2 Data Input Streams in Social Sensing Frameworks

The social sensing is dependent on a wide range of data sources, which may be grouped into vast segments of data about physical signals, digital signals, and behavioral signals. Physical signals comprise of data/information via the fixed and mobile sensors situated in infrastructure, automobiles and wristbands. The sensors detect the real-time data regarding the environmental properties, locomotions, and physiological analysis, which are used in cases such as the monitoring of pollution and evaluation of the population health.

The digital signals rely on online media and internet communications. These are the geotagged tweets, the search queries done, the statistics used in apps, and the contents used in the social network. Digital evidence allows discovering user interests as well as their opinions and patterns of social behavior that can apply to decipher the sentiment of groups and identify emerging concerns. Social media such as twitters and Instagram has been adopted in sensing of disasters, policing and even urbanization (Goyal et al., 2025a).

Behavioral cues are an expression of how individuals deal with surroundings. An example of behavioral indicators is swipes in the public transportation card, location history of mobile users, and energy consumption records. The sources assist cities to know traffic jam patterns, what time is high time utility consumption or crowd traffic during any major event or an emergency.

These mixed heterogeneous sources of data produce a strong dataset; yet one that requires strong fusion techniques. AI is needed to match these signals, clean them, and make sense. Techniques like multi-modal fusion using deep learning can allow the correlation of sensor measurements with social media activity to provide a better situational understanding and more successful actions to take in the city (Chen et al., 2019).

2.3 Challenges in Social Sensing in Urban Environments

In spite of its benefits, social sensing in city environments has numerous challenges and the first challenge is the quality and reliability of data. Misinformation, sarcasm, and ambiguity, which are likely to occur in social media content, may also become the reasons that AI systems fail to be guided by such content. Sensor data is more structured but is prone to problems caused by the calibration, environmental noise or communication latency. Basing accuracy and uniformity of such disparate inputs is a basic challenge.

Misbalance in the sensing sources is also another problem. The implications of digital divides between urban neighbourhoods are that more data might be produced in more affluent or tech-savvy regions of the city whereas disempowered groups are underrepresented. It creates an imbalance that results in partial studies and possibly discriminatory policy-making. AI models must also address these disparities, and doing so critically requires AI models to address it with statistical techniques such as data augmentation, resampling, or even with fairness-aware algorithms (Wang & Wang, 2021).

There are privacy and ethical issues that are looming large. The constant surveillance of citizens even when with good intentions creates the fear of being spied as well as being deprived of liberty. Design practices to allow transparency, consent and data minimisation must be incorporated in the system always to make it trustworthy by the people. Technical solutions like differential privacy, federated learning, and secure multiparty computation look bright, but their use in the real world remains immature.

Lastly, there is the issue of integration with current infrastructure which brings about practical barriers. Many legacy systems are neither interfaced nor compliant to contemporary IoT solutions or AI engines. There have to be cross-agency efforts to institute cross-agency systems, common data standards and common governance organizational structures. Synthesis of this kind will ensure that the prospects of social sensing are not underutilized, and cannot provide the promise of smarter life in the city (Kitchin, 2014).

3. INTERNET OF THINGS (IoT) IN SMART CITIES

3.1 Architecture of IoT-Based Urban Systems

The IoT systems found in smart cities architecture generally consist of three-slided architecture, such as perception, network and application. The perception layer involves inclusion of all the sensors, RFID tags, GPS modules and cameras that capture the real-time data within the physical environment. These gadgets are the first line of defense, in detecting changes in urban conditions, like the air quality or the pedestrian traffic or even traffic congestion.

Figure 2. IoT Architecture for Smart Cities

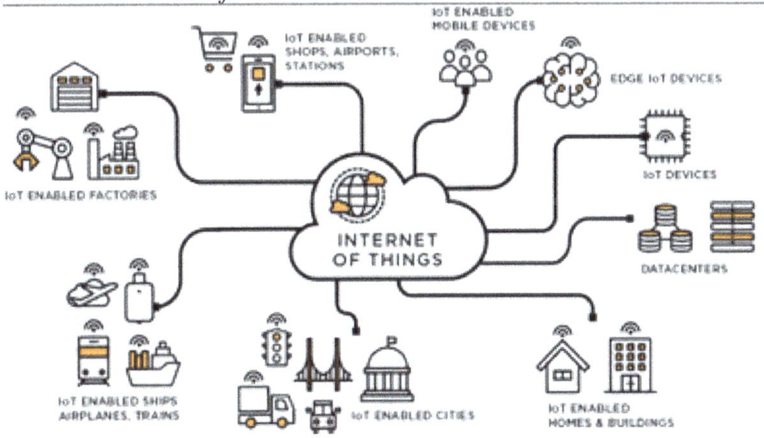

The network layer performs the role of secure and efficient relaying of gathered information. It links the perception devices to the storage and processing nodes along with protocols such as 5G, Wi-Fi, Zigbee or LPWAN. This layer defines the scalability, latency and resilience of the system based on its robustness. In large-scale network system installations, network redundancy and dynamic routing mechanisms are important to maintain the stable data flowv (Satyanarayanan, 2017).

Application layer Cloud and edge platforms, analytical engines and user interfaces constitute the application layer. This layer feeds raw data into actionable insight as an AI model, decision rule, and graphs. As an example, smart parking applications and predictive maintenance systems provide utility managers with real-time availability maps on the basis of sensor input and inform them about possible infrastructure failures in advance. Raw sensing links with the civic action through the application layer.

Fog and edge computing paradigms also find their way in a modern IoT architecture. These allow decentralised processing locally to the source of the data, decreasing load on the network, and speeding up the response time. Edge intelligence is also essential to future-proof smart city IoT because such levels of decentralization are crucial to latency-sensitive applications such as autonomous traffic control or emergency services (Raji & Buolamwini, 2019).

3.2 Sensors, Devices, and Edge Infrastructure

In the environment of a smart city, sensors and IoT devices are the most important tools of a real-time data collection. The examples of such devices are environmental sensors that monitor the air situation and temperature in the area, motion detectors that track the flow of people, cameras to check traffic, and also RFID sensors to track the assets. More developed environments include wearable health monitors and home appliances or smart gadgets that contribute a personal and domestic layer to city (or urban) data collection. Diverse and granular nature of such devices enables cities to understand their operations in a multidimensional manner.

Figure 3. IoT Edge Computing

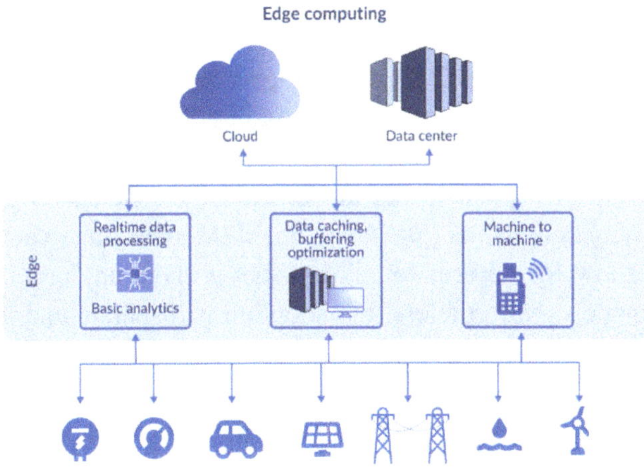

Their application of these sensors should be calculated and comprehensive. Locating sensors in commercial or affluent areas may leave data blind spot in underserved areas, under which bias occurs in decision-making systems. Further on, the sensors need to be adjusted, fixed, and frequently upgraded to reflect precise data acquisition. Redundancy of sensors and self-diagnosis aspects can aid in reducing data loss and downtimes which in turn enhances the reliability of the sensing network.

The edge infrastructure is an essential part of the IoT system efficiency. Instead of using the cloud servers fully, which can cause latency and bandwidth issues, edge computing enables processing the data at a local level, on the gateways, routers or even on the sensors. Such decentralization decreases reliance on high-bandwidth Internet connectivity, minimizing the per-transmission expense and raising responsiveness in real-time (one of the crucial applications being emergency identification, factory automation, or preventative maintenance) (Shi et al., 2016).

Moreover, edge computing guarantees enhanced privacy because the number of raw data going beyond the source is restricted. More sensitive data, including facial images or biometric readings, may be analyzed at the edge and only summary information or metadata are shared. As edge devices are becoming AI-enabled, cities will operate more autonomic, self-governing systems that will decrease human load and increase situational awareness in all sectors.

3.3 Data Collection and Interoperability Challenges

But this large number and variety of IoT devices implemented during the Smarter City is associated with another equally problematic issue of data collection and interoperability. The frequency, structure and format of data differ and varies by type of sensor and the manufacturer of the sensor. Other devices can send data in real-time, whereas others can send a periodic summary. The lack of standardization makes it a challenging process to unify these heterogeneous data streams into a system.

Among the most critical interoperability issues, there are proprietary hardware and software ecosystems. The various vendors have varied communication protocols, storage formats and APIs, which are fragmented and inefficient. Organizations like Open Connectivity Foundation (OCF) and oneM2M are trying to develop common protocols that allow IoT devices to talk together without any problem, regardless of brand and models. Nevertheless, the adoption is disparate and legacy systems are not usually accommodative of new interoperability frameworks (Goyal, 2025a).

Synchronization and consistency of the data are also crucial issues. And when it comes to combining data conveyed by thousands of sensors the time delays, packet loss of communication interference may lead even to incomplete or inconsistent data. They impact the sensible propensities of AI models prepared on this information and diminish belief in computerized ruling frameworks. The protocols of synchronization, databases of time-series, and data buffering should be instantiated in a way that such eventualities are dealt with without affecting performance negatively.

The issue of privacy and security also makes the process of collecting data harder. The attacks that can be launched on the IoT networks include spoofing, data manipulating, and unauthorized access. End-to-end encryption, strong authentication of the device, and audit path via blockchain may serve as additional trust and resilience. To streamline social sensing on smart cities, the functional interoperability of IoT systems is not only necessary, but also secured, scalable, and ethically established.

4. ARTIFICIAL INTELLIGENCE FOR URBAN OPTIMIZATION

4.1 AI Models for Data Analysis and Prediction

The Artificial Intelligence has transformed the process of interpreting the urban data, allowing real-time decisions and long term strategic planning. The use of AI models (especially based on machine learning (ML) and deep learning (DL)) enables the systems of the cities to recognize patterns, predict, and react dynamically. Prediction analytics may also be used in smart cities to predict congestion on the road, energy requirements, outbreaks of disease and the levels of pollution, before measures have to be taken to rectify the situation (D'Andrea, 2020).

Supervised learning models are typically applied in the situation where labeled historical data is provided. As an example, a regression model or neural network can predict eventual traffic congestions during peak hours based on historical data on traffic flow. Conversely, un-supervised models like clustering algorithms can be used to determine anomalous behavior in vast data, like tracking of potential cyber threats in smart grid networks or realization of abnormal water consumption to show leakage.

Dynamic decision making in uncertain environment is also picking up the idea of reinforcement learning (RL). RL agents are known to learn optimum policies with the use of trial and error and are therefore applicable in such areas as adaptive traffic signal control or robotic delivery systems. They make a very good choice when it comes to balancing short-term operations and long-term benefits in terms of resource allocation in smart infrastructures (Yassine et al., 2019).

Also, the simulations with the assistance of AI can be very useful in terms of the urban planning process. They enable city planners to simulate hypothetical situations, he or she may include new routes of transportation or the growth of population; to check how it can affect the area based on AI-driven projections. This futuristic trait is very instrumental in making sure the smart cities develop sustainably and inclusively.

4.2 Machine Learning and Deep Learning in Urban Contexts

Machine learning (ML) also plays the analytical foundation of the largest part of AI applications in smart cities. Examples that have been extensively applied throughout the mobility, energy, and community security sectors include decision trees, support machine classes and squadronization designs. Each system that adopts ML algorithms would be able to learn continuously about the new data as it comes and leverage any changes in urban dynamics and citizen behaviors (Li et al., 2015).

Deep learning (DL), a subdivision of ML, uses several- layer neural networks to extract complicated abstraction in data. In the analysis of visual data, more specifically CCTV or drone footage, CNNs have been found especially useful to carry out, as objects can be detected in real-time, license plates read, or actions be picked out. In the same vein, Recurrent Neural Networks (RNNs) and their more sophisticated versions, e.g., LSTMs, are applied to the temporal sequence predictions, including electricity consumption forecasts or the patterns cognition in the traffic.

Deep learning has been strong in the sense of being able to work with unstructured data, like images, audio, and natural language, a property that makes them suitable in the case of social sensing. Take, as another example, Natural Language Processing (NLP) models that can be used to draw actionable insights out of tweets in the time of emergencies, or audio classifiers identifying aggression or distress in a place of community gathering (Wang et al., 2020).

Nevertheless, the privacy of most deep learning systems leads to disappointing transparency and explainability. Democratization of AI, which aims at making AI decisions in publicly provided services transparent and explainable, is an important effort toward achieving that goal especially under regions where human rights and trust are primarily the concern, such as law enforcement or healthcare.

4.3 AI for Event Detection, Forecasting, and Decision Making

Among the most revolutionary impacts of AI on smart cities, one should note the capability to detect, predict, and react to urban events. Event detection refers to the process of detecting critical events (or abnormal events) in real time real-time accident, fire, or civil turbulence, based on sensor, video camera and social media data. Anomalous movement patterns, aural fingerprint, or emotion peaks can be identified by AI algorithms at a rate higher than human operators.

The proactive management of the city is possible due to forecasting capabilities. As an example, machine learning models learned on weather and infrastructure data can point out potential flood areas or the places in power grid that will be stressed based on priorches numbers. In healthcare, AI can assess the trend in mobility patterns and social demographic information to predict the spread of an illness to give way to precautionary measures like immunization or quarantine (Li et al., 2018).

The other crucial use of AI in urban governance is decision support. Since AI systems make data-driven recommendations, planners and administrators can better design and improve efficiency in operations. AI: Multi-agent systems Multi-agent systems are a form of AI that models interplay between elements of a city, including public transit, utilities and emergency services so that responses to even the most complex situations can be coordinated.

Cities are becoming more complex as they become larger and so the time-sensitive decisions that are made with AI should be a necessity. It must however be supplemented with ethical control, human-in-the-loop, and constant user feedback to promote openness and equity as well as sustainability and utilitarian merit in the long term.

5. AI-IoT INTEGRATION FRAMEWORKS

5.1 Edge and Fog Computing Models

With AI and IoT, the smart city needs effective computing models that would strike a balance with latency, bandwidth, and scalability. The edge and fog computing offer infrastructure where data processing is handled at its origin instead of being carried out to remote cloud servers. This is especially critical in programs that need to be quick in response like autonomous vehicles, emergency detection and industry automation.

Edge computing means that data are processed at the device or gateway itself and it minimizes processing delays and saves bandwidth. Edge-based implementations of AI can detect objects, recognize gestures, or predictive maintenance without requiring access to significant resources. Edge AI also boosts privacy by reducing the amount of raw data transmitted, which is particularly urgent in the case of sensitive data, such as surveillance footage or health indicators (Mehmood et al., 2020).

Fog computing images a mid-layer of the edge and the cloud. It allows preprocessing, aggregation and local storage over a distributed node network. Multilevel design helps distribute the workload, lowers dependency on cloud resources as well as providing interoperability between non-homogeneous IoT systems. Examples of AI use in fog settings are real-time optimisation of transport networks, localised environmental warnings, and energy management in real-time.

Collectively, edge and fog computing enable hierarchical processing paradigm, which can improve the integration of AI and IoT. These will allow smart cities to grow effectively and be resilient, responsive, and user-oriented in data-driven services.

5.2 Real-Time Processing and Context-Aware Analytics

The intelligent cities are dependant on prompt handling of big data flows. It uses real-time analytics powered by AI models to have systems to make sense of the data collected and be able to interpret it in milliseconds that can only display the time-sensitive services being offered including managing traffic lights, detecting crime, and responding to disasters. This necessitates high-throughput data pipelines, low-latency inference engines and event-streaming frameworks capable of low-latency and high-throughput operations.

The characteristic of intelligent urban systems is context-awareness. It is necessary to refer to the possibility of adapting outputs of the system according to situational parameters (location, time, weather, or user profiles). As an example, a smart lighting system can control the visibility according to the adjacent and presence of people. Context-aware analytics relies on metadata, sensor fusion and environmental clues to adapt the outputs of AI to enable more personalized and efficient services (Zhang et al., 2017).

Real-time ingested and processed data are used with streaming AI frameworks like Apache Flink, Kafka Streams and Spark Streaming. These platforms are able to support dynamically updated rules, fault tolerant, and supports scale-out features and thus will be appropriate to roll out at city wide level. When combined with AI models, they allow the development of systems with not only a fast speed but also an intelligent and adaptable one (Goyal et al., 2025b).

The problem is that such technical abilities have to be considered by ethical and policy aspects. To give an example, the context based surveillance should be controlled in order not to come to intrusive monitoring. Real-time intelligence with privacy, transparency can only be realized through collaborative models though the participation of technologists, ethicists and city stakeholders.

5.3 Scalable Architectures for AI-IoT Fusion

AI-IoT systems in smart cities have to guarantee the scaling feature. Cities are becoming more connected, and the proportion of connected devices and data flowing out rise exponentially. The scalable architectures have to enable horizontally, which involves the addition of new nodes and services without interruption of the existing operations, and vertical scaling, which enables more powerful processing over time as requirements increase (Aral, 2020).

With architectures based on microservices, it is possible to adopt modular architecture where each service does a certain task and can be separated, and updated or scaled independently of other services. This is quite consistent with the concepts of AI-IoT fusion, as it is possible to manage separate parts, such as data ingestion, model inference, and visualization shelf. Elastic scaling such as containerization (using such tools as Docker) and resource-efficient allocation with the help of orchestration (with such products as Kubernetes) are supported.

Communication between distributed elements of a smart city system is facilitated by message brokers like MQTT, AMQP and Kafka. These tools provide the reliability, order, and scalability of messages across the devices, the edge nodes, and the cloud platform. They are very significant when it comes to ensuring smooth integration of AI IoT on various levels and regions across the world (Badii et al., 2019).

With the help of scalable architectures, cities will be able to make their smart infrastructure future-proof, cost-effective, and ready to receive a variety of new technologies, devices, and use cases without their performance being impacted in any way.

6. SOCIAL SENSING OPTIMIZATION STRATEGIES

6.1 Data Fusion and Filtering Techniques

The situation is similar in social sensing, where many data sources, including sensor measurements, social media, and any available mobile app log, have to be combined and transformed into a coherent and actionable data collection. Data fusion is a process of combining data obtained by various sources to amplify the malice and sufficientness of the sensing. This may be accomplished on numerous levels: data-level combination combines uncooked knowledge, feature-level combination gathers vital properties, and choice-level combination mixes the conclusions of separate models (Lin et al., 2017).

Filtering aspect is also important to get rid of noise, redundancy and false positives. Sensor data are refined with Kalman filters, particle filters and statistical methods of smoothing, and text data on social media are refined and ranked through NLP-based models. Signal-to-noise ratio is increased through filtering, which is applicable in emergency conditions when it is important to sense quickly but accurately.

Fusion approaches based on machine learning The models involved in the machine learning-based methods may be random forests, SVMs or even ensemble networks to identify those most dependable data sources at given circumstances. The next level of fusion capabilities is brought in by deep learning, where one can observe the attention mechanisms, where we dynamically weigh importance of inputs depending on the context.

Fusion and filtering not only enhance quality of insights, but also ease downstream analytics because of reduction in the burden of computations. This further streamlines the overall system and increases its accurateness and responsiveness to ensure a high level of utilization of both the machine and human-generated data in the work of smart cities (Zanella et al., 2014).

6.2 Redundancy Reduction and Accuracy Enhancement

The problem with the redundancy of data emerging in urban sensing systems is mainly caused by a large number of sensors and overlapping streams of data. Although redundancy may be good in fault tolerance, it also causes wastage of storage, higher transmission fees and difficulty in interpretation. Diminishing redundancy is the action or process of finding duplicate or non-valued information and cutting off that information without jeopardizing important information.

Principal Component Analysis (PCA), t-SNE and autoencoders are dimensionality reduction methods that can be used to remove redundancy in high-dimensional data. These techniques keep the most informative part of the data disposing of noise and redundancy. The Machine Learning models also can be trained to identify and suppress the duplication in our sensor or social media data streams (Goyal et al., 2025c).

It increases accuracy by calibration sensor, outlier detection, and ensemble models. Complex environments with highly unknown environments tend to be better off with ensemble techniques than single model systems. The use of feedback loops in which models learn to reduce errors in the past also adds accuracy to the model as time goes on.

One of the most important considerations is the trade-off that occurs between redundancy and diversity loss. Homogenization of data may occur through over-optimization and take away valuable exceptions. Therefore, the optimization practices should be well-balanced to provide resilience and generalization, in terms of social sensing.

6.3 Crowdsourcing and Incentive Mechanisms

Social sensing also essentializes crowdsourced data when sensors infrastructure is weak or when insights that are personalized to the user are required. An example of the crowdsourcing tasks may include reporting damage or malfunctions to infrastructure, providing location information, or making contributions to real-time warnings of hazards. Nonetheless, it is difficult to encourage or persuade users to input quality data at all times (Tzounis et al., 2017).

The participation in crowdsensing systems is promoted by incentive mechanisms. They may be financial, like micro-payments or vouchers, or non-financial, like reputation systems or gamifications, or on social public recognition. Effective incentive systems need to be designed by knowing what motivates the users and aligning them to civic ends.

Incentives can be personalized via AI that predicts the patterns of user behavior and interests. As an example, reinforcement learning could optimize on incentive strategies, consistently and at low costs. Moreover, smart contract tokens and blockchain-based tokens can bring an open source and secure reward to enable trust and confidence in the systems of data donations (Khan et al., 2020).

It is essential that data authenticity and manipulation prevent crowdsourcing. The verification processes, including a consensus vote, triangulation in data, and anomaly detection, have to be incorporated to ensure data integrity. Performed in a responsible manner, crowdsourcing with AI analytics thus can play an important role in expanding the scale and reach of urban social sensing.

7. APPLICATIONS IN SMART CITIES

7.1 Smart Mobility and Traffic Monitoring

The role of AI and IoT in social sensing is one of the most evident and effective as it is related to the field of smart mobility. With rising traffic volumes and the resulting delays in means of transport and insufficient security in the streets, cities are becoming overpopulated with increasing traffic and an excessive amount of auto vehicles. Sensors installed as IoT, such as traffic cameras, inductive loops, GPS trackers and mobile devices, produce data in real-time regarding vehicle movement, transit occupancy and road conditions. In combination with AI algorithms, such data can be used to model the predictive traffic behavior, dynamic route optimization, and dynamic signal control.

Traffic flow patterns may be analyzed in machine learning models and congestion points may be predicted hours before they occur, allowing preemptive rerouting to optimize commute time. Traffic signal control is undergoing adaptive reinforcement learning, so that traffic signals are controlled in response to changing real-time information, rather than fixed schedules. The predictive maintenance models, which rely on data on usage rates and vehicle diagnostics, also facilitate the work of the public transportation systems and allow them to avoid violating services (Alsaadi & Tubaishat, 2015).

Social sensing furthers these abilities by the means of adding human behavior to the mobility models. As an example, the information on ride-sharing applications, social media check-ins to events, or mobile activity schedules may indicate future traffic or transportation increase. This social layer adds background that could be lacking in pure sensor information so that traffic systems respond more to physical situations.

Overall, AI and IoT when used in mobility are encouraging safe roads, less carbon emissions, and overall customer satisfaction. As we move to mature stages of such systems, the prospect of an autonomous traffic ecosystem with the help of vehicle-to-everything (V2X) communication looks more and more feasible.

7.2 Urban Safety and Emergency Response

The security and safety of urban areas are a priority in any smart city framework and AI-IoT systems greatly improve the capacity to survey, predict, and react to any emergencies. The central sensor is built on IoT-based surveillance cameras, gunshot sensors, air quality sensors, and vibration sensors. This data is studied by AI algorithms to determine anomalies, detect crime, or predict dangerous activity.

In real-life context, deep learning-based object detection and behavioral recognition models to use AI to identify suspicious activity based on CCTV feeds. Mobile data or videos that could be used to analyze some ways to track the density of people in the crowd and possibly avoid stampedes or use in evacuation. Social sensing comes in here, as citizens report incidents using apps or social media and this is processed with the use of natural language processing that acts as a filter and locates the information using geolocation (Goyal et al., 2025d).

In the case of calamities like earthquakes or floods, the AI in smart city systems helps in organising rescue and relief by mapping out the worst hit places depending upon sensor signals and calls by citizens. Cameras installed on drones and sensors to monitor water levels using IoT technology send information to algorithms that evaluate the threat to infrastructure and propose which area should be addressed first.

Such integrated systems make emergency response quicker, more specific and efficient. Nonetheless, it also poses serious ethical questions of surveillance, facial recognition and civil liberties where sound ethical control is demanded as well as technological progress.

7.3 Environmental Monitoring and Public Health

The other essential application of social sensing through the integration of AI and IoT in improving the urban sustainability is on environmental monitoring. Sensors of air and noise pollution installed in cities constantly gather data, which is treated in artificial intelligence models in order to predict the pollution trend, find the hotspots and suggest ways of mitigation. AI-based warnings warn citizens of health threats, especially the sensitive groups of people including children and the aged.

Smart monitoring systems are also useful in water quality and waste management. Water distributions are fitted with sensors to indicate contamination or leakage and smart bin to indicate when a bin is full to reduce collection routes. These systems are more responsive and comprehensive when integrating social information (complaints filed by citizens or health statistics of populations), in other words when placed on top of social data (Guo et al., 2017).

Wearable IoT devices and health apps in the public health can use individualized data streams to aggregate and analyze them to identify outbreaks, trace symptoms or deploy resources in the best way. In the case of pandemics, social sensing is used to model disease spread by using mobility patterns, social distancing patterns, and disease-related keywords across the digital sphere.

It also helps cities use a more holistic and preventative approach through social sensing of environmental and health data. It is a question of finding a balance between the improvement of the public health, the issue of data security and the ethical usage of the person health information.

7.4 Citizen Engagement and Participatory Governance

AI and IoT can also be effectively used as the tool of civic engagement and participatory governance. Citizens provide information via surveys, apps, and their social media usage, and the smart infrastructure would provide facts on their behavior in a passive way. When such inputs are scrutinized together, they give the policy makers a subtle context of the needs and sentiments of communities.

Sentiment analysis, topic modeling and clustering (and therefore, citizen feedback categorization and summarization) can be provided via AI. This will allow the city planners to prioritize issues, identify emerging issues and to customise services accordingly. As an example, an increased number of complaints concerning street lighting in the given district can initiate an automated inspection job run by IoT-managed lighting grids.

Rewards and gamified networks play concepts that motivate citizens to get themselves involved in sustainability activities, including reporting pot holes, recycling or energy conservation. Engagement pattern analysis helps AI models alter such rewarding processes and maximize engagement. Feedback loops provide transparency by keeping the citizens aware of the measures implemented after their participation (Shahzadi et al., 2021).

People-powered AI-IoT-integrated participatory governance does not only enhance democracy but robust urban services and relevance and responsiveness in the form of improved urban services. It is aimed to transform better not just smart and inclusive cities where the voice of everyone will matter towards defining the urban future.

8. SECURITY, PRIVACY, AND ETHICAL CONSIDERATIONS

8.1 Privacy Risks in Social and Sensor Data

Data gathered continuously by the citizens and infrastructure presents a serious risk to privacy. Connected devices tend to collect sensitive data like location, identity, health indicators and behavior patterns. The possibility of intrusion of privacy is quite significant when it is added to social sensing when individuals willingly or unwillingly pass information by interacting digitally.

The issue of informed consent is among the fundamental ones. Data collected about the citizens might be done in large scale, which is not known by the citizens. Yearning or inactive gathering of information via on-street Wi-Fi, surveillance cameras, or the mobile apps commonly has no apparent approach for opting out. Furthermore, re-identification involving anonymized data can occur as well, when additional data is added to such information, which causes the violation of privacy.

These issues are increased by the use of AI, which may make the inference about an individual, like predicting his or her habits, affiliations, or emotional condition. This poses ethical concerns of the use of profiling, surveillance and digital manipulation, especially within sensitive context such as policing or health care.

To offset these risks, principles of privacy-by-design have to be incorporated in each tier of the smart city architecture. These cover data minimization, decentralized data storage, differential privacy, and data governance policies that are transparent to the citizens so that they know more about how their data are utilized and can regulate them.

8.2 Data Anonymization and Access Control

It is essential to ensure data collected does not even jeopardize individual identities and this has to be ensured through strong levels of anonymizing the data and access control. Methods like k-anonymity, l-diversity, and t-closeness are used to mask identifiers of an individual whilst maintaining the analytical value. Nevertheless, the methods are not universal and tend to corrupt data.

More modern techniques such as randomized response and differential privacy involve adding randomness to data, and have mathematically provable privacy guarantees. Federated learning allows learning on decentralised devices with local data, and only model updates are exchanged at scale-this is much safer when compared to direct exposure to raw data. These methods find particular application in healthcare or individual, customized service areas (Nguyen et al., 2021).

The control of access is also essential. Role-based access control, information encryption, and block-chain-ledger registries make it impossible to access sensitive data to unauthorized personalities. It is also possible to have transparent audit trails of all data transactions through blockchain technology at the expense of better accountability and trust.

There is a fine line between the usability of data and privacy. On the one hand, the over-anonymization may make the data useless in terms of decision-making, whereas the underprotection may lead to a wave of popular discontent and compliance breach. Consequently, research and regulatory convergence are more needed than ever to uphold the worth of data as well as the rights of the citizens.

8.3 Governance and Regulatory Frameworks

Regulatory frameworks are important to provide that AI and IoT technologies are utilized in ways that are ethically and fairly. Such frameworks specify the regulations, duty, and systems of accountability to manage the urban data systems. In the absence of this, the chances of biasness, abuse, and disparity get heightened overwhelmingly.

Smart city governance should be inter disciplinary and multi stakeheld, including technologists, policy-makers, lawyers, and civil society representatives. Regulators should stipulate the level of data quality, algorithmic accountability, and moral risk determination. Automatic audit of the algorithms, assessment of impact as well as moral review committees may be employed to act as a milestone prior to the implementation at a large scale (De Filippi & Hassan, 2016).

Smart cities can use international documents like OECD AI Principles, General Data Protection Regulation (GDPR) created by EU, etc., as beneficial guidelines on the national level. Localizing them will provide adherence to the deployment of smart cities by the region in reference to culture, law and ethics.

Finally, governance is not only about mitigating the risk, but also about swaying the trust in the public. Smart cities with an ethical cost of doing business, embracing transparency, and a citizen-centered design will also be more just, resilient, and inclusive.

9. CASE STUDIES AND REAL-WORLD IMPLEMENTATIONS

9.1 Barcelona's Smart Citizen Project

Barcelona has been a trendsetter in smart city innovation process, especially in the diffusion of citizen-focused strategies of the urban development. In a move dubbed as the Smart Citizen, the city rolled out affordable environmental sensors within districts and distributed them to the locals. These gadgets measured air quality, noise, and temperatures and related data was analyzed with the help of AI algorithms and shared publicly through a web platform.

The participatory involvement of the initiative enabled people to work with the data and create awareness and work together in solving the problems. Artificial intelligence models helped detect the areas polluted and were able to predict the changes according to the traffic, weather forecasts. The insights produced affected such urban level decisions as diverting traffic or establishing low-emission areas (Cugurullo, 2018).

The attention Barcelona paid to the use of open data and community participation in the process, as well as the moral use of technologies, has turned it into an example of how to adopt citizen engagement in smart sensing. It shows how the AI-IoT integration may not only be utilized to spy or automate but instead empower and exercise democracy.

9.2 Singapore's Smart Nation Strategy

The Smart Nation project developed by Singapore is considered one of the most ambitious and centralized smart city projects all over the world. It incorporates AI, IoT, and big data in areas such as transportation, the level of health care in society, and public safety. The main features would be the use of thousands of IoT sensors at the street level and the adoption of a surveillance system that works on the aid of AI to manage crowds, organize traffic, and monitor the movements of disease.

One of the most remarkable ones is the so-called Moments of Life platform unifying citizens data in order to provide them with personalized public services that rely on life events. AI-models operate with such information and predict needs, suggest services, and facilitate interactions with governmental agencies. The system both increases efficiency and eliminates bureaucratic friction.

In spite of this success, the surveillance and digital rights of the citizens have been questioned by the approach of Singapore. The model of the city-state exemplifies the possibilities as well as the costs of highly unitary AI-IoT systems and stresses the value of trust, cyber protection, and strong law enforcement.

9.3 Predictive Policing in Tokyo's Public Safety Program

To ensure overall safety of the masses during the Tokyo 2020 Olympics, the city implemented a variety of AI-IoT infrastructures. The historical patterns of crime, social media trends and up-to-date sensor data were used by predictive policing systems to predict what could occur in the future and where potential threats or disturbances were likely to happen. They then dynamically deployed mobile units as well as surveillance systems depending on risk assessments fostered by AI.

This system encompassed both facial recognition on transport platforms, crowd analytics based on CCTV footage, and geofences that alerted user to abnormal activity. These gave a faster rate of response and assisted in curbing some of these events that would otherwise have occurred. There were also the concerns regarding racial profiling, false positives and a lack of transparency though.

The situation in Tokyo provides a good case-study regarding the efficiency of AI-powered security form in high density areas. It further emphasizes the ethical implications of the same by noting the importance of control, human discretion, and constant validity of models.

10. FUTURE DIRECTIONS AND RESEARCH OPPORTUNITIES

10.1 Towards Human-Centric and Ethical AI-IoT Systems

With smart cities becoming increasingly automated and data-driven, human-centric AI-IoT systems become more and more important. Technical performance should not be the only thing that the next generation social sensing systems should care about, but also the area of inclusivity, transparency, and fairness. Human-centric design means that we have systems that are aware of all citizens, the older generation, the disabled, and the underserved in digital realms. This demands intuitiveness, multilingual and accessible user interfaces and services that are receptive to different user inputs in an equally balanced manner.

Designs will also have to incorporate ethics. On the one hand, AI and IoT can empower individuals and their communities by making the latter smarter and more cultured; on the other hand, they threaten to aggravate social disparities unless they are properly controlled. These issues of surveillance, algorithmic discrimination, and misuse of data should be checked by strong principles of ethics. Over-automation and the loss of accountability can be avoided by human-in-the-loop use as the key decisions are handled by people.

The next step is to research directions on how to implement explainable AI in social sensing, as this is beneficial to the user and authorities concerned with the decision-making understandability and what data is being used to draw up such decisions. This accountability can offer familiarity among the people, which is extremely essential to the broad comprehending of smart city practices. Moreover, the models of participatory governance in which communities are introduced to the AI-IoT systems development, the product testing, and the assessment will also contribute to aligning the technological advancement with the values of society.

In the end, what will be achieved is a transformation of technology-driven to values-driven smart cities; where social sensing innovations are no longer developed on behalf of the citizen of smart cities but their co-creation. This change of direction will change the definition of success in the urban implementation of AI related not only to efficiency but to equity and sustainability and also to allowing civic empowerment.

10.2 Federated and Privacy-Preserving Learning Models

The growing levels of granularity of data gathered by social sensing systems means that privacy is of the utmost concern. Convention machine learning models are centralized processes that intend to compile data across various possible devices or places and bring it in one position where it is at risk of being hacked or appropriated thus worthless. The ultimate research will be on federated learning (FL) which is a much more decentralized mechanism and where collaboration in training AI models happens across multiple devices but without the use of raw data. The paradigm improves privacy without missing the advantages of large-scale collective learning (Goyal, 2025b).

Federated learning may empower IoT devices deployed in neighborhoods, buildings, or transit systems in smart city experiments to learn locally using smart city data via locally trained models, e.g. energy forecasting or anomaly detection. The combined updates are thereafter distributed to a central model that has the ability to generalize out of location. This way substantially negates the likelihood of leaked personal information particularly when coupled with such methodologies as secure multi-party processing or homomorphic encryption.

The areas of research are optimizing federated learning in constrained devices, better model convergence, and source data heterogeneity. Special problems of social sensing wherein the quality of the data is uneven, the connectivity sporadic, and the time inconsistent, need to be taken into consideration in federated designs.

The involvement of privacy-preserving machine learning in AI-IoT constructs does not only support compliance with recommendations on data protection acts in different parts of the world, such as GDPR, but also better assures users. With cities becoming more data-intensive, such methods will be the most essential means to make sure that personal freedom and digital innovation do not clash with each other.

10.3 Integration with Next-Generation Technologies: 6G, Digital Twins, and Blockchain

AI-IoT innovation is already taking off and the next generation of AI-IoT will be powered by convergence with 6G communication, digital twins and blockchain. The emerging technologies will increase the scale and performance of social sensing to overcome the limitations in connectivity, the simulation strength, and information security.

It is believed that 6G will be ultra-low latency, massive device connectivity as well as AI-native architectures. This will allow a real-time relationship between billions of edge computing devices and autonomous systems as well as billions of sensors. In the context of social sensing what this implies is that a large volume of data flow can be analyzed in real time as well as the ability to coordinate interventions across, and through, the urban infrastructure. AR/VR in cities and the emergence of drone-based surveillance, autonomous vehicle fleets, and parking, as well as driverless deliveries, should become reality.

Virtual counterparts of real-world elements of the city, so-called digital twins, will be essential to simulating and optimizing city behaviour. These digital twins will be fuelled by social sensing data that will allow predictive models to be applied to everything. Using AI algorithms, simulations of several scenarios can be simulated to determine how the infrastructure will react, and planners can make educated actions instead of putting out fires.

In its turn, blockchain technology provides data sharing frameworks that are tamper-proof and decentralized. It can also give reliable logs of sensor data, citizen contributions, or algorithmic decisions in social sensing. It provides data integrity, increases transparency and allows automatic service management through smart contracts, e.g. congestion pricing or billing energy.

Future studies will target on interoperability of these technologies, and ethical application. Such transdisciplinary collaboration between computer science, urban studies, ethics and the public policy will be vital to develop responsibly and resilient integrated systems that are also innovative.

CONCLUSION

Since smart cities are increasingly becoming highly interactive and data-driven environments, Artificial Intelligence (AI) and the Internet of Things (IoT) integration have become the most promising fields of technology. Combined with other technologies, they allow to have high-level social sensing capabilities, i.e. to collect real-time data via human and sensor sources, analyze it down to action. AI-IoT systems are starting to transform, from 'optimising mobility and public health through all the way to security and participatory governance, how urban environments might be more efficient, equitable and resilient.

This chapter has discussed the building blocks of AI-IoT integration, such as the system architecture, real-time analytics, and optimization to be used in social sensing. The use of these technologies in various fields has proved to be game changers when done with the right intentions. Meanwhile, we have talked about these vital issues: interoperability, data privacy, algorithmic bias, and ethical issues, reminding that the technologic advancement should be combined with responsible and citizen-focused regulation.

Next-generation paradigms like federated learning, edge intelligence, digital twins, and blockchain are on the rise to perpetuate the potential of smart urban systems even more. The technical performance will however not be the only benchmark of success in this area but how well these systems are able to support the value of transparency, inclusiveness, and sustainability. The interaction of technologists, city planners, policymakers and people will be paramount in the formulation of AI-IoT solutions which will ensure that not only will cities be smart- but human as well.

REFERENCES

Alsaadi, H. A., & Tubaishat, A. (2015). Internet of Things: Features, challenges, and vulnerabilities. *International Journal of Advanced Computer Science and Applications*, *6*(11), 128–133.

Aral, S. (2020). *The hype machine: How social media disrupts our elections, our economy, and our health—and how we must adapt*. Currency.

Atzori, L., Iera, A., & Morabito, G. (2010). The internet of things: A survey. *Computer Networks*, *54*(15), 2787–2805. DOI: 10.1016/j.comnet.2010.05.010

Badii, C., Nesi, P., & Paolucci, M. (2019). Predictive models for urban mobility in smart cities. *Sensors (Basel)*, *19*(12), 2781. DOI: 10.3390/s19122781 PMID: 31226869

Chen, X., Ran, X., & Liu, J. (2019). Deep learning with edge computing: A review. *Proceedings of the IEEE*, *107*(8), 1655–1674. DOI: 10.1109/JPROC.2019.2921977

Cugurullo, F. (2018). Exposing smart cities and eco-cities: Frankenstein urbanism and the sustainability challenges of the experimental city. *Environment and Planning A. Environment & Planning A*, *50*(1), 73–92. DOI: 10.1177/0308518X17738535

D'Andrea, A. (2020). The role of digital twins in smart cities: Real-time urban simulation and control. *Smart Cities*, *3*(3), 1055–1070.

De Filippi, P., & Hassan, S. (2016). Blockchain technology as a regulatory technology: From code is law to law is code. *First Monday*, *21*(12). Advance online publication. DOI: 10.5210/fm.v21i12.7113

Goodfellow, I., Pouget-Abadie, J., Mirza, M., Xu, B., Warde-Farley, D., Ozair, S., & Bengio, Y. (2014). Generative adversarial nets. *Advances in Neural Information Processing Systems*, •••, 27.

Goyal, N. K. (2025). Security and Privacy in IoT, Fog, and Blockchain Networks. In Sathio, A., Rind, M., Awan, S., & Junejo, A. (Eds.), *Energy-Efficient Deep Learning Approaches in IoT, Fog, and Green Blockchain Revolution* (pp. 371–398). IGI Global Scientific Publishing., DOI: 10.4018/979-8-3373-0300-0.ch015

Goyal, N. K. (2025). Weaponizing Misinformation and Cyberattacks on Supply Chains. In Bukhari, S., & Zafar, I. (Eds.), *Mass Media and Impact of Fake News on Supply Chains* (pp. 177–210). IGI Global Scientific Publishing., DOI: 10.4018/979-8-3373-2175-2.ch007

Goyal, N. K., Dandotiya, M., Kumari, M., Kaloria, S., & Anushya, A. (2025). AI-Powered Haptic Devices for Patient Rehabilitation. In Mamodiya, U., Kalam, A., Bossoufi, B., Awad, M., & Somwanshi, D. (Eds.), *Integrating AI With Haptic Systems for Smarter Healthcare Solutions* (pp. 143–170). IGI Global Scientific Publishing., DOI: 10.4018/979-8-3373-2307-7.ch007

Goyal, N. K., Dandotiya, M., Kumari, M., Sheikh, M. F., & Sharma, S. (2025). Emerging Job Roles in the Tourism Sector Leveraging Artificial Intelligence. In Silic, M., Luthia, M., Ahmad, F., & Chakraborty, P. (Eds.), *Empowering Sustainable Performance and Competitive Advantage in Tourism* (pp. 89–126). IGI Global Scientific Publishing., DOI: 10.4018/979-8-3373-4546-8.ch004

Goyal, N. K., Dandotiya, M., Qurashi, J. A., Kumari, M., & Sharma, S. (2025). AI-Driven Predictive Analytics for Disease Management. In Ouaissa, M., Ouaissa, M., Imad, M., Qurashi, J., & Farooq, M. (Eds.), *Utilizing AI of Medical Things for Healthcare Security and Sustainability* (pp. 231–276). IGI Global Scientific Publishing., DOI: 10.4018/979-8-3373-0690-2.ch008

Goyal, N. K., Kumari, M., Kushwah, V., Dandotiya, M., & Vijay, A. (2025). Evaluating blockchain security: Trade-offs in consensus mechanisms. *IET Conference Proceedings*, 2024(38), 80–88. DOI: 10.1049/icp.2025.0775

Gubbi, J., Buyya, R., Marusic, S., & Palaniswami, M. (2013). Internet of Things (IoT): A vision, architectural elements, and future directions. *Future Generation Computer Systems*, *29*(7), 1645–1660. DOI: 10.1016/j.future.2013.01.010

Guo, B., Wang, Z., Yu, Z., & Zhou, X. (2017). Mobile crowd sensing and computing: The review of an emerging human-powered sensing paradigm. *ACM Computing Surveys*, *48*(1), 1–31. DOI: 10.1145/2794400

Khan, W. Z., Rehman, M. H. U., Zangoti, H. M., Afzal, M. K., Armi, N., & Salah, K. (2020). Industrial Internet of Things: Recent advances, enabling technologies and open challenges. *Computers & Electrical Engineering*, *81*, 106522. DOI: 10.1016/j.compeleceng.2019.106522

Kingma, D. P., & Welling, M. (2013). Auto-encoding variational Bayes. *arXiv preprint arXiv:1312.6114*.

Kitchin, R. (2014). The real-time city? Big data and smart urbanism. *GeoJournal*, *79*(1), 1–14. DOI: 10.1007/s10708-013-9516-8

Lee, I., & Lee, K. (2015). The Internet of Things (IoT): Applications, investments, and challenges for enterprises. *Business Horizons*, *58*(4), 431–440. DOI: 10.1016/j.bushor.2015.03.008

Li, J., Li, X., Chen, X., Lee, P. P., & Lou, W. (2018). A hybrid cloud approach for secure authorized deduplication. *IEEE Transactions on Parallel and Distributed Systems, 26*(5), 1206–1216. DOI: 10.1109/TPDS.2014.2318320

Li, S., Xu, L. D., & Zhao, S. (2015). The internet of things: A survey. *Information Systems Frontiers, 17*(2), 243–259. DOI: 10.1007/s10796-014-9492-7

Lin, J., Yu, W., Zhang, N., Yang, X., Zhang, H., & Zhao, W. (2017). A survey on Internet of Things: Architecture, enabling technologies, security and privacy, and applications. *IEEE Internet of Things Journal, 4*(5), 1125–1142. DOI: 10.1109/JIOT.2017.2683200

Mehmood, R., Ahmad, F., Yaqoob, I., Imran, M., Guizani, M., & Alturki, B. (2020). Internet of Things (IoT)-based smart cities: Recent advances, requirements, and challenges. *IEEE Communications Surveys and Tutorials, 22*(3), 1656–1686.

Nguyen, D. C., Pathirana, P. N., Ding, M., Seneviratne, A., Li, J., & Niyato, D. (2021). Blockchain for 5G and beyond networks: A state-of-the-art survey. *Journal of Network and Computer Applications, 174*, 102890.

Raji, I. D., & Buolamwini, J. (2019). Actionable auditing: Investigating the impact of publicly naming biased performance results of commercial AI products. *AAAI/ACM Conference on AI, Ethics, and Society*. DOI: 10.1145/3306618.3314244

Satyanarayanan, M. (2017). The emergence of edge computing. *Computer, 50*(1), 30–39. DOI: 10.1109/MC.2017.9

Shahzadi, S., Iqbal, M. Z., Naqvi, S. A. R., & Amin, M. (2021). Role of AI in smart cities: A review. *International Journal of Distributed Sensor Networks, 17*(5). Advance online publication. DOI: 10.1177/15501477211011040

Shi, W., Cao, J., Zhang, Q., Li, Y., & Xu, L. (2016). Edge computing: Vision and challenges. *IEEE Internet of Things Journal, 3*(5), 637–646. DOI: 10.1109/JIOT.2016.2579198

Tzounis, A., Katsoulas, N., Bartzanas, T., & Kittas, C. (2017). Internet of Things in agriculture, recent advances and future challenges. *Biosystems Engineering, 164*, 31–48. DOI: 10.1016/j.biosystemseng.2017.09.007

Wang, S., Zhang, Y., Zhang, Y., Wang, Y., Yang, J., & Wang, Y. (2020). A survey on mobile edge networks: Convergence of computing, caching and communications. *IEEE Access : Practical Innovations, Open Solutions, 5*, 6757–6779. DOI: 10.1109/ACCESS.2017.2685434

Wang, W. Y., & Wang, A. I. (2021). The role of federated learning in smart cities: Privacy-aware AI for decentralized systems. *IEEE Communications Magazine*, *59*(9), 42–48.

Yassine, A., Singh, S., Hossain, M. S., & Muhammad, G. (2019). IoT big data analytics for smart homes with fog computing. *Future Generation Computer Systems*, *91*, 563–573. DOI: 10.1016/j.future.2018.08.040

Zanella, A., Bui, N., Castellani, A., Vangelista, L., & Zorzi, M. (2014). Internet of Things for smart cities. *IEEE Internet of Things Journal*, *1*(1), 22–32. DOI: 10.1109/JIOT.2014.2306328

Zhang, K., Ni, J., Yang, K., Liang, X., Ren, J., & Shen, X. S. (2017). Security and privacy in smart city applications: Challenges and solutions. *IEEE Communications Magazine*, *55*(1), 122–129. DOI: 10.1109/MCOM.2017.1600267CM

Chapter 7
IoT-Based Smart Cradle Baby Monitoring

G. Theivanathan
http://orcid.org/0000-0001-9503-1333
Velammal Engineering College, India

S. Akash
Velammal Engineering College, India

R. Arunvel
Velammal Engineering College, India

N. Akash
Velammal Engineering College, India

ABSTRACT

The rapid advancement of technology has led to the development of innovative solutions in childcare, addressing the challenges faced by modern parents. This paper presents the design and implementation of an IoT-based Smart Baby Cradle that enhances infant safety and comfort through real-time monitoring and automated soothing mechanisms. The system integrates various sensors to detect temperature, humidity, sound, motion, and health parameters. It employs a microcontroller to process sensor data and trigger appropriate actions, such as gentle rocking or playing lullabies, in response to the baby's needs. Parents receive real-time alerts and can remotely monitor and control the cradle using a mobile application. The proposed system not only reduces caregiver stress but also ensures the well-being of the infant, representing a significant advancement in smart parenting solutions.

DOI: 10.4018/979-8-3373-4962-6.ch007

1. INTRODUCTION

The rapid advancement of the Internet of Things (IoT) has enabled the development of intelligent systems capable of real-time monitoring, automation, and remote control across various domains, including healthcare and child safety. Infant care is a critical responsibility that demands continuous attention, as babies are highly vulnerable and unable to communicate their needs effectively. Working parents and caregivers often face challenges in continuously monitoring infants while managing other responsibilities. This necessitates the development of a smart, automated, and reliable baby monitoring system.

An IoT-based smart cradle is an innovative solution designed to enhance infant safety, comfort, and health monitoring. The system integrates sensors, microcontrollers, and wireless communication technologies to monitor key parameters such as baby crying, temperature, humidity, bed wetness, and motion. When abnormal conditions are detected, alerts are sent to parents or caregivers through mobile applications or cloud platforms, enabling immediate response. Additionally, automated features such as cradle swinging and lullaby playback can be activated to soothe the baby without manual intervention.

Traditional baby cradles lack monitoring and alert mechanisms, which can delay caregiver response during emergencies. Existing baby monitoring systems often focus only on video surveillance and may not provide environmental or physiological monitoring. The proposed IoT-based smart cradle addresses these limitations by combining environmental sensing, sound detection, and real-time notification within a single integrated system. By leveraging embedded systems and IoT communication protocols, the system ensures continuous supervision even when parents are physically distant from the infant.

This research aims to design and implement a cost-effective, reliable, and energy-efficient IoT-based smart cradle for real-time baby monitoring. The proposed system enhances infant safety, reduces parental anxiety, and contributes to the advancement of smart healthcare technologies.

Telepatil et al. (2021) viewed that in the world of technology that changes so quickly these days, smart devices have definitely made life better. Childcare is part of this. Parents today often have a hard time making sure their babies are safe and comfortable while also juggling work and family. Traditional cradles are very common, but they're not very useful. Their shape can't adapt to the baby's needs or to changes in the environment, like when the noise, light, or temperature changes. While this is going on, both the baby and the person caring for them may feel stressed if the pain isn't treated quickly. A cool way to solve these issues is to use smart technology, especially the Internet of Things (IoT). Smart Baby Cribs have sensors and a smart control system that let the parents always keep an eye on the baby. we can be sure that they will help the baby right away if they notice any pain or changes in its environment.

Gupta et al. (2021) connected mobile application plays a vital role in modern baby monitoring systems by enabling parents and caregivers to receive real-time alerts and remotely access cradle functions from any location. Through wireless connectivity and cloud-based platforms, the application allows users to continuously monitor the infant's condition even when they are physically away. Instant notifications are generated whenever abnormal situations such as excessive crying, unsafe temperature levels, unusual motion, or wet bedding are detected. This immediate alert mechanism ensures that timely action can be taken to safeguard the baby's health and comfort. In addition to monitoring, the mobile interface provides remote control features, such as activating automatic cradle swinging, playing soothing lullabies, or adjusting environmental settings, thereby enhancing convenience and responsiveness.

Symon et al. (2017) highlighted integration of IoT technology into a traditional baby cradle significantly transforms infant care by combining sensing, automation, and communication within a unified system. Conventional cradles primarily serve as resting equipment without offering any intelligent monitoring capabilities. As a result, caregivers must rely entirely on manual supervision, which may not always be feasible due to professional commitments or household responsibilities. The Smart Baby Cradle based on the Internet of Things addresses these limitations by embedding sensors and microcontrollers that continuously track environmental and behavioral parameters. The collected data is transmitted to a centralized system or cloud server, where it is processed and made accessible through a user-friendly mobile application.

Shreelatha (2017) presented that to enhance neonatal care by integrating modern communication technologies with conventional cradle design. By leveraging IoT architecture, the proposed system ensures continuous connectivity between the baby and the caregiver, thereby reducing anxiety and improving response time during emergencies. Parents can monitor parameters such as room temperature, humidity, crying intensity, and cradle movement in real time. If any parameter crosses predefined safety thresholds, automated alerts are immediately sent to the registered mobile device. This constant supervision strengthens infant safety while offering peace of mind to parents, regardless of their physical location.

Mehta and Rao, (2022) Furthermore, the proposed system contributes to the broader vision of smart healthcare and intelligent home environments. It not only improves convenience but also promotes proactive infant care by preventing potential risks before they escalate into serious concerns. The fusion of traditional cradle functionality with IoT-enabled monitoring creates a reliable, cost-effective, and scalable solution suitable for both residential and clinical settings. By ensuring continuous observation, rapid notification, and remote intervention, the IoT-based Smart Baby Cradle represents a significant step toward creating a safer and more secure environment for newborns. Ultimately, this technological advancement aims to support parents in maintaining constant connection and assurance, thereby fostering improved well-being and protection for infants at all times.

2. LITERATURE SURVEY

Ahmad and Kumar (2021) showed modern baby cradles depend on people to do a lot of the work; they can't do things automatically or keep an eye on things in real time. Many high-end models don't have the necessary environmental sensors to measure noise, temperature, and humidity. They also don't have the connectivity features that would let you monitor and control them from afar, even if they do have basic features like a rocking mechanism. Due to the lack of advanced features, it can be hard for parents, especially those who work, to be there in person to meet their baby's needs. Parents may feel more stressed and anxious because they have to help their kids all the time. When they are not nearby, they might not be able to see or help their child if they are feeling bad. This lack of features makes it clear that we need a better cradle system that can learn what the baby needs and talk to the parents in real time.

Lee, (2021) as technology has changed a lot in the last few decades, it has also had a big impact on how babies are cared for. The way parents deal with the problems that come with raising kids is changing as they use smart devices and other digital tools every day. In their work Innovations in Infant Care: The Role of Technology, Smith goes into great detail about these new technologies and shows how important they are to modern parenting. Smith starts by talking about the problems that modern parents face, like balancing work, home life, and raising a child.

Patel and Sharma, (2023) They says that technological advances like baby monitors with AI and mobile health apps that work with health data provide a lot of relief by making these tasks easier. Kumar, (2022) parents can make better decisions with the real-time feedback these devices provide by tracking their infant's sleep patterns and vital signs. It also notes the proliferation of tech-enhanced feeding solutions, such as bottle warmers and breast pumps, in addition to monitoring systems. Especially for first-time parents, this user-friendliness helps alleviate some of the stress that comes with being a parent. Smith goes into detail about data privacy worries, pointing out that although these devices bring new levels of convenience, they also bring up serious concerns about the safety and storage of private family information.

Singh and Thomas, (2020) the development of smart home technologies that enhance child safety has been on the rise in recent years. According to research,

"Smart Home Technologies for Child Safety," there are a number of IoT devices that can be used to keep an eye on and safeguard children right in their own homes. With these gadgets, parents can rest easy knowing that their children are being watched in real-time and that they will be notified in the event of any potentially dangerous situations. Desai and Kaur, (2022) the book talks about many smart home gadgets, like baby monitors, smart cameras, and door/window sensors, that parents can use to keep an eye on their kids. Wearables with GPS can help parents keep track of where their kids are when they're outside, and smart speakers or hubs can be set up to alert adults when certain conditions are met, like when a child goes into an area they're not supposed to be in or an unexpected door opens. Gonzalez and Lee, (2020) takes a closer look at the application of AI in such technologies. An example would be predictive algorithms that learn the habits of a child, and raise an alarm in case something unusual occurs. Wang and Zhang, (2023) the author recommends users to take other measures to ensure that their home networks are secured to prevent unauthorized access to child-monitoring systems. Overall, the review by Lee provides useful connections that the IoT can be used to enhance child safety by providing parents with modern tools to make their homes and neighborhoods less unsafe.

Chen and Wang, (2019) Examines more recent baby care systems that are automated and smart to enhance caregiving. Kumar and Joshi, (2021) the book explains how there is an increased need to incorporate sensors and technologies to develop baby monitoring and well-being systems. These robotic structures can offer real-time information, insights and alerts in order to decrease the workload of caregivers and also facilitate timely interventions even in their absence. Fernandez and Silva, (2021) analysis includes many modern baby care systems, including smart cribs. Sensors in these cribs monitor the baby's breathing, sleep, and environmental factors like room temperature, humidity, and light. Multiple ways exist for these cribs to detect and respond to abnormalities like breathing issues or dangerous temperature changes. Adjusting settings automatically stabilizes the environment. Second, they can immediately notify caregivers via connected apps.

L. Chen and Patel, (2019) the system responds automatically to keep the baby safe and comfortable, reducing the caregiver's need to watch them. It studied wearable health monitors, another important innovation. These devices track babies' heart rates, oxygen saturation, and movement. Sharma and Gupta, (2022) wearable monitors continuously track these indicators, which can comfort caregivers, especially parents of premature or newborn infants who may need more medical attention. These devices let caregivers know when someone is in pain, which helps solve health problems quickly. Brown and Miller, (2020) our main research goal is to use multiple sensors to make automated baby care systems more accurate and useful. Nguyen and Park, (2023) Reflects that combining information from many sensors, like sound monitors, temperature gauges, motion detectors, and moisture sensors, can give a fuller picture of the baby's health and environment than many devices on the market today. It's hard to make these technologies reliable and simple to use automated baby care systems to boost safety for kids and ease the work of caregivers.

3. PROPOSED METHODOLOGY

It's an innovative system that lets the Smart Baby Cradle understand and respond to an infant's sounds and movements. This makes the cradle safer and more comfortable for the baby. It includes multiple sensors to measure sound, moisture, and temperature to monitor such factors as the room temperature, the cries of the baby and the condition of their wet diapers.

The cradle informs the caregivers immediately via the Blynk app when any of these occurrences occur and it is time to ensure that the caregiver is able to take action immediately using their phones regardless of their location. It is small high-tech innovation, which is known as a smart cradle that takes care of babies in a more active and comprehensive manner than sitting down. In this manner, when the baby cries a system has the potential of playing soft music or rocking the baby. The providers are informed instantly in the event of an occurrence of something bad within the environment, such as a too hot temperature or wet diaper. This is in the interest of the baby and comfort. The Smart Baby Cradle has built-in remote monitoring that allows parents to monitor their child wherever they are. The parents will receive a message immediately in case something is wrong. This is the tool that is significant in the contemporary parenting due to the reduction in the supervision requirements. This would have an easier time and would provide peace of mind to parents.

An IoT smart cradle that has sensors can keep an eye on the baby's breathing, movement, and surroundings at all times. There will be sensors that measure temperature and humidity, health sensors that keep track of temperature and heart rate, and motion detectors that look for any movement that could be dangerous. A computerized system that is powered by AI will scan the data and search using oddities such as strange cries or breathing patterns. In the event that it detects some, it will initiate automatic procedures to relax the individual, including lullabies or gentle rocking. Parents can monitor their babies with the smart cradle by using a smartphone application to make sure that they are safe and well. They are also able to operate functionalities such as music or rocking, receive real-time notifications and have their batteries supported [Figure 1.].

3.1 WORKING PROCEDURE

Step 1: Setup Sensors

Track the infant's whereabouts by installing motion detectors, such as a PIR sensor.

Install a system with sound sensors (a microphone) that can pick up sounds like crying.

Installing sensors for temperature and humidity will help it keep an eye on the environment.

Step 2: Collect Data

The sensors constantly monitor the movements, sounds and any activity around the cradle of the baby.

The motion sensor will be able to detect any movements of the baby and the sound sensor will detect any unusual sound.

Step 3: Data Processing by Microcontroller

The data is sent to a microcontroller, like an ESP8266, Raspberry Pi, or Arduino.

If the microcontroller hears a loud noise or sees a strange movement, it will process this information.

Step 4: Connect to IoT Platform

Connect the device to Wi-Fi. This will let the microcontroller talk to a cloud provider or IoT platform.

An IoT service like Blynk, Firebase, or MQTT can be used to send and store data.

Step 5: Use a mobile app to control and monitor

You can utilize ready-to-use application or create your own to display real-time information on the cradle.

The smartphones allow parents to monitor the vital signs of their babies (i. e., their activity, temperature, and humidity) in real-time.

Step 6: Automated Soothing Actions

A motor is built into the system so that it will rock the cradle on its own if it senses that the baby is crying or moving around too much.

Putting on some calming tunes or lullabies on your speaker.

Modifying environmental factors, such as temperature, as needed.

Step 7: Send Alerts and Notifications

The system quickly alerts the parent's phone of any strange behavior, like a child crying nonstop, breathing irregularly, or being in an unfavorable environment.

Step 8: Parental Control

With the app, parents can: Observe patterns and current information (like past temperature readings, sleep patterns, etc.).

From a distance, you can control the cradle's features, such as turning it on and off and playing music.

You can change the settings to make the cradle's automatic responses more personal.

Step 9: Data Storage and Analysis

People who have children can look back at the data stored on the IoT platform to see how their child's sleeping habits or room arrangement changed over time.

Step 10: Maintenance and Troubleshooting

Make sure that the cradle can connect to the internet well.

It is important to check the sensors often to make sure they are working properly.

Check that the software and app are up to date to stay safe.

4. WORKING PARTICULARS

In this section, the hardware components that cause the Smart Baby Cradle to be working are shown. Various sensors and actuators enable real-time and automatic response. The ESP32 microcontroller that serves as the brains of the system is also a significant component. Taken together these sections form a dependable and efficient modern care system of the baby.

4.1 Hardware Requirements

- ESP32 Microcontroller
- Microphone Sensor
- Blynk-compatible device
- DC Motor
- DHT11 Sensor
- Rain Sensor
- Servo Motor
- Power Supply

The hardware architecture of the IoT-Based Smart Cradle for Baby Monitoring system plays a crucial role in ensuring real-time sensing, processing, communication, and actuation. The proposed system integrates multiple sensors, actuators, and control units to monitor infant conditions and automate cradle operations. The primary hardware components include the ESP32 microcontroller, microphone sensor, Blynk-compatible communication device, DC motor, DHT11 temperature and humidity sensor, rain (moisture) sensor, servo motor, and power supply unit. Each component contributes to achieving a reliable, safe, and intelligent cradle monitoring system.

ESP32 Microcontroller

The ESP32 microcontroller serves as the central processing unit of the smart cradle system. It is a powerful, low-cost, and energy-efficient microcontroller with integrated Wi-Fi and Bluetooth capabilities, making it highly suitable for IoT applications.

- Dual-core 32-bit processor
- Built-in Wi-Fi and Bluetooth
- Multiple GPIO (General Purpose Input/Output) pins
- ADC (Analog-to-Digital Converter) channels
- PWM (Pulse Width Modulation) support
- Low power consumption modes

In the proposed system, the ESP32 performs the following critical functions:

Sensor Data Acquisition: It collects real-time data from the DHT11 sensor (temperature and humidity), microphone sensor (cry detection), and rain sensor (wetness detection).

Data Processing: It processes sensor readings using predefined thresholds to determine abnormal conditions.
Communication: Through built-in Wi-Fi, it transmits data to the Blynk cloud server and receives control commands from the mobile application.
Actuator Control: It controls the DC motor and servo motor for automated cradle movement.

The ESP32 is preferred over traditional microcontrollers due to its superior processing speed, integrated connectivity, and ability to handle multiple tasks simultaneously. Its low-power operation ensures energy efficiency, which is essential for continuous baby monitoring systems.

Microphone Sensor (Sound Detection Module)

The microphone sensor is responsible for detecting the baby's crying sound. It plays a vital role in initiating automated soothing actions and alert notifications.

The microphone module typically consists of:

- An electret microphone
- An amplifier circuit
- A comparator
- Analog and digital output pins

The microphone converts sound waves into electrical signals. These signals are amplified and compared against a threshold value. If the sound intensity exceeds the preset level (indicating crying), the sensor outputs a signal to the ESP32. Role in the Smart Cradle,

- Detects baby crying in real time.
- Triggers automatic cradle swinging.
- Sends instant notification to parents via the Blynk app.
- Activates lullaby playback if integrated.

The sensitivity of the microphone can be adjusted using a potentiometer. This ensures that background noise does not trigger false alarms. Accurate sound detection improves system reliability and reduces unnecessary notifications.

Blynk-Compatible Device (Wi-Fi Enabled Module)

Although the ESP32 has built-in Wi-Fi, it functions as the primary Blynk-compatible device. The Blynk platform requires hardware capable of internet communication to transmit and receive data from the cloud server.

- Connects to local Wi-Fi network.
- Communicates with Blynk cloud server.
- Sends sensor data to mobile application.
- Receives control commands from users.

The integration of Wi-Fi ensures remote accessibility, enabling parents to monitor cradle status from any geographical location. Secure authentication tokens provided by Blynk ensure safe data exchange.

The IoT connectivity enhances:

- Real-time monitoring
- Data logging
- Remote control operations
- Instant push notifications

This hardware component transforms the cradle from a standalone device into an intelligent IoT-enabled system.

DC Motor

The DC motor is used to provide continuous cradle swinging motion to soothe the baby.

A DC motor converts electrical energy into mechanical rotational energy using electromagnetic principles. When current flows through the motor windings, it produces a magnetic field that causes rotation.

- Provides automatic rocking motion.
- Activated when crying is detected.
- Can be controlled remotely via mobile application.
- Operates through motor driver circuits.

Since the ESP32 cannot directly drive a DC motor due to current limitations, a motor driver (such as L298N or L293D) is used to control direction and speed.

The rocking mechanism is designed to mimic natural manual cradle movement, ensuring baby comfort and safety.

DHT11 Temperature and Humidity Sensor

The DHT11 sensor is used to monitor environmental temperature and humidity inside the cradle. The technical specifications are

- Temperature range: 0°C to 50°C
- Humidity range: 20% to 90%
- Digital output
- Low power consumption

Maintaining optimal temperature and humidity is essential for infant health. Extreme conditions may cause discomfort, dehydration, or respiratory issues.

The DHT11 sensor continuously measures:

- Ambient temperature
- Relative humidity

If values exceed safe thresholds, the system:

- Sends alerts to parents
- Displays data on the mobile application

The digital output simplifies interfacing with ESP32, ensuring accurate and stable readings.

Rain Sensor (Moisture Detection Sensor)

The rain sensor in this system is repurposed as a moisture detection sensor to identify bed wetness.

The sensor consists of:

- A sensing plate with conductive traces
- Control module with comparator

When moisture is present, the conductivity between traces increases, altering the output voltage. The ESP32 detects this change and triggers an alert.

This feature enhances hygiene and prevents skin infections such as diaper rash. Quick detection improves baby comfort and health safety.

Servo Motor

The servo motor is used for controlled angular motion in the cradle mechanism. Role in Smart Cradle

- Controls limited-angle rocking motion.
- Provides smoother and controlled swinging.
- Enhances safety compared to continuous DC motor motion.

Unlike DC motors, servo motors offer precise control over angle and speed, making them suitable for gentle rocking operations.

The ESP32 generates PWM signals to determine the servo position. This ensures controlled and safe cradle motion without excessive swinging.

Power Supply Unit

The power supply unit ensures stable and regulated power delivery to all components. The requirements are

- 5V regulated supply for sensors and microcontroller
- 6–12V supply for DC motor
- Adequate current rating
- Protection against voltage fluctuations

A voltage regulator ensures constant voltage levels to prevent damage. Overcurrent protection and filtering capacitors improve system stability.

A reliable power system is essential for uninterrupted baby monitoring, especially during nighttime.

The Smart Baby Cradle's Arduino board is programmed with the Arduino IDE (v1.8.5). The sensors send data to the Arduino board, which then controls things like motors and LEDs. The IDE is flexible and simple to use, even for experienced programmers. Embedded C is used to make certain that sensors and tools for automated baby care can be added to embedded systems quickly and easily.

4.2 Software Requirements

- Arduino Ide – 1.8.5

- Blynk app
- Embedded C
- Libraries

The development of the IoT-Based Smart Cradle for Baby Monitoring system requires a well-defined software environment to ensure reliable data acquisition, processing, communication, and user interaction. The proposed system integrates embedded programming, cloud connectivity, and mobile application interfaces. The primary software requirements include the Arduino Integrated Development Environment (IDE) version 1.8.5, the Blynk mobile application platform, Embedded C programming language, and supporting software libraries. Each of these components plays a critical role in enabling seamless hardware–software integration and real-time monitoring functionality.

Arduino IDE – Version 1.8.5

The Arduino Integrated Development Environment (IDE) version 1.8.5 is used as the primary platform for writing, compiling, and uploading embedded programs to the microcontroller unit (MCU), such as Arduino Uno or NodeMCU (ESP8266/ESP32). The Arduino IDE provides a user-friendly interface that simplifies embedded system development through its built-in editor, compiler, and serial monitor.

Version 1.8.5 is selected due to its stability, compatibility with various board packages, and extensive community support. It supports multiple operating systems, including Windows, macOS, and Linux, making it accessible and flexible for development. The IDE includes a built-in serial communication monitor, which is particularly useful for debugging sensor data, verifying communication protocols, and monitoring real-time system performance.

In the proposed smart cradle system, the Arduino IDE is used to

- Program sensor data acquisition from temperature, humidity, sound, and moisture sensors.
- Implement control logic for cradle swinging motors and alert mechanisms.
- Configure Wi-Fi communication modules (ESP8266/ESP32).
- Manage data transmission to the Blynk cloud server.
- Perform debugging and real-time performance analysis.

The IDE supports uploading firmware directly to the microcontroller via USB or serial communication. Its open-source nature enables easy integration with third-party libraries, which enhances functionality and accelerates system development.

Blynk Application

The Blynk application serves as the IoT platform for remote monitoring and control of the smart cradle system. It provides a cloud-based infrastructure that connects hardware devices to a smartphone application through the internet. Blynk supports real-time data visualization, notifications, and device control through customizable dashboards.

The Blynk ecosystem consists of three primary components:

1. Blynk Mobile App (Android/iOS)
2. Blynk Cloud Server
3. Blynk Library (integrated into the microcontroller firmware)

In the proposed system, the Blynk app enables parents to:

- Monitor real-time temperature and humidity levels.
- Receive instant notifications when the baby cries.
- Get alerts for bed-wetting detection.
- Remotely activate cradle swinging.
- Control lullaby playback features.
- View system status updates.

The application provides virtual widgets such as gauges, graphs, buttons, sliders, and notification elements. These widgets are linked to virtual pins in the microcontroller program, enabling seamless two-way communication between the hardware and the mobile device.

Blynk uses an authentication token to ensure secure communication between the microcontroller and the cloud server. This token is generated during project creation within the app and embedded into the Arduino firmware. Data is transmitted using Wi-Fi protocols, ensuring low-latency communication.

The use of Blynk significantly reduces the complexity of developing a custom mobile application from scratch, thereby minimizing development time and cost. It also provides scalability, allowing additional sensors or functionalities to be incorporated in future system upgrades.

Embedded C Programming

Embedded C is the primary programming language used to develop firmware for the microcontroller in the smart cradle system. It is a specialized extension of the C programming language tailored for embedded systems. Embedded C provides direct access to hardware registers, memory addresses, and input/output ports, making it highly suitable for real-time applications.

In the proposed system, Embedded C is used to implement:

- Sensor interfacing and data acquisition algorithms.
- Threshold-based decision-making logic.
- Motor control operations for cradle movement.
- Interrupt handling for sound detection.
- Wi-Fi communication setup.
- Data formatting and transmission to the cloud.
- Power management routines.

Embedded C enables efficient memory utilization, which is crucial for microcontrollers with limited RAM and storage capacity. The language allows low-level manipulation of hardware components, ensuring precise control over system operations.

Real-time constraints are critical in baby monitoring applications. For example, when a baby cries, the system must immediately detect the sound, process the signal, and send a notification without delay. Embedded C supports deterministic execution, ensuring predictable response times.

Furthermore, modular programming techniques are employed in Embedded C to enhance code readability and maintainability. Functions are structured for sensor reading, alert generation, motor control, and communication handling. This structured approach improves debugging efficiency and supports future enhancements.

Software Libraries

Software libraries play a vital role in simplifying development by providing prewritten functions and modules for hardware communication and data processing. In the IoT-Based Smart Cradle system, various libraries are integrated within the Arduino IDE to support sensors, communication modules, and cloud connectivity.

Commonly used libraries include:

- WiFi or ESP8266WiFi Library: Enables wireless communication between the microcontroller and internet router.
- Blynk Library: Facilitates communication between the microcontroller and Blynk cloud server.
- DHT Library: Used for interfacing with temperature and humidity sensors.
- Servo or Motor Driver Library: Controls cradle swinging mechanism.
- LiquidCrystal or OLED Display Library (if applicable): Displays real-time data locally.

- EEPROM Library: Stores configuration parameters persistently.

These libraries significantly reduce development complexity by abstracting low-level hardware details. Instead of writing lengthy communication protocols, developers can use simple function calls provided by the libraries. This enhances development speed and reduces programming errors.

Additionally, library compatibility with Arduino IDE 1.8.5 ensures stable compilation and efficient execution. The modular structure of libraries allows easy updates and integration of additional features in the future.

Overall Software Architecture Integration

The software components collectively form a layered architecture:

1. Hardware Layer – Sensors, actuators, microcontroller.
2. Firmware Layer – Embedded C program developed in Arduino IDE.
3. Communication Layer – Wi-Fi module and Blynk cloud server.
4. Application Layer – Blynk mobile application interface.

Data flows from sensors to the microcontroller, where it is processed using Embedded C. The processed data is transmitted via Wi-Fi to the Blynk server and displayed on the mobile application. Control commands from the app are sent back to the microcontroller to activate motors or other output devices.

5. EXPERIMENTAL RESULTS AND DISCUSSION

To determine the suitability of the Smart Baby Cradle to the needs of the baby and the individual caring about them, the Smart Baby Cradle was tested in a number of locations. Indicatively, the system detected crying, excessive temperatures and excessive moisture during the tests. These things had been seen by the cradle and now it started to act properly.

It rocked gently to calm the baby, turned on a fan to control the temperature, and notified caregivers instantly through the app. Using the Smart Baby Cradle gave caregivers a lot more peace of mind, according to user feedback. They loved the real-time notifications and the automatic responses it triggered. Caregivers felt a stronger sense of connection to the welfare of their infant. In the future, the Smart Baby Cradle could be upgraded with more sensors to track other important metrics like oxygen levels and heart rate, enhancing its functionality even further. Additionally, capabilities like machine learning algorithms could be implemented to examine data trends over time, enabling the cradle to adjust to the unique likes and dislikes of every infant. The Smart Baby Cradle has the potential to become an even more all-encompassing and effective solution for modern parenting through ongoing innovation and improvement. This could lead to a safer and more caring environment for infants [Figure 2.].

To improve baby care, the baby monitoring system integrates IoT components like the ESP32 microcontroller with environmental detection sensors (e.g., liquid sensors) and actuators (e.g., fan, servo motor). The motor could help with automation, such as adjusting the crib's motion or positioning, and the fan would make sure there's enough airflow. The moisture sensor can detect dampness in the air in real time, immediately notifying those responsible for care. With all of these parts working together, this project gives a smart, real-time way to watch over a baby and make sure they're safe, happy, and healthy. So that the baby doesn't feel any pain or have skin problems, the moisture sensor lets the caretakers know as soon as it detects wetness in the crib. A fan can be set to turn on and off automatically to keep the air moving, and sensors that measure temperature and humidity work together to make the room perfect for the baby. Furthermore, the system can be remotely monitored and alerted in real-time through a mobile app that is linked to it via Wi-Fi. By automating cradle functions and giving caregivers real-time updates, this efficient and responsive system is created through the integration of IoT elements, which enhances infant care [Figure 3.].

To make sure the baby is safe and comfortable, the picture shows the real-time output of a baby monitoring system, which displays important environmental metrics. A temperature reading of 94 degrees Fahrenheit suggests that the air around the baby is very hot, possibly hotter than the ideal temperature range for their comfort, which is usually between 68 and 72 degrees. In the event that this temperature rises too high, the system's built-in fan will sound an alarm to the caregivers, telling them to lower the room's temperature or increase ventilation. The humidity sensor also reads 76%, which is on the high side. Humidity levels between 40% and 60% are considered ideal for a baby's comfort. Babies can feel uncomfortable, overheat, and have breathing problems in places that are very humid. In this case, the system could use a fan or some other method to let air flow [Figure 4.].

Both the temperature and humidity must be kept in check to keep the baby safe and comfortable. The interface of such a monitoring system allows parents to view the information in real time, as such they can monitor the baby remotely and assist in the cases when needed. Besides, the system can be configured to behave in a manner that holds the baby in the optimal position. To be sure that your baby is fine and safe when you are not at home, implementing an IoT baby monitor device is a smart solution. This system is supposed to maintain the health of the baby. It monitors such vital parameters as temperature and humidity in real-time, automatically reacts to the changes, and notifies the caretakers immediately. It is an intelligent, trustworthy, and active manner of raising children in the modern era. It is able to move the cradle, switch the temperature of the room and notify the caregivers as something significant happens in the environment. The sensor has already reached 94-degree Fahrenheit and this is way too hot considering that the optimum baby room temperature is 68 to 72 degrees Fahrenheit (20 to 22 degrees Celsius). When temperatures are this high, babies may feel uncomfortable and are more likely to get heat rash, dehydration, and overheating. The system would send an alert to caregivers, and they would be able to lower the temperature by setting off cooling systems like a fan that is connected. You need to keep the room at the right temperature for the baby's health and good sleep.

It's also above the range of humidity that's best for babies, which is usually between 40% and 60% (76%). Due to high humidity, the room may feel warmer than it really is.

Excessive heat and humidity within an infant's sleeping environment can significantly affect overall health and comfort. Elevated temperatures may cause overheating, which not only disturbs the baby's natural sleep cycle but also increases irritability and restlessness. Poor ventilation combined with high humidity levels can further contribute to respiratory discomfort, making infants more susceptible to breathing difficulties such as nasal congestion, allergies, or mild respiratory distress. Since newborns have delicate immune and respiratory systems, prolonged exposure to unsuitable environmental conditions may lead to more serious health complications over time.

Moist and damp surroundings also create favorable conditions for the growth of mold and microorganisms. Mold spores can spread easily in enclosed spaces and may trigger allergic reactions, skin irritation, or respiratory infections in infants. Preventing excessive moisture accumulation is therefore essential to maintaining a safe and hygienic sleeping area.

By incorporating environmental monitoring and control features, the system functions as a comprehensive baby care solution. It continuously maintains optimal conditions, ensuring that the infant remains comfortable and protected. As a result, caregivers do not need to engage in constant manual supervision, and the baby benefits from a consistently healthy, safe, and well-regulated environment.

6. CONCLUSION

The development of an IoT-enabled Smart Baby Cradle effectively overcomes the limitations of conventional cradles by integrating advanced sensing, automation, and communication technologies. Traditional cradles are designed mainly to provide a comfortable resting space for infants, but they lack continuous monitoring and alert mechanisms. As a result, caregivers must rely entirely on manual supervision to ensure the baby's safety and well-being. In today's fast-paced world, where parents often manage professional responsibilities alongside household duties, constant physical monitoring may not always be possible. The proposed IoT-based smart cradle addresses this challenge by enabling real-time monitoring and automated responses through an interconnected system.

The smart cradle incorporates multiple sensors to continuously observe environmental and behavioral parameters such as temperature, humidity, crying sound, and moisture levels. These parameters are processed by a microcontroller that evaluates whether the readings fall within safe and predefined limits. If any abnormal condition is detected—such as excessive crying, uncomfortable room temperature, or bed wetness—the system immediately generates notifications. These alerts are transmitted to parents through a connected mobile application, allowing them to respond promptly regardless of their physical location. This real-time alert system significantly enhances infant safety and minimizes the risks associated with delayed attention.

Remote accessibility is one of the key strengths of the IoT-enabled cradle. Through a mobile application, parents can monitor live data related to their baby's environment and cradle status from anywhere with internet access. This feature is especially beneficial for working parents, as it ensures continuous awareness of their infant's condition. The application not only displays sensor readings but also enables remote control of certain cradle functions, such as activating gentle rocking or playing soothing music. This combination of monitoring and control reduces parental anxiety and promotes a sense of constant connection.

In addition to monitoring capabilities, the cradle includes automated soothing mechanisms designed to enhance infant comfort. When the system detects crying through the microphone sensor, it can automatically initiate a gentle rocking motion using motors. This motion mimics traditional manual rocking, which is widely recognized for calming babies. Similarly, integrated audio modules can play soft lullabies to soothe the infant. These automated responses provide immediate comfort and reduce the need for constant manual intervention by caregivers.

Overall, the IoT-enabled Smart Baby Cradle represents a significant advancement in baby care technology. By transforming a conventional cradle into an intelligent monitoring and response system, it establishes a safer and more comfortable environment for infants. The integration of real-time alerts, remote accessibility, and automated soothing features not only enhances convenience for caregivers but also promotes improved infant well-being. This innovative approach sets a new standard in smart childcare solutions, offering parents peace of mind and ensuring that babies receive timely attention and optimal comfort at all times.

REFERENCES

Ahmad, N., & Kumar, S. (2021). Real-time infant monitoring system using cloud-enabled IoT devices. *International Journal of Distributed Sensor Networks*, *17*(7), 1–12.

Brown, T., & Miller, J. (2020). Secure IoT frameworks for smart childcare applications. *IEEE Consumer Electronics Magazine*, *9*(5), 56–63.

Chen, L., & Patel, M. (2019). Artificial intelligence in child monitoring systems. *International Journal of Computer Science and Information Security*, *17*(9), 75–82.

Chen, Y., & Wang, L. (2019). Smart home technologies for child safety applications. *Journal of Ambient Intelligence and Humanized Computing*, *10*(8), 3125–3134.

Desai, P., & Kaur, H. (2022). Technology-driven innovations in infant healthcare monitoring. *International Journal of Recent Technology and Engineering*, *11*(3), 201–209.

Fernandez, A., & Silva, M. (2021). Automated baby care systems using intelligent sensing. *Journal of Healthcare Informatics Research*, *5*(3), 211–223.

Gonzalez, R., & Lee, H. (2020). Smart parenting systems and infant care technologies. *IEEE Access : Practical Innovations, Open Solutions*, *8*, 165421–165430.

Gupta, S., Verma, P., & Reddy, K. (2021). Smart cradle automation using IoT and embedded systems. *International Journal of Innovative Technology and Exploring Engineering*, *10*(8), 45–50.

Kumar, P., & Joshi, A. (2021). IoT-enabled child safety monitoring in smart homes. *International Journal of Smart Technology and Learning*, *2*(1), 44–53.

Kumar, R. (2022). Automated infant care systems using IoT technology. *Journal of Embedded Systems and Applications*, *9*(4), 101–108.

Lee, M. (2021). Advanced smart cradle systems for modern childcare. *International Journal of Smart Home*, *15*(2), 55–63.

Mehta, R. and Rao, P. 2022, "IoT-enabled neonatal monitoring and smart cradle architecture," *Journal of Healthcare Engineering*, vol. 2022, no. 3, pp. 1–10.

Nguyen, T., & Park, S. (2023). Sensor-integrated smart cribs for neonatal monitoring. *International Journal of Medical Informatics*, *172*(5), 104–113.

Patel, A., & Sharma, R. (2023). Embedded sensor-based intelligent baby cradle. *International Journal of Electronics and Communication Engineering*, *12*(1), 77–84.

Sharma, R., & Gupta, P. (2022). Predictive analytics for child safety monitoring using AI. *Journal of Artificial Intelligence Research*, *14*(2), 88–97.

Shreelatha, R. P., & Kumar, V. (2017). Wireless baby monitoring system using mobile application. *International Journal of Computer Applications*, *162*(4), 18–23.

Singh, R., & Thomas, J. (2020). Remote baby monitoring using wireless sensor networks. *International Journal of Engineering Research & Technology (Ahmedabad)*, *13*(5), 89–96.

Symon, A. F., Hasan, M., & Hossain, T. (2017). Design and implementation of a smart baby monitoring system using IoT. *IEEE Internet of Things Journal*, *4*(6), 2157–2165.

Telepatil, A. R., Patil, P. P., Yaja, S. S., & Jadhav, S. R. (2021). IoT based smart baby cradle system for infant monitoring. *International Journal of Advanced Research in Computer and Communication Engineering*, *10*(5), 112–117.

Wang, X., & Zhang, Y. (2023). AI-assisted baby care and monitoring systems. *Sensors (Basel)*, *23*(4), 1450–1462.

Chapter 8
Privacy, Security, and Ethical Challenges in AI and IoT Social Sensing Systems

Nathaniel Adeniyi Akande
http://orcid.org/0009-0001-5930-3116
University of Bradford, UK

ABSTRACT

The infrastructure that promises urban revolution of innovations, such as real-time traffic optimization, community surveillance of health metrics, and data-intensive urban governance, does not come without significant risks, including mass surveillance, cybersecurity threats, and discrimination via algorithms. The chapter evaluates the challenges of privacy, security, and ethics, shedding light on governance gaps in legal regulations such as the GDPR and the NIST frameworks by detailing these gaps using examples from a ransomware case in Atlanta and in another case, the racial bias in predictive policing. To address these challenging gaps, the implementation of privacy-by-design system architecture, decentralised data governance systems, and participatory policy systems that bring technological evolution in line with human rights protection and democratic principles was proposed. The conclusion affirms the urgent need to have interdisciplinary cooperation that can promote the production of smart cities that are profitable and fair by developing a thorough culture of trust.

DOI: 10.4018/979-8-3373-4962-6.ch008

1. INTRODUCTION: AI-IOT-DRIVEN SOCIAL SENSING IN SMART CITIES

Artificial Intelligence (AI) and the Internet of Things (IoT) have been developed rapidly, causing the instigation of social sensing, which is a groundbreaking concept in smart-city systems. This system perpetually accumulates and questions raw data on human behaviour, movement patterns, nature, and societal sentiment via the combination of networked gadgets, wearables, smartphones, environmental sensors, and social media platforms (Dhelim et al., 2021). This merging of the two provides municipalities with the ability to maximise infrastructure, improve services with services and progress towards achieving sustainability with predictive analytics and robotised decision-making. At the same time, however, the ubiquity of AI-IoT social sensing raises serious privacy, security, and other ethical issues that demand urgent scholarly and policymaking interest (Alahi et al., 2023).

1.1. Defining AI-IoT Social Sensing

Social sensing is both a socialising process driven by AI-IoT and the socialisation of AI-IoT--the mutual interaction that involves aggregation of distributed sensors and AI used to perceive and describe urban dynamics. Unlike legacy sensing solutions, which rely on a stream of isolated data, social sensing combines heterogeneous data sources, such as GPS traces sourced by smartphones, biometric data collected by personal devices, and sentiment data taken from social media to create actionable intelligence (Rashid et al., 2023). Examples are smart traffic environments that use real-time crowd-sourcing data to reduce traffic congestion and environmental sensors that can optimise air quality to inform and prompt the development of public health policies (Longo et al., 2017). These cases show how ambient datafication, the passive, non-consented data collection, can increase the efficiency of an urban sphere and at the same time threaten individual autonomy.

1.2. The Imperative of Privacy, Security, and Ethics

The advantages of social sensing have serious risks attached to them:

- **Privacy Violations:** There are fears of privacy invasion because the data gathered may be as coarse as location trails or as fine as behavioral biometrics, and can lead to surveillance creep and function creep, where the data used to control traffic is converted to law enforcement or commercial profiling in ways not determined by society (Barkane et al., 2023). Such actions are inconsistent with frameworks such as the GDPR, which speaks about purpose limitation and user consent.

- **Security Vulnerabilities:** IoT systems could be attacked, and it is even possible to attack the machine learning used on the IoT systems (examples include sensor spoofing, intercepting the data being passed between systems, and adversarial AI attacks based on corruption of the AI used predictive models (such as falsifying pollution data or an emergency response attack) (Denis et al. 2025). Inadequate preventive measures are manifested by the devastating effects of the ransomware attack communicated to Atlanta when its smart infrastructure was compromised in 2025 (described below).
- **Ethical Dilemmas:** Algorithms produce an unethical outcome because they are biased and tend to harm marginalised populations, further causing social disparities (Yen & Hung, 2021). The democratic divide also leads to further entrenchment of disparities because the excluded groups in the digital world, like the low-income community, lack a process of appeal against the resulting automated decision-making concerning housing, policing, or social services.

1.3. Chapter Objectives and Structure

This chapter takes a multidisciplinary perspective, which combines knowledge in data ethics, cybersecurity, urban policy, and human rights to:

- Examine and comment critically on tradeoffs about individual rights and civic innovation in the AI-IoT implementation.
- Assess both technical and policy protections, e.g. privacy-by-design designs, and algorithmic responsibility systems.
- Propose inclusive patterns of governance, demonstrating participation-based data-trusts and AI-audits, to make smart city technologies serve democratic purposes.

2. OVERVIEW OF SOCIAL SENSING SYSTEMS IN SMART CITIES

2.1 AI and IoT in Real-Time Social Sensing

AI and IoT in smart cities have transformed the social sensing process through constant real-time data sources that include wearable, environmental sensors, CCTV networks, and social media. Devices that make up the IoT are embedded in the infrastructure of the city, via devices such as fitness trackers, air quality monitors, etc., and AI algorithms interpret that data, deriving consumable insights (Ramani et al., 2025). As an example, the GPS + accelerometer capabilities of smartphones can help to understand traffic and mobility patterns, and CCTV cameras with the functionality of computer vision can reveal unusual phenomena (like accidents or a crowd rush). The social media platforms also surface this ecosystem with sentiment analysis, which has occurred throughout the discovery of the COVID-19 situation, with Twitter information being used to monitor current population situations and misinformation (Osakwe et al., 2020).

But the extent of this data gathering brings with it several very important issues of privacy and consent. The idea of ambient datafication, which involves collecting cross-conceptual data on citizens without their input, can not always be carried out with the knowledge of the citizens, as is the case with the GDPR framework (Cremonini, 2023). Indicatively, facial recognition CCTV systems were widely used to monitor the faces of London residents, resulting in a backlash by the city due to the ability to conduct mass surveillance (Galliven, 2025), which is one of the key issues between innovation and civil liberties. Despite these difficulties, the combination of AI and IoT cannot be supplanted when it comes to urban resilience since it provides detailed insights concerning human behavioural patterns and the state of the environment.

2.2 Applications in Urban Systems

The urban areas are evolving through social sensing systems to offer some of the most important services, which are urban planning, the health of the people, the transport systems, and even disaster management. In urban design, efforts such as the renovation of the old town of Beijing, through social sensing, utilised the data to study the land-use and mobility of residents to optimise the development of infrastructure (Yu et al., 2017). In the same way, wearable data is being used in public health efforts to track epidemics; the TraceTogether app in Singapore used both Bluetooth and GPS to facilitate contact tracing through the COVID-19 pandemic, although issues were raised by critics about the data-retention policy (Asher, 2020).

The traffic management systems, like Barcelona IoT-based smart lights, adapt the signal timing in real-time according to the traffic and pedestrian traffic, decreasing congestion by 21% (Alahi et al., 2023). In the meantime, in emergency response, AI-based systems such as the Smart Dubai initiative in the city of Dubai combine social media feeds and sensor networks to identify disasters (i.e., floods or fires) and to organise first responders (International Telecommunication Union, 2016). However, such applications have their shortcoming. The use of predictive policing technology, including the Chicago system, has been described as perpetuating racial biases because of training data imbalances (Shapiro, 2019).

2.3 Benefits and Societal Impact

Efficiency, sustainability, and data-based governance are some of the promises of the adoption of social sensing systems. The efficiency can be observed in such utilities as smart grids, where AI can analyse IoT-generated data on the use of energy to equalise loads and minimise waste, as in the case of Amsterdam Climate Street (Alghamdi et al., 2025). The advantages of sustainability consist of monitoring pollution; the real-time emissions monitoring in Shenzhen using AI-enabled sensors allows real-time policy changes, reducing PM2.5 levels (Wu et al., 2021).

Participatory frameworks also improve the conduct of governance. The examples of how Saudi Arabia used sentiment analysis on 2 million tweets during the COVID-19 period to inform its education policies (Alotaibi & Nadeem, 2024), which are a combination of online and face-to-face learning, depending on the opinion of the population. Nonetheless, such developments are dependent on transparency and equity. The marginalised populations do not usually have access to digital tools, which further increases the democratic divide. For instance, citizen-centric data trusts developed by Barcelona empower residents, but in low-income areas, such programs often struggle to achieve high participation rates due to digital literacy disparities (Calzada, 2018).

Although AI-IoT social sensing systems have the potential to transform smart cities, their effectiveness will be determined by the ability to strike a balance between innovative use and ethical governance, inclusivity, and compliance with regulations. Privacy-by-design and algorithmic accountability should become the key priorities of policymakers, who must guarantee that such technologies can benefit every citizen in an equal way. Technological expertise is not the only technology that is the future of urban resilience, but also the trust-building method of participatory and transparent structures.

3. PRIVACY RISKS AND SURVEILLANCE CONCERNS IN AI-IOT SOCIAL SENSING SYSTEMS

Social sensing in smart cities integrated with AI and the IoT has brought about new abilities in monitoring and managing cities as never previously seen. The wide-scale accumulation of geolocation, biometrics, behaviour, and ambient data, however, brings privacy and surveillance issues of significant concern, and threatens to upset the balance between civic innovation and personal rights (Syahmi, 2025). These systems can be built under the name of effectiveness and social security, but their data collection activity tends to impinge on individual rights, especially in the form of ambient datafication, functional creep, and risks associated with differential privacy. These tensions are well depicted by the COVID-19 contact tracing apps, which are in the process of transforming into permanent surveillance tools.

3.1 Types of Data Collected: Granular and Intrusive

Social sensing systems collect geolocation data collected on smartphones, biometrics (e.g. facial recognition, heart rate monitors), behavioural patterns (e.g. mobility traces, social media activity) and ambient data (e.g. environmental sensors, audio recordings) (Zhang et al., 2023). As an example, facial recognition and gait tracking, smart CCTV networks in cities such as London and Shenzhen can predict so-called suspicious behaviour, and wearable technologies in healthcare constantly track the vitals of patients without their express permission. This kind of data can be extremely identifying, even after anonymisation, as indicated by research (including studies by Hardesty, 2013; de Montjoye et al., 2013), indicating that merely 4 spatiotemporal points could be used to uniquely identify 95 per cent of people. Such granularity of data collection makes the boundary between the public and the personal so fine that the traditional concept of consent is no longer valid.

3.2 Key Privacy Challenges

- **Ambient Datafication and Lack of Informed Consent:** The central problem is the ambient datafication that implies passive data gathering without involving a user (Kennedy, 2018). As an instance, IoT-powered smart benches in urban green spaces can detect sound or Wi-Fi to calculate the number of people, whereas Bluetooth beacons in retail establishments monitor the movement of the customers. In contrast to active data sharing (e.g., sharing on social media), ambient collection is stealthy (Zhao et al., 2020), so they do not know when and how their data is collected. It is against the principles of informed consent, and this is what the GDPR (2018) emphasises: transparency and control of the user. The operation of ambient intelligence systems in hospitals in the healthcare industry records interactions of staff and patients, but little is revealed about how long the information is stored and by whom.
- **Function Creep: Mission Drift in Data Usage:** Another important issue is that of function creep, or the misuse of data for purposes other than its intended use (Kroops, 2021). Singapore TraceTogether app, which was marketed to be used in COVID-19 contact tracing, was applied at a later point by police in criminal investigations, defying government promises (Asher, 2020). The same has been applied in Australia, where police have used QR code check-in data in bars to detect other crimes not related to the bar (Galloway, 2021), which led to changes in legislation to limit its use. Such instances highlight

the ability of emergency measures to normalise surveillance, which weakens the trust of the populations. Centralised data storage further contributes to the problem of function creep, as the accumulated data is simply too enticing to be used in various ways, whether it is predictive policing or targeted advertisement (Koops, 2021).

- **Differential Privacy Risks:** Privacy violations are usually directed towards marginalised groups. Predictive policing algorithms are based on previous arrest data, which was biased and Black and Hispanic neighbourhoods are over-policed (Brantingham et al., 2018). Equally, women and people of colour also have a better chance of being misidentified by facial recognition systems (Yucer et al., 2024). The issue of the democratic divide also adds to inequalities: low-income communities, which do not have the right to opt out or have a low level of digital literacy, are at a greater risk of data exploitation (Tiffin et al., 2019). As an example, biometric ID systems with increased surveillance of homeless populations in smart cities are likely, and there is no chance to appeal against the misuse of data.

3.3 Case Example: COVID-19 Contact Tracing Apps

The tension between privacy and public health became obvious during the worldwide implementation of COVID-19 contact tracing apps. Whereas apps such as Germany (Luca) (Holl et al., 2022) and Canada (COVID Alert) (Government of Canada, 2020) used decentralized Bluetooth protocols to limit data exposure, other apps were centralized, such as Singapore (TraceTogether) (Asher, 2020) and allowed access to law enforcement. When governments began to use the data without consulting the citizens, a backlash appeared: in Israel, the data on phone location, initially gathered to track people during the quarantine, was later accessed by security agents (Halbfinger et al., 2020; Holmes, 2020). Apps that were even advertised as privacy-preserving were not without criticism; Bluetooth-based systems still kept IP addresses and device identifiers, which could be used to perform linkage attacks. The episode showed a larger trend: when tech adoption is due to a crisis, the issue of accountability is usually avoided, with legacy surveillance systems remaining intact.

Addressing these challenges needs technological protection (e.g. federated studying, differential privacy), coverage reforms (e.g. stringent purpose-limitation issues), and participatory design featuring marginalised populations. The GDPR (2018) and Privacy by Design systems provided by the EU are templates, but there is no even application around the world (European Data Protection Supervisor, 2018). The issue with smart cities is not how to innovate but how to integrate equity and transparency into the data governance process, so that the social sensing is not only an issue of benefiting those being surveilled.

4. SECURITY THREATS IN AI-IOT ECOSYSTEMS: A CRITICAL ANALYSIS

Smart cities with AI and the IoT have completely changed the nature of urban governance, where data analytics live and decisions are often made automatically. Nevertheless, this collision has also created a tricky frame of vulnerability, and AI-IoT ecosystems now face well-crafted cyber-attacks that threaten safety, privacy, and raise a question mark on popular trust. Ranging from the spoofing of individual devices to AI adversarial attacks, the appearance of security risks in such systems is complex and growing. The Atlanta ransomware incident last year that brought the city to its knees and cost millions of dollars to be wiped, is a testament to the disastrous outcome of unmolested weaknesses in the smart city resources.

4.1 The Vulnerability Landscape in AI-IoT Ecosystems

- **Device-Level Threats:** Physical tampering and spoofing of IoT devices can take place, since they are often created with little in terms of computing capabilities and security features. An example of this is that in smart cities, sensors about the environment can be hacked to input inaccurate information, e.g., air quality meters or measures of the traffic, which will result in policymakers making the wrong choice (AlJamal et al., 2024). Wearables and biometrics are no less susceptible either; researchers (see Dou, C., & Huan, 2021; Anthi et al., 2024) have already revealed that it is possible to spoof heart rate trackers with off-the-market radio equipment and gain unauthorised access to potentially sensitive health information. All these risks are further compounded by the absence of strong boot mechanisms on most IoT devices, which enables an attacker to sideload malicious firmware updates.
- **Network-Level Threats:** IoT systems are prone to Man-in-the-Middle (MITM) attacks wherein attackers can exploit the interdependency with the creation of a false two-way transmission among networks and cloud sites (Fereidouni et al., 2025). As an instance, smart traffic lights can be decrypted to cause artificial traffic congestion or even accidents. The 2016 Mirai botnet incident revealed that infected IoT appliances, including CCTV cameras and routers, can be used to carry out Distributed Denial-of-Service (DDoS) attacks that flood infrastructure and block it (Affinito et al., 2023). The improper authentication protocols in the form of default passwords in municipal IoT implementations contribute as well to unauthorised access to networks.
- **AI/ML-Specific Threats:** The IoT systems which use AI are vulnerable to the risk of model poisoning, in which the basic data is injected with corrupt data through the training process to distort the algorithmic results (Wang et al,

2022). An obvious instance is that of the predictive policing systems, whose use is prone to bias, considering that they will reproduce discrimination when the training that shows biased examples of arrests is used. Adversarial attacks ride on the fact that AI depends on pattern recognition inputs- minor manipulations in data will mislead the facial recognition software into assigning wrong labels to people (Chakraborty et al., 2021). In the context of autonomous vehicle networks, which may be doomed in case of manipulated sensor data, these vulnerabilities deserve to be highlighted as especially worrying.

4.2 Consequences of Security Breaches

- **Safety Risks:** Any manipulation of the AI-IoT systems may lead to terrible real-world implications. Smart transportation systems and their satellite-based communications infrastructure are vulnerable to GPS jamming, traffic system hacking and even diversion of emergency vehicles due to spoofed GPS signals, which respond to lost communications or failure to reach their destinations in time (Gupta et al., 2025). On the same note, a hacker can manipulate sensors used in industrial IoT (IIoT) in water treatment plants and cause unnoticed contamination that may result in health emergencies. Even though the Stuxnet worm is not an IoT attack, it exemplified the ability to sabotage cyber-physical systems that might result in damage to physical assets, a trend IoT-enablement critical infrastructure must prevent (Dziwisz, 2023).
- **Erosion of Public Trust**: Recurrent violations will loss of trust of people in smart cities. Not only did the Atlanta ransomware attack cause disruption of municipal services, but it also showed the instance of systemic neglect, as more than a year before the incident, auditors discovered 1500-2000 vulnerabilities that were not fixed on the system (Sebastian et al., 2022; Murciano-Goroff et al., 2024). These failures encourage citizens to avoid engagement in the digitalisation process of governance, occurring when contact tracing apps to prevent COVID-19 penetrated the market, and citizens were quick to reject them due to the fear of using their data inappropriately. The issue of biases in AI-IoT systems, like racial differences in facial recognition, continues, further destroying trust when they are not being solved.

4.3 Case Example: The Atlanta Ransomware Attack

In March 2018, SamSam ransomware was deployed that paralysed the Atlanta outdated IT infrastructure by encrypting the court records, police databases, and utility payment platforms (U.S. Attorney's Office, 2018). This ransomware attack took advantage of servers that had not been patched and poor Remote Desktop

Protocol (RDP) settings, which emphasise how dangerous legacy systems can be in smart city systems (Smyre, 2018). Recovery became very expensive and cost the company a huge sum of 17 million dollars, and some services still required months to be back online (Pascu, 2018; Schwartz, 2025). Three important lessons stood out from the incident:

1. **Proactive Patching is Non-Negotiable**: The breach was aggravated by the inaction of the city concerning pre-attack vulnerability audits.
2. **Decentralised Backups Mitigate Damage**: The use of centralised storage at Atlanta led to a further increase in data loss; backups on the cloud would have speeded up the process.
3. **Cross-Sector Collaboration is Vital**: After the attack, Atlanta collaborated with MS-ISAC and its privately owned companies to redesign its cybersecurity infrastructure, with a focus on the importance of public-private resilience

The safeguarding of AI-IoT security threats needs a layered approach:

- **Zero Trust Architectures**: Lateral movement can also be curtailed by segmenting the networks and having strict access administration control.
- **Explainable AI (XAI)**: Transparent AI models will allow auditors to observe biases and adversarial manipulations.
- **Regulatory Enforcement**: They may force the use of such frameworks as NIST IoT Security Guidelines and GDPR to homogenise the security practices among vendors.

5. ETHICAL DILEMMAS AND GOVERNANCE GAPS IN AI-IOT SOCIAL SENSING SYSTEMS

It has become evident that the realities of AI and IoT-based social sensing are being deployed at a rapid pace in smart cities and are struggling to balance technology creation and social principles with the rigidity of governance frameworks. Such systems that accumulate large volumes of behavioural, biometric and environmental information pose basic questions of data ownership, algorithmic fairness, and consent, a challenge that current legal systems do not fully satisfy. Race prejudice in predictive policing algorithms is an example of the failure of a technocratic answer with no ethical regulation, instead going against it, to create a systemic disadvantage.

5.1 Core Ethical Questions in Social Sensing

- **Ownership of Social Sensing Data**: The most dominant ethical concern is over ownership rights to data in AI-IoT systems. The methodology of social sensing is capable of ambient datafication or data extraction in a passive manner from the privacy of the front of things, wearable devices, and spy eyes without a formal response (Lee, 2019). As an example, instead of sitting on a regular bench, one could sit on a smart bench with environmental sensors which might cut an audio or Wi-Fi signal, whereas a retail store could follow the path of a shopper using Bluetooth beacons, which are practices of having grey areas where there is a junction of a public and a private data there (Torrens, 2022). There is no balance of power as a result of the absence of well-structured ownership arrangements, as companies and states manage flows of data, and citizens are oblivious of how their data is utilised or monetised.
- **Algorithmic Bias and Discrimination:** AI is even more subject to ethical risks because its decision-making is opaque, especially in such high-stakes uses as predictive policing and facial recognition. Research (e.g., Shapiro, 2019; Richardson et al., 2019) indicates that such systems tend to contain historical biases/prejudices; predictive policing systems, to cite but one such example, tend to focus undo attention on Black and Hispanic neighborhoods owing to lopsided training sets based on neighborhoods already disproportionately policed in the past. In the same fashion, facial recognition programs are less accurate when it comes to women and people of colour, thereby causing erroneous identifications during policing (Moy, 2021). The biases weaken the image of justice and fairness, as the ethics guidelines on AI, such as the EU AI Act, define such applications as the ones labelled as high-risk and require transparency.
- **Voluntary Consent in Public Spaces**: The idea of informed consent is vague when it comes to public or semi-public locations where people do not know they are being observed. As an example, the facial recognition CCTV networks implemented by London and the TraceTogether contact-tracing app initially did not have any opt-out provisions, which prompts arguments of forced collection of information in the name of citizen security. These ambient data gathering also tend to ignore the GDPR formula of consent being freely according, specific and unequivocal since the gathering of information in these settings is done on a take-it-or-leave-it system where refusal to receive the information may render one unable to access essential services.

5.2 Limitations of Existing Legal Frameworks

Though the existing regulations, such as GDPR, CCPA, and HIPAA, offer basic security, they do not provide answers regarding the specific issues of AI-IoT social sensing:

- **GDPR's Territorial and Conceptual Gaps**: Although purging and data reduction are required by the GDPR, they are not consistently supported beyond the EU. Besides, it is also grappling with the issue of function creep, in which the information gathered to manage traffic is used by the police, an issue demonstrated by the Singapore TraceTogether debacle.
- **CCPA's Narrow Scope**: CCPA provides Californians with the rights to access and delete personal data, but not the aggregated or anonymised data (Determann & Tam, 2020), which are extensively used in social sensing systems. In this loophole, companies manage to escape the requirements of individual consent.
- **HIPAA's Healthcare Focus**: HIPAA protects health information, but disregards biometric data accumulated by wearables or environmental monitors (Sivakumar et al, 2024), leaving the health information adjacent unprotected.

5.3 Case Example: Racial Bias in Predictive Policing

The COMPAS algorithm, which is popular in the courts of the United States in determining the likelihood of recidivism, represents the case of governance and ethical failures of AI-driven social sensing. In 2016, a Pro Publica investigation showed that the algorithm was incorrectly labelling black defendants three times more likely than white defendants to get a false high-risk rating, and white defendants were evenly likely to be defined as a low-risk one (16 per cent vs. 5 per cent) (Larson et al., 2016). This bias was the compound of training data, which is based on systemic disparities within the policing, and not on objective crime rates. COMPAS is still used, despite various accusations, exposing the absence of an accountability system in AI regulation.

The case of COMPAS highlights three very important gaps:

1. **Transparency**: The algorithm was proprietary software and, therefore, was not auditable, which also disregards the GDPR right to explanation.
2. **Redress**: No appeal mechanism was available to affected people fighting against the decisions made by algorithms, and the EU AI Act aims to provide such protection by requiring a provision of a human supervisor in decision-making.

3. **Contextual Ignorance**: The algorithm disregarded socio-economic concern and focused on past arrest data as indifferent data inputs- an issue, which is accentuated by a democratic gap in smart cities and relegates marginalised citizens to their role in the development of the system.

The solutions to these dilemmas need a multi-stakeholder partnership and dynamic regulatory systems:

- **Participatory Design**: Making marginalised groups part of the development of AI systems, as is being done in Barcelona with the citizen-centric data trusts, allows for reducing bias and trust.
- **Explainable AI (XAI)**: Such methods as interpretable machine learning and algorithmic impact assessment (required by the Canadian Directive on Automated Decision-Making) can increase transparency.
- **Dynamic Consent Models**: Inspired by healthcare ethics, real-time opt-out interfaces may enable opt-outs in public spaces through smartphone or physical opt-out kiosks.

6. STRATEGIES FOR ETHICAL AND SECURE SOCIAL SENSING IN SMART CITIES

The application of social sensing through AI in the context of smart cities necessitates efficient mechanisms to balance the technological innovations, ethical imperatives, and security concerns. The concern of privacy-preserving technical measures, adaptive governance systems, and accountability mechanisms centred on citizens is paramount because, as these systems continue to infiltrate our urban lifestyle, they become capable of capturing all manner of mobility data, physical biology, heart rate, images, biometrics, and so on. The example of the data sovereignty project in Barcelona shows the way cities could become the first ones to implement ethical social sensing through the incorporation of decentralised technologies, dynamic consent frameworks, and participatory governance.

6.1 Technical Safeguards: Embedding Ethics into Design

- **Privacy-by-Design and Security-by-Default:** Privacy-by-design requires that the systems, including their data protection, are designed with privacy in mind and not as an afterthought, which can be added after the system has been designed. Among these, we may consider such techniques as data minimisation (the minimal amount is collected), pseudonymization (the database is disassociated with any identifiers), and end-to-end encryption of both data at rest and data in transit (Alauthman et al., 2025). An example is how Singapore in its TraceTogether COVID-19 app also began as centralised user data, but, since then, it transferred to decentralised Bluetooth protocols to prevent privacy risks as much as possible, which also complies with the data protection-by-default provision of GDPR. In the same manner, security-by-default hardens systems so that they are not easily hacked as soon as they are deployed to the environment (Simoiu, 2020). Incidents like the Mirai botnet attack, where default passwords in the IoT devices were used, demonstrate the importance of a required security setup on the piece of devices, including multi-factor keys and periodic firmware revisit.

- **Decentralised Solutions: Blockchain and Beyond:** Social sensing systems benefit through transparency and auditability through decentralised architectures, like blockchain. The DECODE project in Barcelona uses blockchain to enable citizens to grant or revoke controls over where their data goes by preventing commercial exploitation of personal data, but only allowing it to be used in approved applications (e.g., to optimise traffic) (Monge et al., 2022). Further enhance privacy using Zero-knowledge proofs (ZKPs) that certify transactions without disclosing structured information (Bhattacharya et al., 2024), which tried to implement a healthcare blockchain-controlled to verify prescriptions without disclosing patient records. Nevertheless, decentralisation is not a silver bullet: scalability becomes a problem with public blockchains (e.g., Ethereum), whereas privacy requirements can be at the cost of transparency in the case of the private/permissioned ones (e.g., Hyperledger). Hybrids, that is, the combination of decentralised storage and federated learning (where AI model training is done on devices on the edge), offer a viable middle way, as envisioned in the cases of Urban Digital Twins, where real-time analytics and sustainability planning over long horizons are balanced (Ferr

6.2 Policy and Governance: From Consent to Oversight

- **Dynamic Consent Models**: Social sensing involves a granular data collection that makes classic broad consent systems, in which the user grants broad permission on the future use of the data, much less sufficient (Lee & Lee, 2022). Sometimes the process of obtaining consent can be dynamic, as it is already implemented in the biomedical research field (Ivanova & Katsaounis, 2021; Merlec et al., 2021) and enables a user to adjust settings in real-time with the help of digital tools. To illustrate, the EnCoRe project allows participants of the biobank to be in/out of any study, which overcomes the arguments of dissatisfaction with the silent inclusion of biobank participants into research in passive consent regimes. In the context of smart cities, a dynamic consent would enable citizens to directly switch the right to utilise their smart meter or CCTV information task, which remains difficult among audiences with low levels of digital literacy.

- **Independent Oversight Bodies**: One of the solutions to reduce risks of function creep and algorithmic bias-related concerns is independent, multidisciplinary oversight organisations, such as technologists, ethicists, and representatives of the community, etc. The attempted implementation of such surveillance in the project of the existing Smart City in Toronto (which led to the popular opposition to some actions implementing the idea of the Sidewalk Labs to monetise collected data) (Wang et al., 2021) is an example of the significant consequences of the lack of such regulation. In contrast, the Algorithmic Transparency Register in Amsterdam requires AI systems to be transparent with their purpose and impact, to be accountable (GPAI, 2024). These organs must implement algorithmic impact assessments (AIAs) that serve functions similar to environmental audits to determine the fairness of the systems and the potential unintended damages before they go into operation.

6.3 Case Example: Barcelona's Citizen-Centric Data Governance

The DECODE project of Barcelona is the perfect embodiment of ethical social sensing by doing three innovative things:

1. **Data Sovereignty**: With the help of blockchain-based licenses, citizens can control who gets their data and can choose to share it or not with the government or advertisers in cases of a public service (e.g., to understand pollution).
2. **Participatory Design**: The city partners with the community members in co-designing policies, including how sensor data will be used to fight housing shortage caused by tourism: the process has been recorded in its Digital City Roadmap.
3. **Transparent Infrastructure**: Information flows can be audited when sensor data is logged on open-source platforms such as Sentilo.

Nevertheless, there are also some problems behind the model used in Barcelona, such as scalability and the necessity to rely on civil involvement as a part of the model, which is reflected in the dynamic consent system, where the level of participation is unequal in each case.

Technology and governance must have synergistic properties to be able to perform ethical social sensing:

- Technically, privacy-preserving solutions such as ZKPs and hybrid blockchains will also need to change to support large-scale urban data.
- Politically, the introduction of dynamic consent and policing bodies should be established upon legislation, as it would be suggested in the AI Act in the EU.
- Socially, cities need to invest in digital literacy to benefit the marginalised groups, with equal advantages.

7. POLICY AND REGULATORY RECOMMENDATIONS FOR ETHICAL AI-IOT SOCIAL SENSING

The explosive development of AI-IoT-enabled social sensing in the context of smart cities has entailed sound policy and regulatory measures to advance a balance amid innovation and privacy, security and equity. In existence are the global best practices or standards, such as the GDPR, NIST Privacy Framework, and ISO/IEC 27701, that offer basic guidance, yet there remains a gap in how to deal with the various issues that only ambient data gathering, algorithmic bias, and transnational data flow have specific solutions to. Mandatory privacy impact assessments (PIAs), participatory governance, and digital literacy programs are only some ways policymakers need to close these gaps, as illustrated by the EU Digital Education Action Plan (20212027). These are important to allow smart cities to foster transparency, accountability and inclusivity in their technological implementations.

7.1 Global Frameworks: GDPR, NIST, and ISO 27701

At present, the gold standard in data privacy is the General Data Protection Regulation (GDPR), which focuses on the importance of principles such as purpose limitation, data minimisation and the idea of accountability (Hjerppe et al., 2019). Nevertheless, its territorial limits and enforcement capacities have failed to focus on the modern-time social sensing data, including biometrics of public CCTV or mobility histories. Another example is that GDPR stipulates the necessity of explicit consent, which is usually bypassed in smart cities, where datafication of the ambient is an involuntary process (Rubinstein & Petkova, 2019).

Complementing GDPR, the NIST Privacy Framework provides a risk-based, flexible framework to deal with privacy risks, especially in IoT ecosystems. It can also preserve privacy as its basic functions, Identify, Govern, Control, Communicate, and Protect, can allow organisations to adapt precision to the scenario, including health-monitoring wearables or the smart traffic system (Force, 2018). In the meantime, ISO/IEC 27701 shall expand the ISO/IEC 27001 security controls to cover privacy equivalent requirements, giving organisations that manage personally identifiable information (PII), such as a Privacy Information Management System (PIMS). Although ISO 27701 supports the goals of GDPR, it does not serve as an alternative to legal compliance, but, on the contrary, it helps to bring the concept of privacy-by-design into practice, including pseudonymization and data retention restrictions.

7.2 Proposed Interventions

Mandatory Privacy Impact Assessments (PIAs): The PIAs play an important role in the identification and management of privacy threats in the AI-IoT projects. In another example, the U.S. Department of Homeland Security (DHS) requires the PIAs on any system that processes identifiable data, and their deployment should be transparent and accountable, such as facial recognition at the border (Homeland Security, 2024). In the case of AI-driven systems, the AI can review algorithm bias, creep in functions and provenance of data. The Guide to PIAs on AI Systems, produced by New South Wales, suggests evaluating whether such implications of AI (e.g., predictive policing scores) should be considered as personal information under the privacy laws, since it may continue the existing discrimination even when anonymised (Information and Privacy Commission, 2024). Integrating PIAs in procurement processes, as is the case with the EU with its AI Act, can be done to block potential causes of harm before deployment, like biased job applicant recruitment algorithms or unjust health benefits eligibility systems.

Public Consultations and Digital Literacy Programs: Policymaking that is inclusive necessitates involvement of the people in how they contribute to aligning technologies with community principles. An example of a participatory design that allows a citizen to regulate data-sharing rights is the DECODE project of Barcelona. But it is important that the marginalised groups have sufficient digital literacy to interact in a meaningful way. This is covered by the Digital Education Action Plan (20212027) of the European Union (2022) that targets blended learning, AI ethics training of educators, and wider access for women to STEM. Critics (such as Castaneeda et al., 2023; Gottschalk & Weise, 2023) claim that the plan pays too much attention to formal education; however, such non-governmental organisations as the European Association for Adult Education (EAEA) urge to increase their investments in non-formal processes of education to captivate vulnerable groups, such as rural communities or low-income families with no access to broadband.

7.3 Case Example: EU's Digital Education Action Plan

An example of such a policy is the Digital Education Action Plan of the EU, which enhances ethical adoption of technology that serves to resolve capacity problems. Among its 14 actions, it involves:

1. **Ethical AI Guidelines for Educators**: Educating teachers to confront the AI tool bias in the interface.
2. **European Digital Skills Certificate (EDSC)**: Competencies of newly created jobs in the IoT and analytics sectors are being standardised.
3. **Public-Private Partnerships**: Investing in tech companies to invest in infrastructure, but EAEA cautions against the commercial interests that might isolate marginalised students.

The technological ambitions of smart cities can be seen in the focus of the plan on integrating 5G and AI in education, but emphasise that such integration must be coupled with corresponding investments in online security and critical thinking to fight disinformation. As an example, the national digital literacy program in Finland educates residents to question the AI-based algorithms used in news production that can be implemented in smart cities management.

AI-IoT social sensing requires progressive adaptation to control it successfully. Policymakers should:

- **Harmonise Standards**: Essentially, national smart city policy should include the integration of GDPR, NIST, and ISO 27701 requirements, as suggested in a Model AI Governance Framework proposed in Singapore.
- **Enforce Dynamic Oversight**: Set up independent oversight mechanisms that check algorithmic systems, as they have done in Amsterdam with the Algorithmic Transparency Register.
- **Scale Digital Empowerment**: Encourage an EU model of education that builds on community-facilitated workshops in data literacy so that citizens in every country of the EU can learn how to go around and contest surveillance technologies.

The EU way of doing things shows that technological development and human rights are not mutually independent of each other. Smart cities can innovate and transform data into public goods and not venues of exclusion, especially by incorporating ethical frameworks within the economy of education and regulation.

8. FUTURE OUTLOOK: THE EVOLUTION OF AI-IOT SOCIAL SENSING IN SMART CITIES

The fast development of AI and IoT-enabled social sensing is on the verge of transforming the very concept of smart cities, providing the ability to sense emotions and behaviour and making the digital twin and metaverse even more tightly connected to smart cities. Nevertheless, in light of the development of such technologies, urban administrators are faced with the difficult task of balancing between innovation and security of rights, so that the technological advancement is not accompanied by a loss in the sphere of privacy, fairness, or democratic responsibility.

8.1. Emerging Trends: Emotional and Behavioural Sensing

The next wave is emotion AI (affective computing) that helps machines to identify and act on human feelings using facial expressions, vocal tone, and physiological signs. As an example, AI-enhanced wearables and ambient sensors can track stress levels at the workplace or indicate distress in the places of aggregation and thus allow timely mental health treatment (Roy et al., 2025). But this poses an ethical dilemma mainly in terms of consent and misuse of data. Emotion recognition is considered to be of high risk in the proposed AI Act in the EU, and the EU wants rigorous regulation to guard against abusive use of such AI, e.g. discriminatory hiring bots or exploitative advertisement.

Advanced behavioural sensing with Artificial Behaviour Intelligence (ABI) even interpolates intent, based on posture, gait, and interaction patterns, and can be used in autonomous driving (e.g. in anticipating pedestrian movements), or security (e.g. in detecting suspicious practices) (Jo et al., 2025). But, cultural prejudices within the training data may result in discriminatory behaviours such as identifying ethnic minorities using facial recognition systems.

8.2. Integration with Digital Twins and the Metaverse

The metaverse is also combining with digital twins, the virtual replicas of physical assets, to provide immersive urban simulations. To illustrate, Singapore has already implemented the virtual city model, which functions on the current IoT data to manage the traffic and energy consumption (Walker, 2023), and Dubai plans to offer its citizens access to the digital twins of government services through the metaverse project (UAE, 2023). In medicine, organ digital twins can be used to train surgeons in complex surgery to conduct operations virtually.

Nevertheless, such integration increases risk. At the same time, the central question in data sovereignty is whether metaverse platforms, which are commonly privately owned, should host the critical city infrastructure. As a counter-model, the DECODE project in Barcelona provides citizens with a blockchain to make decisions regarding who can access their data. Moreover, the risks of cybersecurity are likely to increase since hackers might use digital twins to sabotage the power lines or transportation systems in smart cities, as shown in ransomware attacks like those in Atlanta.

8.3. Balancing Innovation with Rights Protection

To manage these technologies in a responsible, ethical manner, cities need to implement right-orientedness:

- **Dynamic Consent Models**: Let the citizen tinker with data-sharing options in real-time, modelled after the healthcare biobanks pilot.
- **Algorithmic Transparency**: The Algorithmic Register introduced in Amsterdam requires transparency about the aim and the effects of AI systems.
- **Ethical AI Audits**: Discrimination will be reduced by conducting regular bias assessments that are proposed by the EU AI Act in prediction policing or emotion AI.

The Responsible Smart Cities Toolkit, prepared by the UNDP, emphasises that technologies and smart cities should be co-designed with marginalised communities so that, similar to the unsuccessful Sidewalk Labs project in Toronto, cities do not impose their solutions based on corporate interests rather than those of inhabitants.

The future of smart cities is associated with active governance. Emotional AI and metaverse-enabled digital twins are only more unfathomable as the policymakers will need to implement privacy-by-design, cross-cultural fairness, and social control. It is only through entrenching these tenets that cities will be able to guide social sensing as an instrument of encompassing development, rather than a mechanism of surveillance and isolation. It is not just a challenge in technology but one in the most human sense: to create more just cities that are also smarter.

CONCLUSION

The combination of AI and IoT technologies in smart cities is the equivalent of an unfamiliar opportunity and a fortifying weakness to the current local government. Social sensing systems based on AI-IoT are revolutionary in the optimisation of the traffic streams, monitoring of the health of the population, and decision-making based on the data. Nevertheless, these functionalities that guarantee performance and promptness present convoluted networks of privacy and data abuse, Internet security issues and moral dilemmas that have since become a matter of emergency. The identified key challenges demonstrate underlying conflicts between democratic values and technological innovation. The issue of privacy arises as a result of ambient datafication, in which citizens unsuspectedly lose their data in their daily encounters with smart city infrastructure. The main reason is the security weaknesses of critical systems in the urban environment that can turn into cyberattacks, resulting in failures in delivering necessary services, as happened with the ransomware attack against Atlanta. The most concerning of those gaps is when ethics enables algorithmic bias to be used as a systemic discriminatory mechanism, especially against groups of people who have less access to other resources to protest against cases of automated decisions that affect their access to housing, jobs, and justice.

Nonetheless, there are workable solutions ahead. Ethics can be directly incorporated into the structure of a system through technical solutions: privacy-by-design architectures, decentralised blockchain ecosystems, and federated learning solutions. Such policy frameworks as the mandatory privacy impact assessment, the dynamic consent model, and the algorithmic transparency requirements offer accountability in the form of governance to both the public and private actors. But most importantly, participatory design approaches where communities are at the centre of technology development, like it is done in Barcelona during the DECODE project, are the best way to show how cities can build trust and enable innovation.

These challenges will be aggravated further in the future development of emotional AI and the metaverse-enhanced digital twins. With machines becoming capable of reading and reacting to human emotions, and with digital simulations of urban infrastructure becoming places of democratic action, the reward of successful governance is even greater. The new environment does not require only the reactive control but also the proactive frameworks that may adjust to the change in technology and also accommodate the basic human rights.

This examination alludes to the expediency of the interdisciplinary cooperation of technologists, ethicists, policymakers, and most perilously, the communities afflicted. The social scientists need to collaborate with computer scientists so that they are aware of the wider implications of their invention. To create a system that serves the interests of the public without undermining personal freedom, urban planners need to cooperate with privacy activists so that joint efforts can be used to develop systems or mechanisms in which the interests of both parties would be achieved. Law scholars are compelled to partake in community organising because the law regulations must accommodate the multiplicity of views and safeguard the weak strata of society. Equitable smart city solutions are not entirely about technological solutions or interventions. It requires the radical rethinking of our approach to urban innovation, where ethical concerns would be at the centre instead of the margins of development practices. This implies the creation of systems that are not only technically sound, but also socially equitable, not only effective but also clear, not only novel but also responsible to, and answerable by, the communities in which they profess to operate.

At this critical point, the decisions we make about AI-IoT social sensing during this time will determine urban living far in the future. It is our job to make sure that tomorrow we have smarter cities and that these cities are not only technologically smarter than the ones we have today, but above all, more human, more equal, and more democratic.

REFERENCES

Affinito, A., Zinno, S., Stanco, G., Botta, A., & Ventre, G. (2023). The evolution of Mirai botnet scans over six years. *Journal of Information Security and Applications*, *79*, 103629. DOI: 10.1016/j.jisa.2023.103629

Alahi, M. E. E., Sukkuea, A., Tina, F. W., Nag, A., Kurdthongmee, W., Suwannarat, K., & Mukhopadhyay, S. C. (2023). Integration of IoT-Enabled Technologies and Artificial Intelligence (AI) for Smart City Scenario: Recent Advancements and Future Trends. *Sensors, 23*(11), 5206. MDPI. https://www.mdpi.com/1424-8220/23/11/5206

Alauthman, M., Alkasassbeh, M. S., Al-Qerem, A., Alateef, S., & Almomani, A. (2025). Privacy-Preserving Security: Balancing Protection and Data Utility in Digital Systems. In *Complexities and Challenges for Securing Digital Assets and Infrastructure* (pp. 483-506). IGI Global Scientific Publishing.

Alghamdi, W. Y., Alshamrani, R. M., Aloufi, R. K., Ba Lhamar, S. O., Altwirqi, R. A., Alotaibi, F. S., Althobaiti, S. M., Altalhi, H. M., Alshamrani, S. A., & Alazwari, A. S. (2025). Exploring the Synergy Between Digital Twin Technology and Artificial Intelligence: A Comprehensive Survey. *International Journal of Advanced Computer Science and Applications*, *16*(3). Advance online publication. DOI: 10.14569/IJACSA.2025.0160399

AlJamal, M., Mughaid, A., Bani-Salameh, H., Alzubi, S., & Abualigah, L. (2024). Optimizing risk mitigation: A simulation-based model for detecting fake IoT clients in smart city environments. *Sustainable Computing : Informatics and Systems*, *43*, 101019. DOI: 10.1016/j.suscom.2024.101019

Alotaibi, A., & Nadeem, F. (2024). Leveraging social media and deep learning for sentiment analysis for smart governance: A case study of public reactions to educational reforms in Saudi Arabia. *Computers*, *13*(11), 280. DOI: 10.3390/computers13110280

Anthi, E., Williams, L., Ieropoulos, V., & Spyridopoulos, T. (2024). Investigating radio frequency vulnerabilities in the Internet of Things (IoT). *IoT*, *5*(2), 356–380. DOI: 10.3390/iot5020018

Asher, S. (2020, July 4). TraceTogether: Singapore turns to wearable contact-tracing Covid tech. *BBC News*. Retrieved from https://www.bbc.com/news/technology-53146360

Barkane, I., Adamsone-Fiskovica, A., & Kilis, E. (2023). Creeping road traffic surveillance in Latvia: Social and legal implications of digital policing tools. *Surveillance & Society, 21*(4), 432–446. DOI: 10.24908/ss.v21i4.15812

Bhattacharya, S., Seth, D., Panyam, S., & Gangrade, P. (2024). Enhancing Digital Privacy: The Application of Zero-Knowledge Proofs in Authentication Systems. [IJCTT]. *International Journal of Computer Trends and Technology, 72*(4), 34–41. DOI: 10.14445/22312803/IJCTT-V72I4P104

Brantingham, P. J., Valasik, M., & Mohler, G. O. (2018). Does predictive policing lead to biased arrests? Results from a randomized controlled trial. *Statistics and Public Policy (Philadelphia, Pa.), 5*(1), 1–6. DOI: 10.1080/2330443X.2018.1438940

Calzada, I. (2018). (Smart) citizens from data providers to decision-makers? The case study of Barcelona. *Sustainability (Basel), 10*(9), 3252. DOI: 10.3390/su10093252

Chakraborty, A., Alam, M., Dey, V., Chattopadhyay, A., & Mukhopadhyay, D. (2021). A survey on adversarial attacks and defences. *CAAI Transactions on Intelligence Technology, 6*(1), 25–45. DOI: 10.1049/cit2.12028

Cremonini, M. (2023). A Critical Take on Privacy in a Datafied Society. *arXiv preprint arXiv:2308.02573.*

de Montjoye, Y.-A., Hidalgo, C. A., Verleysen, M., & Blondel, V. D. (2013). Unique in the Crowd: The privacy bounds of human mobility. *Scientific Reports, 3*(1), 1376. Advance online publication. DOI: 10.1038/srep01376 PMID: 23524645

Denis, A., Thomas, A., Robert, W., Samuel, A., Kabiito, S. P., Morish, Z., Sallam, M., Ali, G., & Mijwil, M. M. (2025). A Survey on Artificial Intelligence and Blockchain Applications in Cybersecurity for Smart Cities. *SHIFRA, 2025*, 1–45. DOI: 10.70470/SHIFRA/2025/001

Determann, L., & Tam, J. (2020). The California Privacy Rights Act of 2020: A broad and complex data processing regulation that applies to businesses worldwide. *Journal of Data Protection & Privacy, 4*(1), 7–21. DOI: 10.69554/GCLK6627

Dhelim, S., Ning, H., Farha, F., Chen, L., Atzori, L., & Daneshmand, M. (2021). IoT-enabled social relationships meet artificial social intelligence. *IEEE Internet of Things Journal, 8*(24), 17817–17828. DOI: 10.1109/JIOT.2021.3081556

Dou, C., & Huan, H. (2021). Full respiration rate monitoring exploiting Doppler information with commodity Wi-Fi devices. *Sensors (Basel), 21*(10), 3505. DOI: 10.3390/s21103505 PMID: 34069847

Dziwisz, D. (2023). Stuxnet. In *The Handbook of Homeland Security* (pp. 289–293). CRC Press.

European Data Protection Supervisor. (2018). *Privacy by Design.* https://www.edps.europa.eu/data-protection/our-work/subjects/privacy-design_en?page=5

European Union (2022, October 6). *Digital education action plan (2021–2027).* Europa.eu. https://eur-lex.europa.eu/legal-content/EN/TXT/?uri=legissum:4617905

Fereidouni, H., Fadeitcheva, O., & Zalai, M. (2025). IoT and man-in-the-middle attacks. *Security and Privacy*, *8*(2), e70016. DOI: 10.1002/spy2.70016

Force, J. T. (2018). Risk management framework for information systems and organizations. *NIST Special Publication, 800*, 37.

Galliven, H. (2025, July 12). *Croydon's fixed facial recognition cameras spark debate*. BBC News. Retrieved from https://www.bbc.com/news/articles/cy0w5egz91no

Galloway, A. (2021, September 5). *Police using QR check-in data to solve crimes.* The Sydney Morning Herald. https://www.smh.com.au/politics/federal/breach-of-trust-police-using-qr-check-in-data-to-solve-crimes-20210903-p58om8.html

GDPR. (2018). *General Data Protection Regulation (GDPR).* General Data Protection Regulation (GDPR). https://gdpr-info.eu/issues/consent/

Gottschalk, F., & Weise, C. (2023). Digital equity and inclusion in education: An overview of practice and policy in OECD countries. *OECD Education Working Papers*, (299), 0_1-75.

Government of Canada. (2020, September 15). *Download COVID Alert: Canada's exposure notification app*. Aem. https://www.canada.ca/en/public-health/services/diseases/coronavirus-disease-covid-19/covid-alert.html

GPAI. (2024). "Algorithmic Transparency in the Public Sector: A state-of-the-art report of algorithmic transparency instruments". Report, May 2024, Global Partnership on Artificial Intelligence. https://wp.oecd.ai/app/uploads/2025/05/algorithmic-transparency-in-the-public-sector.pdf

Gupta, J. K., Negi, G. S., & Krishna, G. (2025). Smart Transportation Systems for Vehicle Geographical Tracking. In *Federated Learning Based Intelligent Systems to Handle Issues and Challenges in IoVs (Part 2)* (pp. 160-199). Bentham Science Publishers. DOI: 10.2174/9789815322224125030008

Halbfinger, D. M., Kershner, I., & Bergman, R. (2020, March 16). To Track Coronavirus, Israel Moves to Tap Secret Trove of Cellphone Data. *The New York Times*. https://www.nytimes.com/2020/03/16/world/middleeast/israel-coronavirus-cellphone-tracking.html

Hardesty, L. (2013). *How hard is it to "de-anonymize" cellphone data?* MIT News. https://news.mit.edu/2013/how-hard-it-de-anonymize-cellphone-data

Hjerppe, K., Ruohonen, J., & Leppänen, V. (2019, September). The general data protection regulation: requirements, architectures, and constraints. In *2019 IEEE 27th International Requirements Engineering Conference (RE)* (pp. 265-275). IEEE. DOI: 10.1109/RE.2019.00036

Holl, F., Flemisch, F., Swoboda, W., & Schobel, J. (2022). Apps for Covid-19 in Germany: Assessment Using the German Mobile App Rating Scale. *JAMIA Open*, *5*(4), ooac082. Advance online publication. DOI: 10.1093/jamiaopen/ooac082 PMID: 36225895

Holmes, O. (2020, March 17). Israel to track mobile phones of suspected coronavirus cases. *The Guardian*. https://www.theguardian.com/world/2020/mar/17/israel-to-track-mobile-phones-of-suspected-coronavirus-cases

Homeland Security. (2024, September 7). *Privacy Impact Assessment for the Customer Profile Management System*. https://www.dhs.gov/sites/default/files/2024-11/24_0930_priv_pia-dhs-uscis-cpms-060d.pdf

Information and Privacy Commission. (2024, August). *A guide to undertaking Privacy Impact Assessments on AI systems and projects*. Retrieved from https://www.ipc.nsw.gov.au/sites/default/files/2024-08/Guide_A_guide_to_undertaking_Privacy_Impact_Assessments_on_AI_systems_and_projects_August_2024_CONSULTATION_DRAFT.pdf

International Telecommunication Union. (2016). *Implementing ITU-T International Standards to Shape Smart Sustainable Cities: The Case of Dubai*. https://www.itu.int/en/publications/Documents/tsb/2016-DubaiCase/files/downloads/Dubai%20Case%20Study-410059-FINAL.pdf

Ivanova, D., & Katsaounis, P. (2021). Real-time dynamic tiered e-consent: A novel tool for patients' engagement and common ontology system for the management of medical data. *Innovations in Digital Health, Diagnostics, and Biomarkers*, *1*(2), 45–49. DOI: 10.36401/IDDB-21-01

Jo, K., Choi, J., Kim, K., Kim, S., Nguyen, D.-L., Vo, X.-T., Priadana, A., & Tran, T.-D. (2025). *Artificial Behavior Intelligence: Technology, Challenges, and Future Directions*. ArXiv.org. https://arxiv.org/abs/2505.03315 DOI: 10.1109/HSI66212.2025.11142417

Kennedy, H. (2018). Living with data: Aligning data studies and data activism through a focus on everyday experiences of datafication. *Krisis: Journal for Contemporary Philosophy*, *2018*(1), 18–30. DOI: 10.21827/krisis.38.1.37184

Koops, B. J. (2021). The concept of function creep. *Law, Innovation and Technology*, *13*(1), 29–56. DOI: 10.1080/17579961.2021.1898299

Larson, J., Mattu, S., Kirchner, L., & Angwin, J. (2016, May 23). *How We Analyzed the COMPAS Recidivism Algorithm*. ProPublica. https://www.propublica.org/article/how-we-analyzed-the-compas-recidivism-algorithm

Lee, C. S. (2019). Datafication, dataveillance, and the social credit system as China's new normal. *Online Information Review*, *43*(6), 952–970. DOI: 10.1108/OIR-08-2018-0231

Lee, H., & Lee, U. (2022). Toward dynamic consent for privacy-aware pervasive health and well-being: A scoping review and research directions. *IEEE Pervasive Computing*, *21*(4), 25–32. DOI: 10.1109/MPRV.2022.3210747

Longo, A., Zappatore, M., Bochicchio, M., & Navathe, S. B. (2017). Crowd-sourced data collection for urban monitoring via mobile sensors. *ACM Transactions on Internet Technology*, *18*(1), 1–21. DOI: 10.1145/3093895

Lu, J., & Merritt, J. (2024). *How equitable data practices can help shape urban planning*. [online] World Economic Forum. Available at: https://www.weforum.org/stories/2024/10/equitable-data-practices-urban-planning/

Merlec, M. M., Lee, Y. K., Hong, S. P., & In, H. P. (2021). A smart contract-based dynamic consent management system for personal data usage under GDPR. *Sensors (Basel)*, *21*(23), 7994. DOI: 10.3390/s21237994 PMID: 34883997

Monge, F., Barns, S., Kattel, R., & Bria, F. (2022). A new data deal: the case of Barcelona - UCL Discovery. *Ucl.ac.uk*. https://discovery.ucl.ac.uk/id/eprint/10196041/1/Kattel_new_data_deal_barcelona_fernando_barns_kattel_and_bria.pdf

Moy, L. (2021). Facing injustice: How face recognition technology may increase the incidence of misidentifications and wrongful convictions. *SSRN*, *30*, 337. DOI: 10.2139/ssrn.4101826

Murciano-Goroff, R., Zhuo, R., & Greenstein, S. (2024). *Navigating software vulnerabilities: Eighteen years of evidence from medium and large us organizations* (No. w32696). National Bureau of Economic Research.

Osakwe, Z. T., Ikhapoh, I., Arora, B. K., & Bubu, O. M. (2020). Identifying public concerns and reactions during the COVID-19 pandemic on Twitter: A text-mining analysis. *Public Health Nursing (Boston, Mass.)*, *38*(2), 145–151. DOI: 10.1111/phn.12843 PMID: 33258149

Pascu, L. (2018). *At $17 million, Atlanta network recovery six times more expensive than estimated.* Hot for Security. https://www.bitdefender.com/en-us/blog/hotforsecurity/at-17-million-atlanta-network-recovery-six-times-more-expensive-than-estimated

Ramani, D. R., Sujitha, B. B., & Tangade, S. (2025). Smart Environmental Monitoring Systems: IoT and Sensor-Based Advancements. *Environmental Monitoring Using Artificial Intelligence*, 45-60.

Rashid, M. T., Wei, N., & Wang, D. (2023). A survey on social-physical sensing: An emerging sensing paradigm that explores the collective intelligence of humans and machines. *Collective Intelligence*, *2*(2), 26339137231170825. DOI: 10.1177/26339137231170825

Richardson, R., Schultz, J. M., & Crawford, K. (2019). Dirty data, bad predictions: How civil rights violations impact police data, predictive policing systems, and justice. *NYUL Rev. Online*, *94*, 15.

Roy, P., Ghosh, S., Sarkar, S. K., Podder, A., & Paul, S. (2025). AI-Driven Ambient Intelligence Systems for Mental Health Monitoring and Proactive Intervention. *Cuestiones de Fisioterapia*, *54*(3), 2743–2752.

Rubinstein, I. S., & Petkova, B. (2019). Governing privacy in the datafied city. *The Fordham Urban Law Journal*, *47*, 755.

Schwartz, M. J. (2025, August 6). *Atlanta's Reported Ransomware Bill: Up to $17 Million.* https://www.bankinfosecurity.com/atlantas-reported-ransomware-bill-up-to-17-million-a-11281

Sebastian, G. (2022). Could incorporating cybersecurity reporting into SOX have prevented most data breaches at US publicly traded companies? An exploratory study. *International Cybersecurity Law Review*, *3*(2), 367–383. DOI: 10.1365/s43439-022-00062-x

Shapiro, A. (2019). Predictive policing for reform? Indeterminacy and intervention in big data policing. *Surveillance & Society*, *17*(3/4), 456–472. DOI: 10.24908/ss.v17i3/4.10410
Shapiro, A. (2019). Predictive policing for reform? Indeterminacy and intervention in big data policing. *Surveillance & Society*, *17*(3/4), 456–472. DOI: 10.24908/ss.v17i3/4.10410
Simoiu, C. (2020). *Secure by Default: A Behavioral Approach to Cyber Security.* Stanford University.

Sivakumar, C. L. V., Mone, V., & Abdumukhtor, R. (2024). Addressing privacy concerns with wearable health monitoring technology. *Wiley Interdisciplinary Reviews. Data Mining and Knowledge Discovery*, *14*(3), e1535. DOI: 10.1002/widm.1535

Smyre, B. (2018, March 27). *City of Atlanta 2018 Ransomware Hack: What We Know and What You Can Learn From It | Raxis*. Raxis.com. https://raxis.com/blog/city-of-atlanta-2018-ransomware-hack-what-we-know-and-what-you-can-learn-from-it/

Syahmi, H. (2025). Privacy and ethical implications of big data utilization in public transportation surveillance. *International Journal of Advanced Cybersecurity Systems, Technologies, and Applications*, *9*(1), 1–10.

Tiffin, N., George, A., & LeFevre, A. E. (2019). How to use relevant data for maximal benefit with minimal risk: Digital health data governance to protect vulnerable populations in low-income and middle-income countries. *BMJ Global Health*, *4*(2), e001395. DOI: 10.1136/bmjgh-2019-001395 PMID: 31139457

Torrens, P. M. (2022). Smart and sentient retail high streets. *Smart Cities*, *5*(4), 1670–1720. DOI: 10.3390/smartcities5040085

UAE. (2023, November 7). *Dubai Metaverse Strategy - The Official Portal of the UAE Government.* https://u.ae/en/about-the-uae/strategies-initiatives-and-awards/strategies-plans-and-visions/government-services-and-digital-transformation/dubai-metaverse-strategy

U.S. Attorney's Office. (2018, December 5). *Atlanta U.S. Attorney Charges Iranian nationals for City Of Atlanta ransomware attack*. https://www.justice.gov/usao-ndga/pr/atlanta-us-attorney-charges-iranian-nationals-city-atlanta-ransomware-attack

Walker, A. (2023, May 4). *Singapore's digital twin – from science fiction to hi-tech reality*. Infrastructure Global. https://infra.global/singapores-digital-twin-from-science-fiction-to-hi-tech-reality

Wang, Z., Ma, J., Wang, X., Hu, J., Qin, Z., & Ren, K. (2022). Threats to training: A survey of poisoning attacks and defenses on machine learning systems. *ACM Computing Surveys*, *55*(7), 1–36. DOI: 10.1145/3538707

Wu, J., Luo, K., Wang, Y., & Wang, Z. (2021). Urban road greenbelt configuration: The perspective of PM2.5 removal and air quality regulation. *Environment International*, *157*, 106786. DOI: 10.1016/j.envint.2021.106786 PMID: 34314980

Yen, C. P., & Hung, T. W. (2021). Achieving equity with predictive policing algorithms: A social safety net perspective. *Science and Engineering Ethics*, *27*(3), 36. DOI: 10.1007/s11948-021-00312-x PMID: 34075448

Yu, W., Mao, Q., Yang, S., Zhang, S., & Rong, Y. (2017). Social Sensing: The Necessary Component of Planning Support System for Smart City in the Era of Big Data. *Lecture Notes in Geoinformation and Cartography*, 231–244. DOI: 10.1007/978-3-319-57819-4_13

Yucer, S., Tektas, F., Al Moubayed, N., & Breckon, T. (2024). Racial bias within face recognition: A survey. *ACM Computing Surveys*, *57*(4), 1–39. DOI: 10.1145/3705295

Zhang, H., Ibrahim, A., Parsia, B., Poliakoff, E., & Harper, S. (2023). Passive social sensing with smartphones: A systematic review. *Computing*, *105*(1), 29–51. DOI: 10.1007/s00607-022-01112-2

Zhao, Y., Zhang, X., Wang, J., Zhang, K., & Ordonez de Pablos, P. (2020). How do features of social media influence knowledge sharing? An ambient awareness perspective. *Journal of Knowledge Management*, *24*(2), 439–462. DOI: 10.1108/JKM-10-2019-0543

Chapter 9
Smartphone Detection With Social Sensor Systems

Binastya Anggara Sekti
http://orcid.org/0000-0001-5489-4888
Universitas Esa Unggul, Indonesia

ABSTRACT

Smartphones have evolved into powerful social sensors, offering a rich ecosystem of data through GPS, cameras, microphones, and connectivity modules. In the context of smart cities, these devices serve as ubiquitous nodes that continuously interact with the environment, enabling real-time insights into mobility, public safety, environmental quality, and resource management. We explore how smartphone-based detection systems integrate with AI and IoT frameworks to deliver actionable intelligence for city authorities. It highlights participatory and opportunistic sensing approaches, advanced data collection techniques, and the role of cloud-edge integration in achieving scalability. Case studies about crowd density monitoring and smart tourism showcase the transformative potential of smartphones in urban governance. By combining technical, social, and moral perspectives, it provides a roadmap for harnessing smartphones as integral components of future-ready, citizen-centric smart city ecosystems.

1. INTRODUCTION

Smart cities represent a transformative paradigm where digital technologies, IoT devices, and artificial intelligence converge to optimize urban services, enhance quality of life, and promote sustainable development. In this ecosystem, smartphones

DOI: 10.4018/979-8-3373-4962-6.ch009

have emerged as critical nodes because of their widespread adoption and integration into everyday life. With global smartphone penetration surpassing 6.8 billion users by 2023, these devices provide unprecedented opportunities for large-scale, real-time data collection. Unlike static sensors embedded in infrastructure, smartphones travel with their users, enabling a more dynamic and personalized form of sensing in urban contexts. With the rapid growth of smartphone penetration around the world, the role of navigation cues has become more important in using smartphones for consumers in online business activities (Liu, Huang, & Yen, 2022).

The proliferation of smartphone sensors—GPS, accelerometers, gyroscopes, cameras, microphones, Wi-Fi, and Bluetooth—has transformed each device into a portable "social sensor". These sensors generate vast streams of data that reflect mobility patterns, environmental conditions, and social interactions. Such data, when aggregated and anonymized, can support city planners in understanding crowd flows, predicting congestion, monitoring air quality, and even detecting emergencies. This shift from fixed infrastructure sensing to citizen-driven smartphone sensing significantly reduces deployment costs while increasing coverage and flexibility. Smartphones are personal communication tools that have many sensors, capable of monitoring social behavior. Calls, messages, app usage, and even face-to-face conversations can be disrupted with smartphones (Zhang, Ibrahim, Parsia, Poliakoff, & Harper, 2023).

Moreover, smartphones act as bridges between individuals and IoT ecosystems, enabling two-way interaction: not only do they collect data, but they also serve as feedback channels where citizens can receive alerts, guidance, or recommendations. For instance, location-aware apps can inform users of safe evacuation routes during floods, or notify them about air pollution hotspots in real time. This citizen–system loop is fundamental to building participatory smart cities where residents are not passive recipients of services, but active contributors to urban intelligence. Despite these advantages, the integration of smartphones into social sensing frameworks is not without challenges. Issues of data privacy, security, interoperability, and digital divides present substantial barriers. Ethical questions emerge regarding consent, data ownership, and algorithmic biases when personal information is leveraged for collective benefits. Hence, while smartphones offer a powerful tool for sensing urban dynamics, their role must be carefully balanced with considerations of inclusivity, transparency, and accountability. In the field of smart healthcare, video consultations have become the primary way to provide medical services for remote patients. In this process, it is crucial to ensure the legal identity of both parties in communication and to protect patient privacy data from unlawful interception and interference (He, Fan, & Zhang, 2025).

This chapter aims to explore the methodologies, applications, and implications of smartphone detection within social sensing systems. It provides a technical and socio-ethical overview of how smartphones contribute to the smart city vision, supported by real-world case studies and future-oriented perspectives. Table 1 summarizes the key characteristics of smartphones as social sensors, while Figure 1 illustrates their integration into the broader smart city ecosystem.

Table 1. Key Characteristics of Smartphones as Social Sensors

Feature	Description	Example Application
Ubiquity	Billions of users worldwide provide wide population coverage	Crowd mobility analysis
Multi-sensor capability	GPS, accelerometer, gyroscope, microphone, camera, Wi-Fi, Bluetooth	Activity recognition, pollution
Mobility	Devices move with users, capturing real-time urban dynamics	Traffic congestion detection
User interaction	Allows participatory sensing via apps and surveys	Reporting waste or infrastructure
Connectivity	Interfaces with IoT networks and cloud platforms	Smart home integration

Figure 1. Smartphones as nodes in the social sensing ecosystem

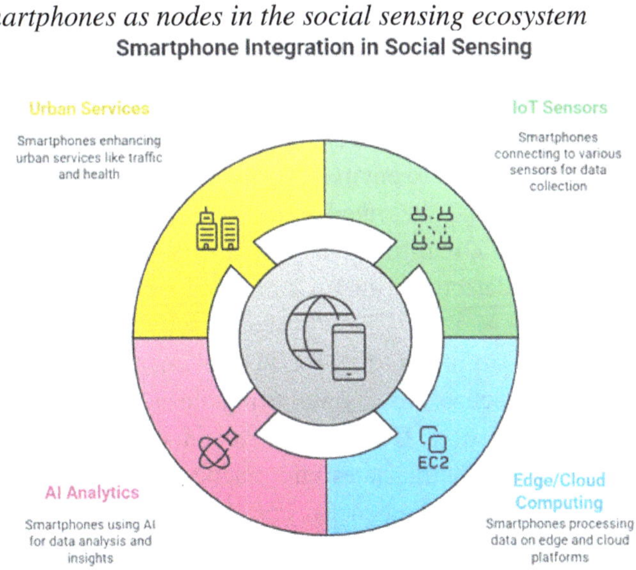

2. OBJECTIVES OF THE CHAPTER

The overarching objective of this chapter is to explore the role of smartphones as powerful agents in social sensing systems within smart cities. In an era where urban spaces are increasingly defined by digital infrastructure, the smartphone provides a unique lens to observe, analyze, and improve city services. Setting clear objectives ensures that the discussion not only highlights technical aspects but also addresses socio-ethical dimensions that shape the practical deployment of smartphone-based sensing frameworks. The emergence of new wireless technologies, such as the smartphones and Internet of Things, allows digitalizing new and diverse urban activities (Benseny, Lahteenmaki, Toyli, & Hammainen, 2023).

The first objective is to explain the conceptual and technical foundations of smartphone-based social sensing. This involves defining what constitutes a social sensor, describing the types of smartphone-generated data, and outlining models of participatory and opportunistic sensing. By clarifying these fundamentals, the chapter establishes a common ground for researchers, practitioners, and policymakers to understand the unique capabilities and limitations of smartphones in urban contexts. With various applications on smartphones, smartphones can be used to collect sensor data and gather basic truth information using various activities such as questionnaires to be crossed to certain respondents as an object of research (Erdelyi et al., 2025).

The second objective is to explore methods of data collection and processing in urban environments. Smartphones generate heterogeneous data streams—ranging from geospatial location (GPS) to environmental audio and image data—that must be collected, anonymized, and integrated with IoT platforms. Establishing effective frameworks for data fusion, real-time analytics, and scalable storage is critical to ensure that smartphone detection systems can operate efficiently and securely in smart city applications.

The third objective is to present real-world implementations and case studies. By analyzing examples such as crowd monitoring during festivals, contact tracing during public health crises (e.g., COVID-19), and environmental monitoring, this chapter highlights how theory translates into practice. The inclusion of case studies offers tangible evidence of the benefits and challenges faced in diverse socio-urban contexts, demonstrating both the technical feasibility and social impact of smartphone-based sensing systems. Impaired social functioning is a symptom of mental illness (e.g., depression, schizophrenia) and a wide range of other conditions (e.g., cognitive decline in the elderly, dementia). Today, assessing social functioning relies on subjective evaluations and self assessments (Wang et al., 2020).

The fourth objective centers on evaluating challenges, privacy concerns, and ethical considerations. As smartphones collect data that may reveal sensitive personal behaviors, robust safeguards are required to ensure compliance with global privacy standards such as GDPR. Ethical objectives also include promoting inclusivity, ensuring transparency in data usage, and avoiding algorithmic biases that could marginalize specific groups within the urban population. With the General Data Protection Regulation (GDPR), the European Union established a global standard for the protection of personal data and digital privacy ("Beyond GDPR: Employing AI to Make Personal Data Useful to Consumers," 2020).

Finally, this chapter aims to propose integration strategies with AI algorithms and IoT frameworks that maximize utility while minimizing risks. This includes the exploration of advanced techniques such as federated learning, blockchain-enabled decentralized sensing, and edge AI for on-device computation. Together, these strategies form a vision of citizen-centric, adaptive, and resilient urban intelligence, where smartphones act not just as data collection tools, but as enablers of a more participatory and sustainable smart city future. Table 2 shows the Objectives of the Chapter and Expected Outcomes. Figure 2 shows the Flow of Objectives in Smartphone-Based Social Sensing.

Table 2. Objectives of the Chapter and Expected Outcomes

Objective	**Expected Outcome**
Explain conceptual and technical foundations of smartphone sensing	Establish theoretical framework and definitions
Outline data collection and processing methods	Provide models for scalable, real-time, and secure smartphone-based sensing
Present case studies and implementations	Demonstrate practical applications and lessons learned
Evaluate challenges, privacy, and ethical considerations	Promote trust, inclusivity, and responsible data governance
Explore integration with AI and IoT frameworks	Offer future-oriented strategies for sustainable and adaptive smart city systems

Figure 2. Flow of Objectives in Smartphone-Based Social Sensing

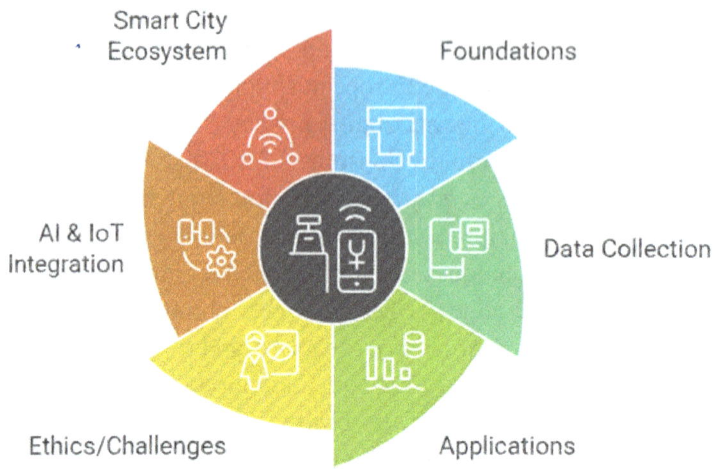

3. THE ROLE OF SMARTPHONES IN SOCIAL SENSING

Smartphones play an increasingly central role in the ecosystem of smart cities. Unlike specialized IoT devices that serve specific functions, smartphones are carried by billions of people every day, acting as portable, sensor-rich platforms. With GPS, accelerometers, gyroscopes, cameras, microphones, Bluetooth, and Wi-Fi, smartphones serve as powerful gateways to transform citizens into "human sensors". This ubiquity allows cities to capture real-time, context-aware, and multimodal data on mobility, environmental conditions, social behaviors, and even emergency incidents.

The growing importance of smartphones in social sensing reflects not just their technical capabilities but also their ability to facilitate participatory governance and citizen engagement. Through dedicated apps and mobile-based platforms, citizens can contribute data, report issues, and receive feedback in near real time. This dual role—data provider and service beneficiary—positions smartphones as bridges between people and the smart city infrastructure, ensuring inclusivity and responsiveness.

3.1 The Characteristics of Smartphones as Social Sensors

Smartphones are uniquely positioned to act as both active and passive sensors. They provide ubiquitous connectivity (cellular, Wi-Fi, Bluetooth), multimodal sensing capabilities (location, motion, audio, image), and computational power to process data locally. Furthermore, the penetration rate of smartphones is significantly higher than most other IoT devices, making them an affordable and scalable tool for mass sensing. By exploring the use of applications to detect littering activities by vehicles and pedestrians, a visual detection model will be able to identify the individuals responsible through facial recognition and license plate detection. Collecting the right data for the detection of littering will bring very high value to the community (Alharbi, Alsulami, Aljohani, Alharbi, & Albaradei, 2025).

Three major characteristics highlight the role of smartphones in social sensing:

1. Portability – carried everywhere, providing continuous spatiotemporal coverage.
2. Diversity of Sensors – enabling multimodal insights, from geospatial mobility to environmental noise.
3. Bidirectional Communication – citizens can both send data and receive service-related information.

Table 3 shows the Characteristics of Smartphones as Social Sensors. Figure 3 shows the Smartphone as a Multi-Sensor Node in Smart Cities.

Table 3. Characteristics of Smartphones as Social Sensors

Characteristic	Description	Example Application
Portability	Smartphones accompany users throughout daily life	Crowd mobility monitoring
Sensor Diversity	Equipped with GPS, accelerometers, gyroscope, microphone, camera	Environmental monitoring, AR services
Bidirectional Communication	Data sharing combined with service feedback loops	Traffic updates, health notifications

Figure 3. Smartphone as a Multi-Sensor Node in Smart Cities

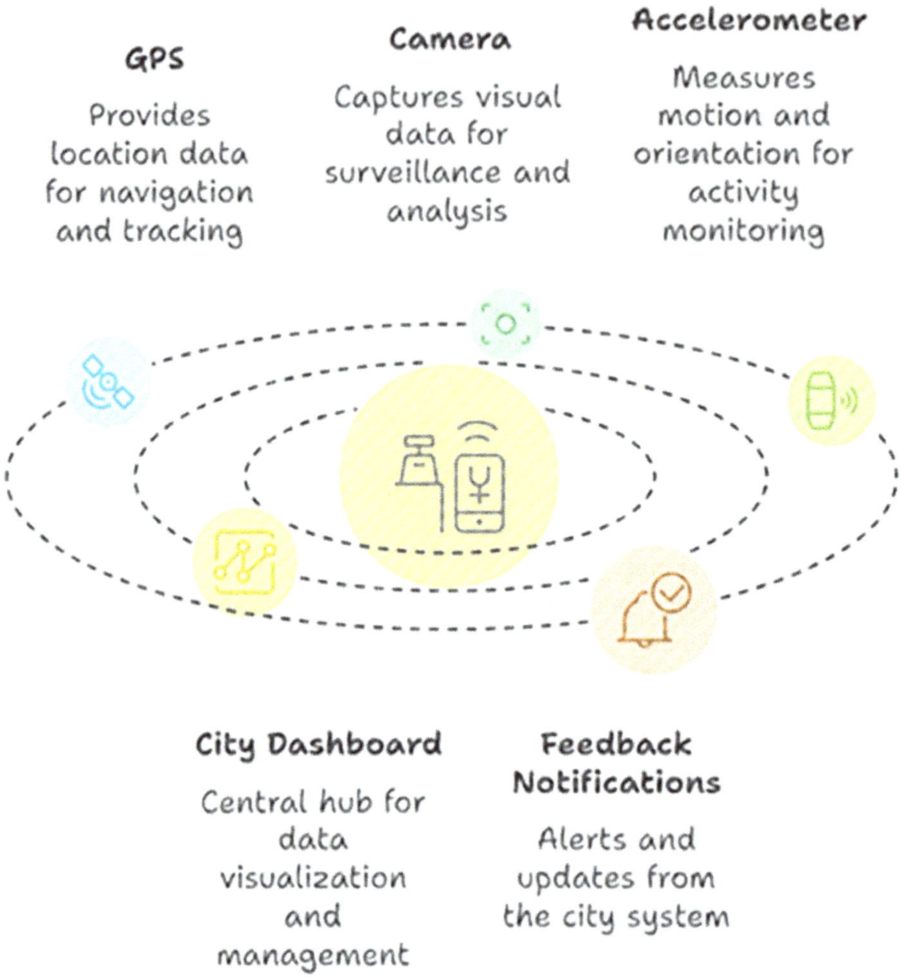

3.2 Data Collection Techniques

Human activity recognition (HAR) techniques are playing a significant role in monitoring the daily activities of human life such as elderly care, investigation activities, healthcare, sports, and smart homes. Smartphones incorporated with varieties of motion sensors like accelerometers and gyroscopes are widely used inertial sensors that can identify different physical conditions of human (Ahmed, Rafiq, & Islam, 2020). Smartphones enable two primary modes of social sensing: participatory sensing and opportunistic sensing.

- Participatory sensing involves citizens actively contributing data through surveys, apps, or direct reporting. For example, reporting potholes, garbage piles, or water leaks via mobile apps. This model encourages community involvement and civic responsibility.
- Opportunistic sensing is passive and automated, collecting data without active user involvement. For instance, GPS traces, Wi-Fi logs, and accelerometer data can reveal traffic congestion or pedestrian flows without requiring manual input.

In modern smart cities, hybrid approaches are often adopted—combining active citizen input with passive sensing to achieve both richness of data and scale of coverage. Table 4 shows the Smartphone Data Collection Approaches.

Table 4. Smartphone Data Collection Approaches

Approach	Description	Example Use Cases
Participatory	Citizens manually submit or report data	Garbage reporting, crowd safety alerts
Opportunistic	Passive background collection via smartphone sensors	GPS for traffic density, Bluetooth for proximity
Hybrid	Combination of both active and passive models	Disaster early warning, event monitoring

3.3 Integration with Social Sensor Systems

The data collected by smartphones does not exist in isolation—it must be integrated into larger social sensor systems that involve cloud computing, edge computing, and AI-driven analytics. Smartphone data streams are transmitted through secure APIs and data pipelines to be aggregated with other IoT inputs such as environmental sensors, surveillance cameras, or smart meters. Wearables are wireless devices that we "wear" on our bodies. The proliferation of devices embedded in clothing, medical smart clothing, leather patches, smartwatches, and wristbands, along with conventional smartphones, has expanded techniques in data collection (Torres-Sospedra & Ometov, 2021).

The integration framework typically follows a layered architecture:

1. Data Collection Layer – smartphones and IoT devices.
2. Data Aggregation Layer – cloud/edge nodes that preprocess and standardize data.
3. Analytics Layer – AI/ML models for recognition, prediction, and anomaly detection.

4. Decision & Feedback Layer – dashboards, alerts, and services for citizens and city managers.

This architecture ensures that smartphone-based sensing contributes effectively to smart city management, enabling scalable solutions in mobility monitoring, disaster management, environmental health, and public safety. Table 5 shows the Integration of Smartphones in Social Sensing Systems. Figure 4 shows the Architecture of Smartphone-Based Social Sensing.

Table 5. Integration of Smartphones in Social Sensing Systems

Layer	Function	Example Tools/Technologies
Data Collection Layer	Capture data from sensors and apps	GPS, Wi-Fi, camera, microphone
Data Aggregation Layer	Preprocess and integrate heterogeneous data	Cloud gateways, edge servers
Analytics Layer	Apply AI/ML models for insights	Anomaly detection, predictive analytics
Decision & Feedback Layer	Provide actionable outputs and user interaction	City dashboards, citizen alert systems

Figure 4. Architecture of Smartphone-Based Social Sensing

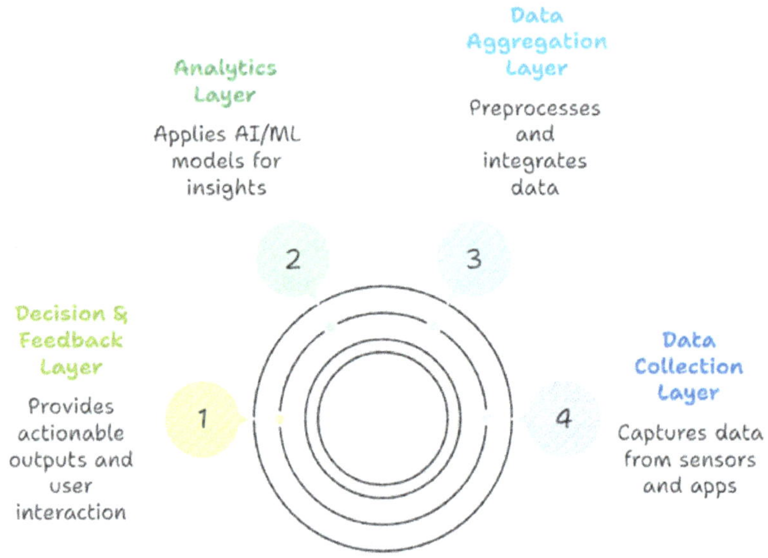

4. DATA COLLECTION TECHNIQUES

Smartphone-based social sensing in smart cities relies heavily on robust and adaptable data collection techniques. The variety of sensors embedded in smartphones enables both user-driven and system-driven mechanisms for gathering information about urban environments. These approaches, when properly designed, can deliver insights on traffic, pollution, safety, and public health in real time. However, successful implementation requires addressing key challenges such as data quality, scalability, and citizen engagement. The concept of smart cities and the role of the Internet of Things (IoT) and machine learning (ML) in realizing smart environments have always been data-centric. Smart cities leverage technology and data to improve the quality of life of citizens and improve the efficiency of urban services. IoT and machine learning have emerged as key technologies to enable smart city solutions that rely on large-scale data collection, analysis, and decision-making (Ullah et al., 2024).

In this section, three primary techniques are explored: crowdsourcing and participatory sensing, opportunistic sensing, and data fusion from multiple sensors. Together, these methods illustrate how smartphones can be utilized as multi-layered data contributors to the broader social sensing ecosystem.

4.1 Crowdsourcing and Participatory Sensing

Crowdsourcing and participatory sensing leverage the active involvement of citizens in data collection. Mobile applications act as gateways for people to report real-time incidents such as traffic congestion, flooding, infrastructure breakdowns, or even emergency hazards. This user-driven model empowers individuals to become active contributors in smart city governance. Effective incident detection is essential for emergency response and transportation management. Traditional methods relying on stationary technologies are often costly and provide limited coverage, prompting the exploration of crowdsourced data such as Waze (Hossain, Lee, Besenski, Dimitrijevic, & Spasovic, 2025).

To increase participation, incentive mechanisms such as gamification, reward points, or social recognition can be applied. For example, platforms like *Waze* encourage drivers to report road conditions and traffic jams, while citizens' apps in Jakarta allow users to report flooding events. Table 6 shows the Architecture of Smartphone-Based Social Sensing. Figure 5 shows the Crowdsourcing via Smartphone Apps.

Table 6. Crowdsourcing/Participatory Sensing Examples

Application Domain	Example Case	Incentive Mechanism
Traffic Monitoring	Waze crowd-based traffic reports	Gamification, social points
Infrastructure Failures	Reporting potholes or power outages	Public recognition, tokens
Public Safety	Flood or fire alerts through citizen apps	Social responsibility, badges

Figure 5. Crowdsourcing via Smartphone Apps

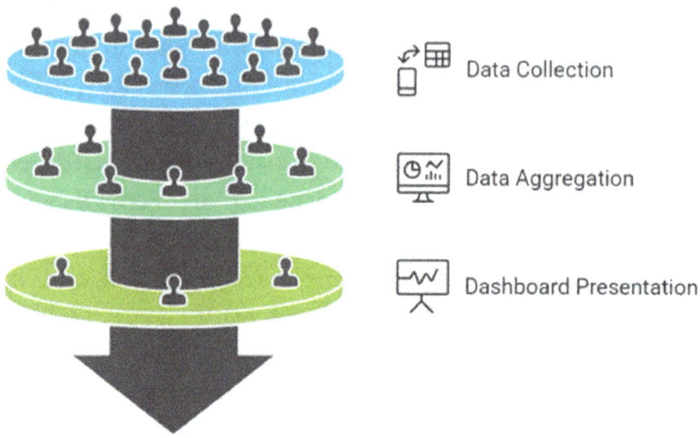

4.2 Opportunistic Sensing

Opportunistic sensing involves the automatic collection of data without explicit user input. Smartphones continuously log GPS locations, Wi-Fi signals, accelerometer data, and Bluetooth encounters in the background. This data can then be aggregated to infer patterns of human mobility, transportation bottlenecks, or pedestrian density. With the surging of smart device sensing and mobile networking, GPS data has been widely available for identifying vehicle position and route on the road map (Luo, Hou, Cai, & Guo, 2020).

For example:

- GPS traces can estimate congestion levels in urban roads.
- Wi-Fi hotspots provide insights into crowd density in malls, stations, or tourist areas.
- Accelerometer and gyroscope data can detect whether a person is walking, cycling, or driving.

Opportunistic sensing reduces the burden on citizens, enabling scalable sensing. However, it also introduces privacy and consent concerns since data collection occurs passively. Table 7 shows the Opportunistic Sensing Examples. Figure 6 shows the Opportunistic Data Collection Flow.

Table 7. Opportunistic Sensing Examples

Sensor Used	Data Collected	Urban Application
GPS	Location, trajectory	Traffic congestion mapping
Wi-Fi/Bluetooth	Device proximity, density	Crowd management, retail analytics
Accelerometer	Motion patterns	Mobility behavior, public transport usage

Figure 6. Opportunistic Data Collection Flow

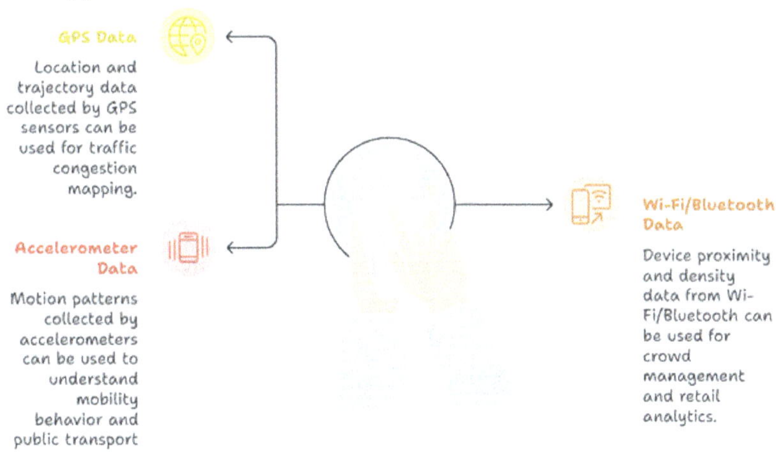

4.3 Data Fusion from Multiple Sensors

While single-sensor data provides valuable insights, data fusion is essential to enhance accuracy and reliability in urban sensing. By combining multiple smartphone sensors, systems can overcome limitations of individual modalities. For example:

- GPS + Wi-Fi fusion improves location accuracy in dense urban environments where GPS signals are obstructed by high-rise buildings.
- Microphone + air quality sensors allow cities to assess overall environmental quality, correlating noise pollution with air particulate levels.

- Camera + accelerometer can detect road surface quality (e.g., potholes) by correlating visual data with vibration patterns.

Data fusion ensures multimodal intelligence, making urban sensing more robust and less prone to errors. However, challenges include synchronization across data streams and computational overhead in real-time environments. Table 8 shows the Examples of Smartphone Data Fusion.

Table 8. Examples of Smartphone Data Fusion

Fusion Approach	Description	Application Case
GPS + Wi-Fi	Improves positioning in urban canyons	Navigation, indoor-outdoor localization
Microphone + Air Pollution Data	Correlates sound levels with pollutants	Environmental health monitoring
Camera + Accelerometer	Detects surface vibrations + visual images	Road quality, smart infrastructure alerts

5. INTEGRATION WITH SOCIAL SENSOR SYSTEMS

Integrating smartphone-based data into larger social sensing systems is crucial for enabling real-time insights and decision-making in smart cities. Smartphones act as distributed sensing nodes, but without a coherent architecture and integration strategy, the raw data remains fragmented and underutilized. A robust integration framework ensures that the massive data streams generated by citizens' devices are collected, pre-processed, transmitted, stored, and analyzed efficiently while maintaining privacy, security, and interoperability.

This section outlines the layered system architecture, the role of cloud and edge computing integration, and the importance of standardized APIs and data pipelines in ensuring smooth and scalable deployments.

5.1 System Architecture

Smartphone sensing systems generally adopt a layered architecture, which structures data flow from raw acquisition to actionable intelligence. This layered approach ensures modularity, scalability, and flexibility, allowing cities to adapt technologies for different applications such as traffic management, environmental monitoring, and public safety. Table 9 shows the Smartphone-Based Social Sensing Architecture. Figure 7 shows the Smartphone Social Sensing Architecture.

Table 9. Smartphone-Based Social Sensing Architecture

Layer	Functionality	Example Use Case
Sensing Layer	Smartphones capture data (GPS, Wi-Fi, accelerometer)	Collecting pedestrian mobility data
Edge Layer	Local pre-processing, anonymization, event filtering	Removing duplicate GPS signals
Cloud Layer	Large-scale storage, AI-driven analytics, fusion	Real-time traffic flow prediction
Application Layer	Dashboards, alerts, visualization for city planners	Smart traffic light adjustments, pollution maps

Figure 7. Smartphone Social Sensing Architecture

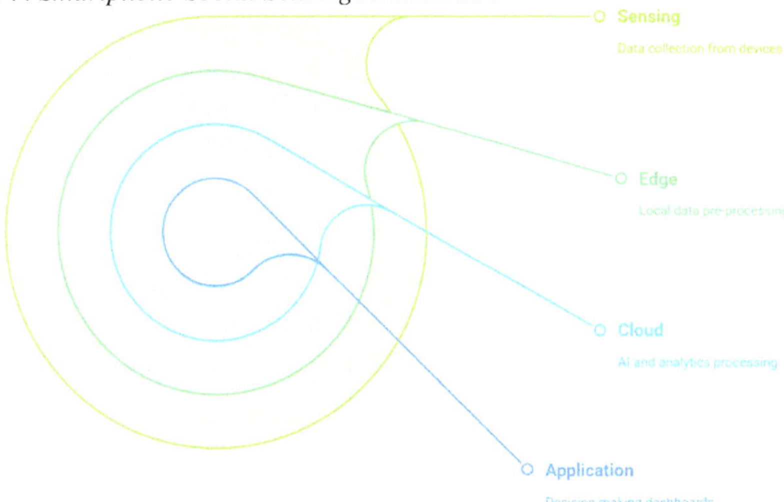

This architecture ensures that data flows seamlessly while minimizing bandwidth, optimizing computation, and protecting user privacy.

5.2 Cloud and Edge Computing Integration

The widespread adoption of cloud computing, edge computing, and the Internet of Things across various domains has enhanced data management and automation capabilities. However, large-scale IoT implementations face considerable challenges, including concerns about data quality, security risks, and the need to identify anomalies (Baimukhanov, Ali, & Yazici, 2025). The combination of edge computing and cloud computing represents a hybrid model that balances latency reduction, scalability, and computational efficiency.

- Edge Computing: Data is pre-processed locally (e.g., on the smartphone itself or nearby edge servers). This reduces latency, minimizes bandwidth consumption, and supports real-time applications such as crowd management or traffic light adjustments.
- Cloud Computing: Heavy computations (e.g., deep learning model training, predictive simulations) are performed in scalable cloud infrastructures such as AWS IoT, Google Cloud IoT, or Azure IoT. Cloud systems also provide centralized data storage and long-term analytics.

For example, in crowd density monitoring, edge servers filter irrelevant data (like duplicate GPS points), while cloud AI systems generate predictive models for mobility patterns across the entire city. Table 10 shows the Cloud vs. Edge in Smartphone Social Sensing. Figure 8 shows the Hybrid Edge-Cloud Integration for Social Sensing.

Table 10. Cloud vs. Edge in Smartphone Social Sensing

Feature	Edge Computing	Cloud Computing
Latency	Ultra-low (real-time response)	Higher (depends on network connectivity)
Computation	Limited, lightweight models	Heavy, large-scale AI/ML
Storage	Local, temporary	Centralized, long-term
Use Case	Real-time alerts (traffic, crowding)	Predictive analytics, city planning

Figure 8. Hybrid Edge-Cloud Integration for Social Sensing

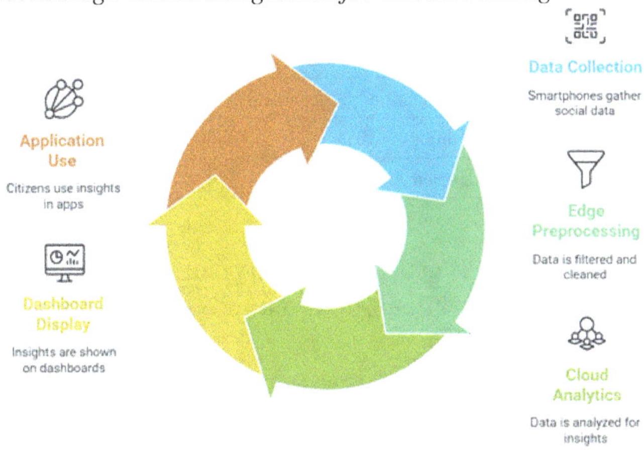

5.3 API Standards and Data Pipelines

To ensure interoperability between diverse devices, platforms, and city systems, standardized APIs and data pipelines are critical. APIs (Application Programming Interfaces) allow different systems to communicate, while data pipelines ensure that the massive inflow of smartphone data is properly ingested, cleaned, anonymized, and analyzed.

- API Standards: Organizations such as the Open Geospatial Consortium (OGC) and FIWARE define interoperability protocols, enabling seamless integration across heterogeneous urban systems.
- Data Pipelines: A typical pipeline includes (1) ingestion (from smartphone sensors), (2) cleaning (removing errors, duplicates), (3) anonymization (user privacy protection), and (4) real-time analytics (feeding dashboards).

Table 11 shows the Components of Smartphone Data Pipeline.

Table 11. Components of Smartphone Data Pipeline

Stage	Description	Example Tool/Framework
Ingestion	Collecting raw sensor data	Kafka, MQTT, REST APIs
Cleaning	Removing noise, handling missing data	Spark, TensorFlow Data API
Anonymization	Protecting user privacy	Differential Privacy, k-anonymity
Analytics	Real-time event detection, AI modeling	Apache Flink, Edge AI models

Through standardized APIs and automated pipelines, cities can scale up smartphone social sensing systems while ensuring data privacy, reliability, and actionable intelligence.

6. AI AND REAL-TIME ANALYTICS

The integration of AI-driven analytics into smartphone-based social sensing systems is a cornerstone for building responsive, intelligent, and adaptive smart cities. Smartphones generate a continuous stream of data that, when combined with AI models, can uncover patterns, detect anomalies, and trigger real-time responses to urban challenges. This section elaborates on the three main areas of AI application: activity recognition, anomaly detection, and real-time event detection.

6.1 Machine Learning for Activity Recognition

Smartphones, equipped with accelerometers, gyroscopes, GPS, and magnetometers, provide rich data sources to infer user activities. Machine learning algorithms such as Support Vector Machines (SVMs), Random Forests, and increasingly deep learning models (e.g., CNNs and LSTMs) have been widely adopted for activity recognition.

These systems can distinguish between activities such as walking, running, cycling, driving, or staying idle, which are crucial inputs for urban mobility analysis, public transport planning, and pedestrian safety. For instance, a city can estimate real-time pedestrian flow in public areas to adjust traffic light timing or allocate public transport resources dynamically. Table 12 shows the Machine Learning Models for Smartphone-Based Activity Recognition. Figure 9 shows the Smartphone Sensors Feeding ML Models for Activity Recognition.

Table 12. Machine Learning Models for Smartphone-Based Activity Recognition

Model/Approach	Features Used (Sensors)	Accuracy (Benchmark Studies)	Application Example
SVM	Accelerometer + Gyroscope	80–85%	Detecting walking vs. driving
Random Forest	GPS + Accelerometer + Magnetometer	85–90%	Cycling detection
CNN (Deep Learning)	Raw accelerometer data	92–95%	Complex gesture recognition
LSTM (Deep Learning)	Sequential sensor time-series	95%+	Driving behavior monitoring

Figure 9. Smartphone Sensors Feeding ML Models for Activity Recognition

By enabling cities to understand mobility at a granular level, activity recognition paves the way for smarter transportation systems and adaptive infrastructure.

6.2 Anomaly Detection

Anomaly detection refers to the identification of unexpected mobility or behavioral patterns. In the context of social sensing, anomalies may include:

- Sudden crowd surges → signaling possible protests, panic, or emergencies.
- Unusual vehicle flows → indicating accidents, traffic jams, or road closures.
- Deviations in pedestrian movement → signaling unsafe conditions or public disturbances.

AI algorithms such as Isolation Forests, Autoencoders, and Graph Neural Networks (GNNs) are increasingly applied to detect such irregularities in smartphone-sensed data (Chandola et al., 2009). For instance, autoencoder models can learn the "normal" mobility patterns of a city and flag any deviation as an anomaly. Table 13 shows the Applications of Anomaly Detection in Social Sensing. Figure 10 shows the Anomaly Detection Workflow in Social Sensing.

Table 13. Applications of Anomaly Detection in Social Sensing

Context	Anomaly Indicator	AI Model/Approach Used	Outcome for City Management
Traffic Mobility	Sudden congestion on arterial road	Isolation Forest	Immediate rerouting suggestions
Public Safety	Crowd panic or protest	LSTM Autoencoder	Alerts law enforcement units
Health Monitoring	Sudden clustering of symptoms	Clustering + Anomaly Detection	Early epidemic warning

Figure 10. Anomaly Detection Workflow in Social Sensing

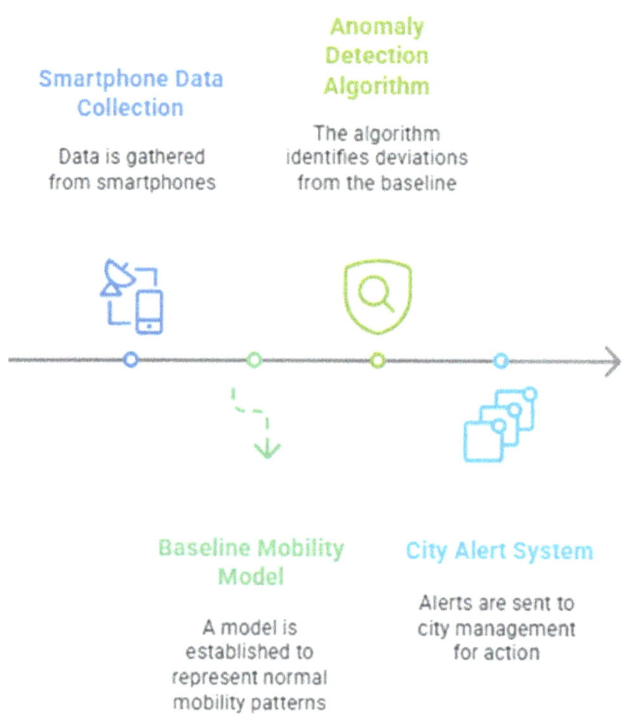

With anomaly detection, cities can move from reactive to proactive governance, anticipating risks before they escalate.

6.3 Real-Time Event Detection

Perhaps the most impactful application of smartphone-based social sensing is real-time event detection. By integrating live sensor data with predictive AI models, urban systems can detect and respond to emergencies as they unfold.

Examples include:

- Flood Warnings: Smartphones detect unusual humidity, rainfall patterns, or user reports of flooding, triggering automated city alerts.
- Traffic Accidents: Accelerometer data from smartphones in vehicles can identify collision patterns and automatically notify emergency services.
- Public Safety Threats: Sudden abnormal acoustic signals (explosions, gunshots) or large-scale social media spikes can trigger instant alerts to authorities.

Table 14 shows the Real-Time Event Detection Applications. Figure 11 shows the Real-Time Event Detection Pipeline.

Table 14. Real-Time Event Detection Applications

Event Type	Data Source (Smartphones)	AI Approach	City Response
Flooding	User reports + GPS + IoT sensors	Predictive ML models	Evacuation & resource dispatch
Traffic Accident	Accelerometer + Gyroscope	Rule-based + ML hybrid	Automatic ambulance dispatch
Public Safety Threat	Microphone + Social Media + Mobility Data	NLP + Acoustic AI models	Police alert & containment

Figure 11. Real-Time Event Detection Pipeline

Event Type	Data Source (Smartphones)	AI Approach	City Response
Flooding	User reports + GPS + IoT sensors	Predictive ML models	Evacuation & resource dispatch
Traffic Accident	Accelerometer + Gyroscope	Rule-based + ML hybrid	Automatic ambulance dispatch
Public Safety Threat	Microphone + Social Media + Mobility Data	NLP + Acoustic AI models	Police alert & containment

This approach transforms smartphones into early warning systems, bridging the gap between citizens' experiences and institutional responses.

7. CASE STUDIES AND APPLICATIONS

The adoption of smartphone-based social sensing has moved beyond theoretical frameworks and pilot projects to concrete, real-world implementations across global cities. By turning citizens' devices into active and passive data contributors, governments and urban managers have been able to address complex challenges in public safety, health, environmental management, and tourism. This section highlights four case studies that demonstrate the breadth of smartphone detection in smart city applications.

7.1 Crowd Density Monitoring in South Korea

As COVID-19 hounds the world, the common cause of finding a swift solution to manage the pandemic has brought together researchers, institutions, governments, and society at large. The Internet of Things (IoT), artificial intelligence (AI) - including machine learning (ML) and Big Data analytics - as well as Robotics and Blockchain, are the four decisive areas of technological innovation that have been ingenuity harnessed to fight this pandemic and future ones (Firouzi et al., 2021). In the wake of the COVID-19 pandemic, South Korea implemented smartphone-based crowd density monitoring systems to prevent infection clusters, particularly in nightlife and entertainment districts. Using anonymized GPS traces from smartphones, city authorities were able to identify hotspots where large numbers of people gathered at specific times. Alerts were then sent to both citizens and local government units to encourage social distancing or deploy enforcement resources when needed.

This proactive use of smartphone data helped significantly reduce the risk of viral spread while preserving citizens' mobility freedoms. It also illustrates the potential of social sensing in maintaining a balance between public health safety and economic activity in densely populated urban environments.

7.2 Contact Tracing in Singapore

Singapore is world's 5th smartest city, top in Asia (Dawar, Srivastava, Singal, Dhyani, & Rastogi, 2025). Singapore pioneered digital contact tracing through the TraceTogether app, which used Bluetooth Low Energy (BLE) technology to detect close contact events between smartphones. When a user tested positive for COVID-19, health authorities could quickly identify individuals at risk of exposure based on logged interactions.

This case shows how smartphone sensing, when combined with data anonymization and strong regulatory frameworks, can scale rapidly in a crisis. Beyond pandemic response, the success of TraceTogether demonstrates how Bluetooth-based proximity sensing could be adapted for other urban safety purposes, such as tracking exposure to environmental hazards or monitoring crowd movements during emergencies.

7.3 Environmental Mapping with NoiseTube

Smartphone applications have emerged as powerful tools in the realm of environmental monitoring, particularly in the context of smart cities. As urbanization continues to advance at an unprecedented pace, the need for efficient and real-time monitoring of environmental parameters has become increasingly crucial. Smartphones, equipped with a myriad of sensors and connectivity options, have proven to be invaluable instruments in this regard (Wolniak, Jonek-Kowalska, & Grebski, 2023). Environmental monitoring is another field where smartphones have proven transformative. The NoiseTube project, first deployed in Europe, allows citizens to use their smartphones as noise sensors, collecting data on sound levels across urban spaces. Uploaded data is then aggregated into interactive maps that reveal noise pollution hotspots.

This participatory sensing approach democratizes environmental science, empowering citizens to contribute actively to urban sustainability goals. It also provides municipal authorities with cost-effective monitoring tools, avoiding the need for large-scale deployment of expensive sensors. Similar methods can be extended to air quality, urban heat mapping, and even micro-level environmental hazards, showcasing the versatility of smartphone-based sensing systems.

7.4 Smart Tourism in Barcelona

From the results of a study on the impact of smart city technology on urban tourism, in the cities of Amsterdam, Barcelona, and Vienna, and considered the implications for the development of smart tourism in Belgrade and other surrounding Serbian cities. This research shows how smart city technology contributes to improving the efficiency, digital engagement, and sustainability of urban tourism (Vujko, Knežević, & Arsić, 2025). Barcelona, as one of Europe's leading smart cities, has integrated smartphone sensing into its smart tourism strategy. Through location-based services and mobile applications, tourists receive personalized guidance to cultural attractions, while the system simultaneously monitors crowd flows across popular destinations.

When overcrowding is detected in a particular landmark, the system suggests alternative attractions, effectively distributing tourist density and reducing pressure on fragile cultural sites. This not only enhances the visitor experience but also supports sustainable tourism practices by mitigating overtourism challenges.

7.5 Insights from Case Studies

Taken together, these case studies highlight the multi-dimensional potential of smartphone-based social sensing in smart city ecosystems. From public health management in South Korea and Singapore, to environmental stewardship in Europe, and sustainable tourism in Barcelona, smartphones act as bridges between individual behaviors and collective urban outcomes.

The central lesson is that while the technology is available, its effectiveness depends heavily on citizen trust, strong governance, and ethical frameworks that ensure data is used responsibly. These insights provide critical guidance for cities seeking to scale smartphone-based sensing solutions for future challenges.

8. PRIVACY, SECURITY, AND ETHICAL ISSUES

As smartphone-based social sensing continues to expand, privacy, security, and ethics become the most critical challenges to ensuring both citizen trust and sustainable deployment. While the technological benefits are evident, the unintended risks associated with large-scale data collection—particularly data generated from personal devices—cannot be ignored. Addressing these challenges requires a careful balance between innovation and the protection of fundamental human rights. Current practices and challenges in data management, offer a roadmap for strengthening the governance, security, and ethical use of AI in genomics and microscopy. While strong governance frameworks and ethical practices have laid the groundwork for data integrity and transparency, there is still an urgent need for collaborative efforts to develop interoperable platforms and transparent data sharing policies (Taddese, Addis, & Tam, 2025).

8.1 Data Anonymization

Sharing health data holds great potential for advancing medical research but also poses many challenges, including the need to protect people's privacy (Halilovic, Meurers, Otte, & Prasser, 2025). Smartphone sensing systems generate highly sensitive datasets, including location traces, mobility patterns, and behavioral indicators. If left unprotected, such data could expose personal routines, home addresses, or even political affiliations. To mitigate these risks, data anonymization techniques such as differential privacy, k-anonymity, and data perturbation are increasingly being adopted.

For instance, Apple's differential privacy framework injects statistical noise into aggregated data, making it nearly impossible to trace contributions back to individual users while still maintaining the analytical value of datasets. In the context of smart cities, such approaches allow municipalities to monitor urban mobility or environmental conditions without compromising personal privacy.

8.2 Consent and Transparency

Urbanization is reshaping landscapes and poses unprecedented sustainability challenges, necessitating a more integrative approach to urban observation. Recent advances in traditional remote sensing technologies and emerging social sensing technologies, emphasize their convergence in urban science (Yu, 2025). A core principle of ethical social sensing is ensuring that citizens maintain agency over their own data. This requires transparent consent mechanisms, where individuals explicitly agree to share specific categories of data for defined purposes. Unfortunately, many existing applications bury consent clauses in lengthy privacy agreements that citizens rarely read.

A better model is the use of granular consent dashboards, where users can enable or disable data-sharing preferences dynamically. Singapore's TraceTogether app demonstrated this by allowing users to see how their data was stored, who could access it, and under what circumstances it would be shared. Such transparency fosters trust and increases citizen willingness to participate in smart city initiatives.

8.3 Legal Compliance

Smartphone-based sensing must also align with global and local regulatory frameworks. In the European Union, the General Data Protection Regulation (GDPR) mandates strict requirements around personal data usage, including the right to be forgotten, informed consent, and limitations on cross-border data transfer. Similarly, countries like Indonesia are implementing Personal Data Protection Acts (PDPAs) that define rules for digital ecosystems, including smart cities.

Failure to comply can result not only in legal penalties but also in erosion of public trust, undermining the very foundation of citizen-driven sensing. Ensuring compliance means embedding privacy-by-design principles into smartphone sensing architectures, from initial data collection to final analytics.

8.4 Ethical Considerations Beyond Privacy

While privacy and security are often the most discussed, ethical challenges go further. Key questions include: Who owns the data? Who benefits from it? How do we prevent bias in AI-driven analytics of social sensing data? For example, crowd monitoring systems might inadvertently over-police certain districts while under-monitoring others, reproducing existing social inequities.

Addressing these concerns requires embedding fairness, accountability, and inclusivity into design processes. Emerging technologies like federated learning allow AI models to be trained locally on devices, ensuring sensitive data never leaves users' smartphones, while blockchain can provide transparent, tamper-proof logs of how data is collected and shared.

8.5 Towards Trustworthy Social Sensing

Ultimately, the success of smartphone-based social sensing in smart cities hinges not on technology alone but on public trust. Citizens must believe that their contributions are safeguarded, used responsibly, and provide tangible benefits in return. Cities that fail to uphold privacy and ethical standards risk facing, as seen in Toronto's Sidewalk Labs project, which was abandoned in 2020 after public concerns about mass surveillance.

Therefore, the future of social sensing public backlash lies in systems that are not only technologically advanced but also transparent, secure, and ethically grounded. Embedding these safeguards ensures that social sensing remains a tool for empowering communities rather than exploiting them.

9. FUTURE DIRECTIONS

In the future, the use of smartphones as a social sensing system in smart cities will be further strengthened by the integration of new, more sophisticated technologies. One of them is the implementation of federated learning that allows training of artificial intelligence models directly on the device without having to move raw data to a central server. This approach not only maintains user privacy, but also improves computing efficiency by leveraging processing power at the edge of the network.

In addition, it is predictable to play a central role in real-time decision-making. Edge AI systems will be able to recognize patterns, detect anomalies, or provide alerts directly at the location of events without relying on a cloud network. This will be especially important in the context of critical services such as disaster response, public health, and traffic management. Meanwhile, blockchain has the potential to be the primary infrastructure for building transparency and trust. By recording each data flow in a decentralized manner, people can be confident that the information they provide cannot be manipulated. This mechanism also opens up opportunities for a new participatory model, where citizens get direct incentives for their data contributions.

Another trend worth noting is the move towards greater inclusivity and participation. Not all citizens have the same level of digital literacy, so it is necessary to design a system that is user-friendly, multi-lingual, and easily accessible. Thus, the future of social sensing is not only about technology, but also about how it is used to build cities that are more equitable, responsive, and adaptive to the needs of all levels of society.

10. CONCLUSION

Smartphones have evolved from just a means of communication to an important element in smart city infrastructure through the concept of social sensing. With sensors embedded in it—such as GPS, accelerometers, cameras, and network modules—smartphones are able to provide real-time data that describes behavior, mobility, and environmental conditions. This data, when combined with artificial intelligence and the Internet of Things, presents an incredible opportunity to build cities that are more responsive, efficient, and oriented to the needs of its citizens. This marks a major shift in the way cities are managed, from a traditional reactive approach to a proactive, data-driven system.

In addition to the technical aspect, the success of smartphones as a social sensor also depends on citizen participation. The participatory approach, in which people voluntarily report on their surroundings, is combined with an opportunistic approach that leverages automated data from devices. The synergy between the two creates a richer, more accurate, and layered database. This proves that residents are not only recipients of services, but also active partners in the urban governance process. However, this engagement requires trust between citizens and the government, which can only be achieved if the privacy and security aspects are properly maintained.

Despite the great benefits, serious challenges remain. Privacy, security, and ethics issues are critical points that must be handled carefully. Without strong protection mechanisms, citizens may be reluctant to participate for fear that their personal data will be misused. Therefore, the application of data anonymization techniques, compliance with privacy regulations, and the provision of transparent controls for users are essential. Furthermore, ethical discourse on how data is used for public purposes needs to be continuously developed so that there is no bias or discrimination in data-based decision-making.

To answer these challenges, cutting-edge technologies such as federated learning, edge AI, and blockchain provide a new direction. Federated learning allows data processing to be carried out on the user's device, so that privacy is better maintained. Edge AI supports real-time analysis without having to rely on the cloud all the time, so decisions can be made faster in the field. Blockchain, on the other hand, ensures transparency and integrity of data so that people can verify that their contributions are being used honestly. The integration of these technologies will strengthen the reliability of the system, while fostering public trust in smart cities based on social sensing.

In the end, smartphone-based social sensing is not just a technological innovation, but also a social transformation. He taught that the future of smart cities must be built on the basis of collaboration between technology, government, and society. With a framework that balances technical efficiency and the protection of citizens' rights, social sensing can be the foundation of a more inclusive, equitable, and sustainable city. If challenges can be addressed wisely, then future cities will not only be smarter in managing resources, but also more humane in responding to the needs of their citizens.

REFERENCES

Ahmed, N., Rafiq, J. I., & Islam, M. R. (2020). Enhanced human activity recognition based on smartphone sensor data using hybrid feature selection model. *Sensors (Basel)*, *20*(1), 317. Advance online publication. DOI: 10.3390/s20010317 PMID: 31935943

Alharbi, E., Alsulami, G., Aljohani, S., Alharbi, W., & Albaradei, S. (2025). Real-time detection and monitoring of public littering behavior using deep learning for a sustainable environment. *Scientific Reports*, *15*(1), 3000. Advance online publication. DOI: 10.1038/s41598-024-77118-x PMID: 39848984

Baimukhanov, S., Ali, H., & Yazici, A. (2025). Enhancing ML-based anomaly detection in data management for security through integration of IoT, cloud, and edge computing. *Expert Systems with Applications*, *293*, 128700. Advance online publication. DOI: 10.1016/j.eswa.2025.128700

Benseny, J., Lahteenmaki, J., Toyli, J., & Hammainen, H. (2023). Urban wireless traffic evolution: The role of new devices and the effect of policy. *Telecommunications Policy*, *47*(7), 102595. Advance online publication. DOI: 10.1016/j.telpol.2023.102595

Beyond GDPR: Employing AI to Make Personal Data Useful to Consumers. (2020). *Trusted Data*. The MIT Press. DOI: 10.7551/mitpress/12439.003.0014

Dawar, I., Srivastava, A., Singal, M., Dhyani, N., & Rastogi, S. (2025). A systematic literature review on municipal solid waste management using machine learning and deep learning. *Artificial Intelligence Review*, *58*(6), 183. Advance online publication. DOI: 10.1007/s10462-025-11196-9

Erdelyi, V., Mizumoto, T., Kitai, Y., Ishimaru, D., Adachi, H., Higashino, T., & Ikeda, M. (2025). Detecting Subtle Signs of School Attendance Issues Using Smartphone-Based Sensing. *IEEE Access : Practical Innovations, Open Solutions*, *13*, 4652–4669. DOI: 10.1109/ACCESS.2024.3523108

Firouzi, F., Farahani, B., Daneshmand, M., Grise, K., Song, J., Saracco, R., Wang, L. L., Lo, K., Angelov, P., Soares, E., Loh, P.-S., Talebpour, Z., Moradi, R., Goodarzi, M., Ashraf, H., Talebpour, M., Talebpour, A., Romeo, L., Das, R., & Luo, A. (2021). Harnessing the Power of Smart and Connected Health to Tackle COVID-19: IoT, AI, Robotics, and Blockchain for a Better World. *IEEE Internet of Things Journal*, *8*(16), 12826–12846. DOI: 10.1109/JIOT.2021.3073904 PMID: 35782886

Halilovic, M., Meurers, T., Otte, K., & Prasser, F. (2025). Parallel privacy preservation through partitioning (P4): A scalable data anonymization algorithm for health data. *BMC Medical Informatics and Decision Making*, *25*(1), 129. Advance online publication. DOI: 10.1186/s12911-025-02959-z PMID: 40075355

He, Y., Fan, J., & Zhang, Y. (2025). Controlled quantum authentication confidential communication protocol for smart healthcare. *Scientific Reports*, *15*(1), 18094. Advance online publication. DOI: 10.1038/s41598-025-02448-3 PMID: 40413245

Hossain, M. T., Lee, J., Besenski, D., Dimitrijevic, B., & Spasovic, L. (2025). Developing a Prediction Model for Real-Time Incident Detection Leveraging User-Oriented Participatory Sensing Data. *Information (Basel)*, *16*(6), 423. Advance online publication. DOI: 10.3390/info16060423

Liu, C.-L., Huang, H.-H., & Yen, D. C. (2022). The effects of navigation cues and involvement of highly innovative consumers on attitudes toward mobile application stores. *Journal of Electronic Commerce in Organizations*, *20*(1), 1–20. Advance online publication. DOI: 10.4018/JECO.305736

Luo, L., Hou, X., Cai, W., & Guo, B. (2020). Incremental route inference from low-sampling GPS data: An opportunistic approach to online map matching. *Information Sciences*, *512*, 1407–1423. DOI: 10.1016/j.ins.2019.10.060

Taddese, A. A., Addis, A. C., & Tam, B. T. (2025). Data stewardship and curation practices in AI-based genomics and automated microscopy image analysis for high-throughput screening studies: Promoting robust and ethical AI applications. *Human Genomics*, *19*(1), 16. Advance online publication. DOI: 10.1186/s40246-025-00716-x PMID: 39988670

Torres-Sospedra, J., & Ometov, A. (2021). *Data from Smartphones and Wearables. Data*. MDPI AG., DOI: 10.3390/data6050045

Ullah, A., Anwar, S. M., Li, J., Nadeem, L., Mahmood, T., Rehman, A., & Saba, T. (2024). Smart cities: The role of Internet of Things and machine learning in realizing a data-centric smart environment. *Complex & Intelligent Systems*, *10*(1), 1607–1637. DOI: 10.1007/s40747-023-01175-4

Vujko, A., Knežević, M., & Arsić, M. (2025). The Future Is in Sustainable Urban Tourism: Technological Innovations, Emerging Mobility Systems and Their Role in Shaping Smart Cities. *Urban Science (Basel, Switzerland)*, *9*(5), 169. Advance online publication. DOI: 10.3390/urbansci9050169

Wang, W., Mirjafari, S., Harari, G., Ben-Zeev, D., Brian, R., Choudhury, T., & Campbell, A. (2020). Social Sensing: Assessing Social Functioning of Patients Living with Schizophrenia using Mobile Phone Sensing. In *Conference on Human Factors in Computing Systems - Proceedings*. DOI: 10.1145/3313831.3376855

Wolniak, R., Jonek-Kowalska, I., & Grebski, W. (2023). Environmental monitoring in smart city–smartphone applications aspects. *Zeszyty Naukowe Politechniki Śląskiej. Organizacja i Zarządzanie*, *185*, 605–615.

Yu, D. (2025). Toward Integrated Urban Observatories: Synthesizing Remote and Social Sensing in Urban Science. *Remote Sensing (Basel)*, *17*(12), 2041. Advance online publication. DOI: 10.3390/rs17122041

Zhang, H., Ibrahim, A., Parsia, B., Poliakoff, E., & Harper, S. (2023). Passive social sensing with smartphones: A systematic review. *Computing*, *105*(1), 29–51. DOI: 10.1007/s00607-022-01112-2

Chapter 10
Social Sensing as a Catalyst for Effective Public Utilities in Urban Areas

Lokesh Jain
http://orcid.org/0000-0001-5764-3822
School of Engineering and Technology, Vivekananda Institute of Professional Studies- Technical Campus, Delhi, India

ABSTRACT

Social sensing is reinventing smart areas by combining human-created information on mobile devices, social media, and crowdsourcing with conventional Internet of Things (IoT) systems to improve urban governance. It helps to facilitate intelligent transport, water and energy management, waste management, and emergency response using real-time insights of citizens. Although the reliability, privacy, and data integration have been challenging, the latest tools in the form of AI, blockchain, and interoperability standards will enhance trust, inclusivity, and resiliency in cities, as well as make them more sustainable and citizen-centric.

1. INTRODUCTION OF SOCIAL SENSING IN SMART ZONES

In the modern era characterized by high rates of urbanization and technological growth, the cities of the world are becoming increasingly complex systems, which have dynamic requirements. Conventional methods of urban governance are no longer enough to contain the strains of growth in populations, vehicular movement, and the generation of wastes, energy usage, and escalating demands of civil amenities. In

DOI: 10.4018/979-8-3373-4962-6.ch010

response to such challenges, the idea of smart zones is brought up more often, as it is a promising solution that allows urban planners and governments to implement. Smart zones are identified areas in cities that drastically deploy digital technologies to improve city systems in terms of efficiency, security, and sustainability. These are the spheres of the convergence of innovation and infrastructure, in which living data, automation, analytics, and connectivity are all used with the aim of enhancing the management and coordination of urban functions. The effectiveness of smart zones depends not only on the implementation of physical infrastructure, including sensors and cameras, but also on citizen participation using what can be described as social sensing (Alvarez et al., 2023; Okoli & Kabaso, 2024; Taher et al., 2018).

1.1 Introduction of Smart Zones

A smart zone is a specific area in cities that is implanted with several interdependent digital technologies that are meant to maximize the performance of daily city activities toward efficiency, safety, and sustainability. Such areas are no longer confined to geographical locations, i.e., they may extend as small as a single street or a complex of buildings, or as large as a sector of a city. Smart zones, in essence, involve the use of real-time data mining and smart analytics to observe, regulate, and streamline diverse municipal operations such as traffic management systems, energy consumption, waste management, environmental control measures, and community safety. Such a data-driven and real-time action will allow governments and city planners to make data-driven decisions within a short period, to act before challenges in the cities occur, and to positively impact the lives of city residents. For example, the traffic lights in an intelligent business area can interact with each other to change the traffic flow depending on the time and decrease peak-hour traffic jams. Parking sensors are possible in these areas that may inform users about empty parking slots in applications and reduce unneeded transportation and air emissions. Equally, in the residential smart zones, smart waste bins could be installed in designated areas, which are fitted with fill level sensors to indicate to the sanitation services when the waste bins require emptying, thereby ensuring sanitation is done effectively and there is no build-up of waste. The real-time energy monitoring systems in industrial zones assist factories in optimum power usage at reduced costs of operation and environmentally friendly carbon footprints (Castrellon et al., 2024; Schmidtke et al., 2022).

In general, smart zones are a bright example of the way to turn the city areas into smart ecosystems, predicting, anticipating, and reacting instantly, leading to sustainable urban development as shown in Figure 1. These make an enormous difference in urban planning and the delivery of viable services to the people, as they are able to interlink systems and involve both machine and human inputs.

Figure 1. Smart Zone Infrastructure

These are examples of the concentrated smart zone infrastructures:

- **Intelligent trash bins:** These bins have fixtures where the fill-level is gauged, and where the bins are solar-powered compactors that notify waste management teams only when they need to be cleaned. San Francisco has such bins in which the frequency of collection is reduced, thereby minimizing overflow and the sanitary problems caused by it.
- **Smart Street Lighting:** The applications of adaptive streetlights use motion sensors and light sensors to automatically adjust the brightness of the lights. Cities such as Paris and Miami use these systems more to provide greater safety to people and reduce their energy bills considerably. There are also environmental sensors in the network to monitor the quality of the air.
- **IoT-enabled Transit and Traffic Controls:** Barcelona also uses a suite of sensors, CCTV cameras, and linked traffic lights to manage traffic on the fly. The real-time information is fed to the control central centers, where programming of the signal timing and automatic incident detection is made immediately.
- **Unified Smart Infrastructures (Living Labs):** Aveiro Tech City in Portugal is a modern smart zone where 44 multimodal access (WiFi, 5G, LoRa-WAN)

points connect sensors in the streets and the public transport. The nodes of edge computing perform data processing at the edge, enabling applications such as mobility, environmental monitoring, and emergency use.

1.2 The Working Combinations of the Components

A smart zone combines and links different related items together into a single blueprint to work in harmony, streetlights, bins, traffic systems, environmental monitors, city Wi-Fi, and kiosks as an integrated urban health system.

- **Traffic and Safety:** Smart urban traffic also promotes smart zones, where CCTV cameras, sensors, and smart bounded streets are incorporated to promote the management of traffic and public safety. All these technologies constantly make a surveillance of the movement of vehicles as well as people, and the authorities are able to control the positioning of vehicles on the road and the reaction to an incident. Digital displays in streetlights can also give real-time messages and other information concerning public safety in cases of emergency.
- **Waste and Sanitation:** The smart bins have sensors to monitor the fullness of the bins and send a command to the cleaning services to visit and get rid of the waste. Such a system eliminates overflow and encourages a cleaner environment. Moreover, smart kiosks enable citizens to have a platform through which they can leave a complaint or report sanitation problems of a service offered to the people (Debnath et al., 2016).
- **Environmental Monitoring:** Smart zones are also equipped with sensors in the streetlights and infrastructure to analyze the ongoing environmental quality. Such sensors check the values such as the quality of the air, the temperature, the humidity, and the footfall, which are essential data to make decisions on public health and the environment. This into detects pollution routines among the city planners, the climate conditions, and the pattern of movement of the people who use their feet, thus this data lets the city planners know how to get actual optimization of the city design and the allocation of resources, enabling people to live sustainably.
- **Energy Management and Connectivity:** Smart zones enhance the use of green energy and digital inclusion by revamping smart infrastructures such as solar-powered benches that offer device charging to citizens and access to Wi-Fi. These green alternatives utilize solar energy, which is renewable and limits the use of conventional forms of energy. In addition, intelligent zones use high-speed fiber or mm Wave networks to support high-performance connectivity and enable real-time transfer of data and coordination among

different sectors of the public services by using the optimized living lab environments.

1.3 Transition from traditional sensing to social sensing

The use of IoT technologies, a network of connected devices and sensors physically installed throughout the urban environment, has been the cornerstone of smart city infrastructures. Such devices constantly gather quantitative data elements connected with the environmental conditions immediately around them (temperature, humidity, or air pollution), the operations (traffic, electricity consumption, etc.), and the state of the infrastructure itself (structural stress in bridges or occupancy in public buildings, etc.). They serve as the origin of their contributions in urban analytics, facilitating early developments of automation, predictive maintenance, and energy optimization. Nonetheless, having an IoT-only approach has limitations. Although physical sensors are accurate, they are limited by cost, coverage, and context. They have to be installed, maintained, and calibrated, which makes it costly to carry out a massive deployment. More importantly, such things are passive reporters: they gather information and are unable to interpret the context or provide semantic information. To illustrate, a traffic sensor would tell the density of the cars; however cannot give a reason as to why the traffic is experienced, whether because of a protest, accident, or a parade. Moreover, IoT systems do not usually capture the unexpected aberrations or social-cultural processes, e.g., the shift in population sentiments, partial panic, or civil riots, taking the form of an event that occurs outside the same physical instrumentation area. That is where social sensing comes in, a revolutionary shift that allows human experience to enter into the data landscape of the smart city. Social sensing is the act or service of acquiring information created by a person via the use of mobile phones, wearable gadgets, open-air social media networks, crowdsourcing applications, messaging apps, and even geolocated pictures or videos. The model is a reimaging of the role of the urban resident not only as a service receiver, but also as an actor and sensor within the city data environment.

Whenever one tweets about the existence of a pothole, provides an uploaded video of a rally, posts a report of a power cut on a civic app, or writes a review of a journey on public transportation, they are serving a centrally dispersed, collaborative sense network. In contrast to IoT sensors, individuals may add contexts, emotions, opinions, and descriptions, which ascribe meaning to the raw data. The streams of data generated by humans are probably more dynamic, subtle, and geographically spread out, as compared to the static IoT devices. The concept of social sensing, combined with the legacy IoT infrastructure, makes the urban management model more sustainable and open. For example, when the government sponsors a large event, IoT can give you density analytics of the crowd, real-time social media posts can tell authorities about the satisfaction of the people, any safety threats, or that there is an improper flow of traffic. On one edge, in environmental monitoring, sensor information may validate air pollution; on the other edge, citizen reports may indicate signs of health grievances, bad odors, or even point toward an inside cause such as unlawful burning or industry spills. This two-purpose system is the precision of machines via the IoT layer and the human intuition via the social sensing layer, both in terms of the situational awareness of the city administrators (Gubbi et al., 2013).

1.4 Concept and working of social sensing

The concept that underlies social sensing can be explained as follows: citizens are the most active urban actors and, therefore, they can spot and report on the conditions of their environment more than anyone. In the course of their daily use of infrastructure and their even more common direct contact with the city material, landscapes, and services, people readily learn about the phenomenon of vandalism or facilities in bad shape, street riots or environmental oddities, or the risks to their safety. In contrast to traditional sensors based on passive monitoring of pre-determined indicators, humans bring along a contextually shaped and flexible type of observation that imparts the uncertainty complexities of the real world. The social sensing process is organized into several steps, transforming individual observations into valuable actionable urban intelligence.

- **Data Generation:** The initial and most important phase of social sensing is data generation. Data are produced by persons using a broad range of individual devices and online platforms. Smartphones are specifically potent instruments, as they allow taking pictures, recording video, exchanging voice messages, sending text reports, or permit location tracking with the help of GPS. The wearable devices can also be used to add physiological data in the case of events such as heat waves or pollution outbreaks. Social media such as Twitter, Facebook, and Instagram offer unstructured and rich information under status updates, hashtags, geo-tagged posts, or live streaming that give a real-time impression of sentiment and events in the city. These may be either spontaneous or activated by targeted campaigns or rewards-driven reporting tools introduced by cities.
- **Data Aggregation:** After the generation of the data, it must be gathered in a methodical and practical format. This is achieved by the use of a blend between civic participation applications, reporting portals provided by governments, data fusion platforms using the sensors, and social monitoring of third-party social media tools. Simultaneously, social media is tapped into

through automated instruments by searching keywords and location to retrieve social media posts that are relevant to urban conditions. Aggregation combines information of different types, text, images, videos, voice, and metadata, among other things, which are gathered by various means and in a variety of forms.

- **Data Processing:** In this step, the raw data is refined further, filtered, and analyzed to derive useful inferences. Machine learning (ML), AI and natural language processing (NLP) are important here. The NLP algorithms can parse social media posts or textual descriptions to extract themes, sentiment, or urgency. This can be done by implementing image recognition models to validate the presence of infrastructure problems or crowd density by reviewing photos. The audio analysis may recognize the level of background noise, sirens, or announcements of people. The spatial and temporal data are used to network motifs both spatially and over time. The step in processing also has also systems to eliminate false positives, copies, or spam-only valuable and credible information is passed ahead.
- **Actionable Insights:** The last step is the transformation of the studied information into knowledge that the city authorities can use to act. These findings are presented in dashboards or GIS or can be transmitted directly to the current municipal systems, such as the traffic control offices, emergency response centers, or the maintenance departments. An example is when there is an influx of alerts in a particular area regarding a power cut, the electricity board can then be alerted to look into the matter. Likewise, topics such as citizen allegations of waterlogging during the rains could facilitate the adjustment of the drainage schedules in the public works departments. Predictive analytics are also supported by actionable insights where the recurrent patterns of data may be used to predict infrastructure breakdown, traffic spikes, or air quality deterioration, and beat them in advance.

Overall, this four-step model renovates the daily interactions of suffering to form a decentralized network of sensors and give municipalities a higher level of agility, scale, and human-level interactions to drive their mode of urban governance. Across the board, social sensing helps to increase the efficiency of operations, as well as promoting both civic participation and confidence among residents and the city systems through a continuous feedback loop (Wang et al., 2015).

2. USE OF SOCIAL SENSING BY PUBLIC UTILITIES

Social sensing is one of the groundbreaking ways of collecting, interpreting, and acting on real-time environmental and human data. The availability of mobile technologies, GPS, and social media as a platform makes it possible to use individuals as sensors in a city, with or without their knowledge. Individuals rely on smartphones to conduct their day-to-day activities and wearable devices to help improve their overall experiences, which creates volumes of data on which solutions can be derived to increase the value in public services. Within the domain of public utilities, this crowdsourced and ambient data paradigm is transformative, altering operations, enhancing efficiency, and maximizing sustainability.

2.1 Intelligent Transportation Systems

Over the last few years, intelligent transportation systems (ITS) have undergone considerable development, largely due to the incorporation of social sensing technologies, in particular. ITS uses sensing, communication, and data processing to enhance the safety, efficiency, and sustainability of the transportation system, as shown in Figure 2 (Ouallane et al., 2021).

- **Real-Time Traffic Tracking and Navigation:** Social sensing applications, such as Google Maps and Waze, use real-time user input to capture traffic flow, identify areas of congestion, and offer the best route out of congestion problems. These apps gather information involving users' mobile devices, e.g., the location of the user via GPS, speed, and direction of traffic flow, and combine them to create a whole picture of the traffic trends. As many users on a given section of the road start to slow down, this is perceived by the system as a traffic jam, and so they immediately alert the rest of the drivers and provide them with an alternative route. It saves time and fuel for the commuters. To the city planners and traffic authorities, it provides useful information about the traffic congestion areas, the busiest time period of the day, and risky corners, and hence, the targeted development of infrastructure and traffic control measures can be provided.
- **Traffic signals and adaptive infrastructure planning:** In urban areas with sophisticated ITS, information gathered through social sensing can be used to adaptively alter the timing of traffic lights. An example of this would be when a spike in traffic is identified on a particular route as a result of a sports event or road closure, it is possible to adjust traffic lights in real time to alleviate the traffic. Such systems not only utilize fixed sensors but also use the input of mobile users, in a way generating a kind of half-fixed half half-floating sensor

type system. Also, the gathered information assists in long-term planning. The history of traffic movements can be determined using social sensing, and this information can be used by the developers and transportation authorities to forecast future infrastructure requirements. It may result in the building of new transit systems, roads, or even pedestrian pathways in places of maximum demand.

- **Improvement of Public Transit:** Social sensing is also beneficial in the area of the public transportation system. The peak commuters can use social networking sites or transit apps to report issues such as delays, hygiene, etc. This feedback, when added to the ITS framework, will help the transit authorities react quickly to service interruptions and learn to expand the services based on the needs of users. Transit prediction systems also use crowdsourced data, as well, to tell prospective passengers when to expect a particular wait length, where the nearest vehicle is, and service announcements. This builds more faith in state transportation and can become a change modality to move out of personal cars, which decongests cities and improves emissions.

Figure 2. Social Sensing-based Intelligent Transportation System

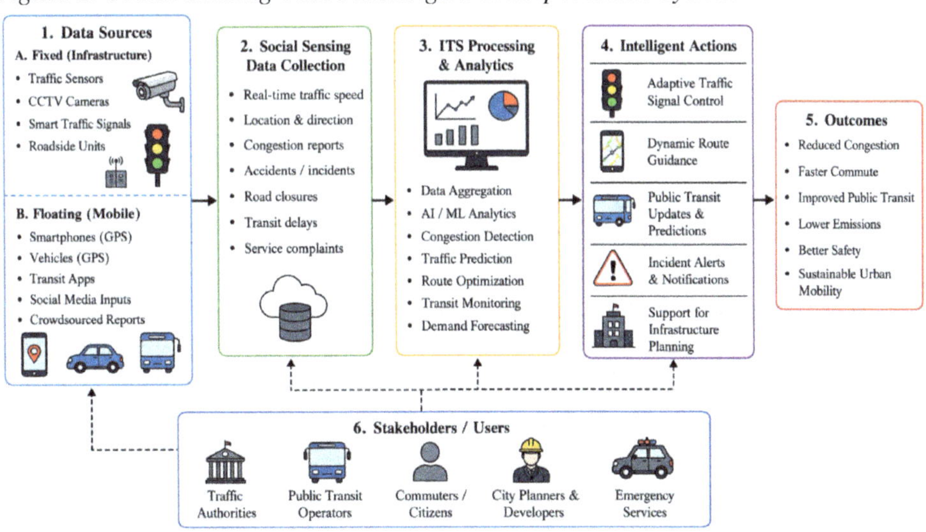

2.2 Energy and Water Control

Water and energy are two of the most fundamental and yet insecure public utilities. The effective use of such resources is not only economically important but also crucial to maintain environmental sustainability and the well-being of the associated population. Social sensing introduces an initiative level of insight and command of these systems (Jan et al., 2021; Ullo & Sinha, 2020).

- **Fault reporting and leak detection:** Conventionally, water leakage or an electric fault was realized through regular checking or customer complaints after significant damage had been caused. Social sensing disrupts this by allowing real-time, community-based monitoring. Mobile applications tend to include applications to geolocate, capture images, and categorize leaks, broken pipelines, power outages, or unusual consumption trends to easily report to users. This crowdsourced information assists the utility companies in locating and addressing the issues much faster than ever before. Early detection will help avoid wasting water, will be cheaper to repair, and there will be a decreased threat of damaging a property or loss of service.
- **Behavioral Insights and Demand Management:** Utilities can perform analytics on anonymized data gathered through smart meters and mobile applications, with the user's consent, and receive the pattern of usage. This allows creating individualized suggestions that will lead to more sustainable behaviors. Residents, as an example, may be alerted that their patterns are above the community average or when spikes in usage may show a broken pipe, etc. In addition, using behavioral data, utilities could predict demand and supply accordingly because they accumulate data, making it possible to make predictions.
- **Infrastructure Planning and Policy using:** Social sensing in the management of the utilities also acts as a guiding factor in the creation of the policy and planning of its infrastructure. Information about the problems that were most frequently reported, gaps in the services, and areas with significant consumption rates can be utilized to prioritize the areas of infrastructure improvement, e.g., the change of old pipelines, construction of new substations, or initiation of incentives for solar panel installation. Social sensing can also produce more inclusive policymaking.

2.3 Organized Waste Management

Urban governance plays a key role in managing waste, including the collection, transportation, processing, recycling, and disposal of waste products. This process has received a substantial boost through social sensing in enhancing efficiency and transparency in the process, as shown in Figure 3 (Wijayaningsih et al., 2020).

2.3.1 Citizen Reporting and Geo-tagged Feedback

The mobile apps have enabled residents to report cases of overflowing garbage cans, illegal dumping, and missed collections. These reports can also be geotagged, i.e., the place where the incident occurs is automatically registered. One can also upload pictures to give a visual representation, and others even have provisions for classifying the kind of waste or giving them urgency levels. The system decentralizes the process of monitoring and allows the residents to participate in the process of urban cleanliness.

2.3.2 Dynamic Scheduling of the Waste Collection

Historically, schedules were determined and waste was collected based on predetermined collections that did not consider the fluctuating generation of waste by different locations or days of the week. Information about social sensing can be used to adopt dynamic scheduling and fertilizer truck routes depending on real needs, and does not need any specific schedule. As an example, when residents call to complain about overflowing bins on their street and no one is on the next street, routes can be optimized to deal with this problem. It not only makes the service more efficient but also minimizes the use of fuel and greenhouse gas emissions due to unnecessary runs of trucks.

2.3.3 Participation and Awareness in the Community

In addition to the delivery of services, social sensing creates a culture of civic responsibility. Gamified applications or point-based reward systems can be used to reward waste reporting and recycling. Citizens may find themselves more likely to repeat the activity when they find actual results of their work that can be observed by people around them, and which include clean streets or any form of public acknowledgment. Further, the information about the state of affairs visualized on the basis of social sensing reports can be provided to the population to increase awareness.

Figure 3. Social Sensing-based Smart Waste Management System

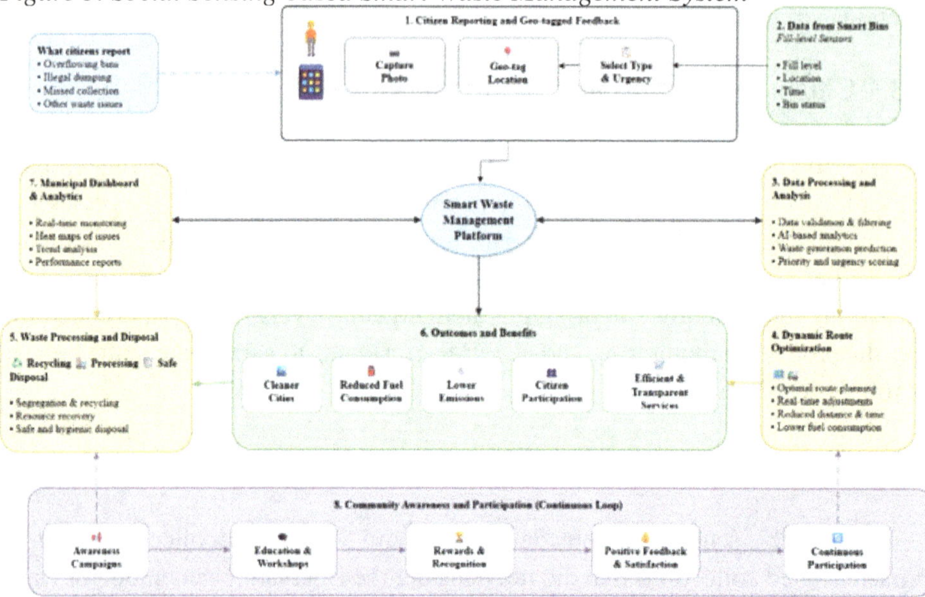

2.4 Citizen and emergency safety

Social sensing should enhance public safety and develop means of responding to emergencies. With the spread of smartphones and social media, there is more of the concept that citizens are becoming first responders, where emergencies are being reported live and more frequently than old systems could even identify them.

2.4.1 Incident Detection and Quick Response

In the case of emergency reporting, such as fires, floods, road accidents, or criminal activities, social media or specific applications are usually sought out by the people for reporting. Such reports are usually accompanied by photos, videos, and location details as well as descriptions. This real-time information flow can be accessed by the emergency services to get an idea of the situation at hand and to act accordingly to improve it.

2.4.2 Situational responsiveness and coordination

The real-time citizen reports enhance the awareness of a situation in the case of crises. Emergency operation centers can form a clear image of the magnitude and dispersion of an event by examining the amount and contents of the social sensing inputs. For example, in the case of a flood, crowdsourced pictures would be able to demonstrate the level of water in different places, thus providing an idea of possible ways of evacuation and relief distribution planning. A shared, real-time information environment also helps agencies, such as firefighters, police, and medical teams, work together much more effectively. This will reduce repetition in the efforts and make resources be used effectively.

3. SOCIAL AND TECHNICAL OBSTACLES TO SOCIAL SENSING

The concept of social sensing or participatory sensing has become a powerful concept in contemporary digital and smart city environments. It allows people, usually through mobile phones or built-in sensors, to add data on the environment as well. These data may be related to traffic information, pollution rates, emergencies, and infrastructure issues. Although the potential merits of social sensing in areas such as transportation, health, managing energy, and public security are universally recognized, large-scale deployment of social sensing is still intensively challenged by technical and social issues. Such difficulties need to be overcome to release its full potential and ensure that the systems founded on social sensing are ethical, correct, safe, and long-lasting as shown in Figure 4 (Rashid et al., 2023; Wang et al., 2019).

3.1 Establishing Data Credibility and Viability

A major issue in social sensing is the credibility and validity of the information gathered by people. In contrast to industrial sensor-type data or institutional monitoring based on data, social sensing takes human users as its input. Such data are usually subjective and unstructured and may be based on perceptions or prejudices. In addition, it may be subjected to conscious falsification, misinformation, or duplication. An example is that a given individual may report a traffic jam that is not true at all due to frustration or may overstate a problem prevailing in the streets as a way of catching attention. It is also possible to have malicious users who exploit social sensing systems by influencing them for personal gain. The lack of a robust system to validate and authenticate such information may compromise the whole sensing system.

3.1.1 Techniques of Data Standardization and Validation

Data generated by users is in different forms, including text messages, photos, videos, geographical positioning data, hashtags, and voice commands. These formats lack uniformity, and hence, they are challenging to integrate into larger systems. To face this, there needs to be technical solutions on several levels:

- ***Preprocessing and Cleaning:*** Shortcuts Raw data require preprocessing to eliminate all irrelevant data, normalize language (e.g., eliminate slang, make corrections, and eliminate bad grammar), and detect duplicates.
- ***NLP:*** NLP can be applied to obtaining meaning that lies in free unstructured textual data and aims to convert the qualitative number of the input data into quantifiable information.
- ***Crowd Credibility Scoring:*** The credibility of each user can be evaluated by systems using a historical behavior-based scoring and reputation-based scoring. As an example, a user whose previous issues could be presented in line with the facts might be granted a higher credibility score.
- ***Redundancy Filtering***: The system can process how many users shared the same information about an event. The higher the degree of redundancy in independent sources, the more confident one can be of the data.
- ***Geospatial and Temporal Correlation:*** The geospatial and temporal patterns of the data entries can be cross-checked. For example, heavy rainfall in a report of a dry area with no previous weather reports is given a red flag.

3.1.2 AI-based anomaly detection

ML-based models are created to find patterns and discover outliers in big data. The irregular data may be identified to be manual review or automatically discarded in these models in case they are incorrect. Further solidification of data validation can be achieved using deep-learning methods, such as Convolutional Neural Networks (CNNs) on image data and Recurrent Neural Networks (RNNs) on time-series input, where data can be supplemented with context from external databases when fused. Nevertheless, they need labeled data and need to keep on being retrained to fit the new user patterns and city conditions; that imposes further levels of difficulty and computational expense.

Figure 4. Overall Challenges in Social Sensing

3.2 Ethics and the Right to Privacy

Basic Issues of Privacy Another critical issue of social sensing that needs to be addressed is the preservation of the privacy of individuals when the information about them is gathered, processed, and perhaps broadcast. Social sensing systems have the potential to collect very sensitive data, such as the positions of people, their health conditions, behavior, and even political views, either directly by receiving it in the form of inputs or indirectly by making inferences. Such data can be used inappropriately to spy, discriminate, or commercially exploit in the wrong hands. These dangers have led to discussions on the issue of digital privacy on the global scene, as well as on the development of regulatory standards, which require rigid restrictions of consent, data minimization, transparency, and user control.

3.2.1 User awareness and conversant consent

Informed consent acquisition is not a checkbox exercise but a process through which users get assurance that users:

- What information is being gathered?
- What will the usage be and how long?
- Who shall be in a position to access it?
- What are the potential risks and benefits?

Unfortunately, these details are buried under the long and technical terms of service used by many applications that many users have never read. This is an act of undermining consent and damages the user's trust. In theory, one possible solution is floated by many researchers; the concept of privacy by design is a principle in which systems are designed with the idea of securing user information and communicating effectively.

3.2.2 Anonymization and Differential Privacy

To protect identities and enable the utilization of data for the public benefit, social sensing systems implement anonymity techniques, procedures by which data may be deleted or obscured to reveal no Personally Identifiable Information (PII). But it has also been demonstrated that even anonymized data can be traced by sophisticated re-identification attacks when used in combination with auxiliary datasets. This has ushered in the use of differential privacy, which is a method of adding mathematical noise to the data to guarantee that individual records cannot be de-anonymised. Apple and Google already have data analytics frameworks in terms of differential privacy.

3.3 Consolidation with the Current Infrastructure

3.3.1 Incompatibility of legacy systems

Most of the public utility and urban administration systems were created decades ago, and little consideration was given to the citizen-generated information. These legacy systems commonly involve antiquated protocols, an immutable structure, and closed databases. They are inefficient in absorbing the social sensing-generated dynamic, unstructured, and high-volume data. As an example, a traffic system in a city may be configured to receive signals that are measured by loop detectors or traffic cameras, but not easily scaled to include coordinated tweets or citizen reporting via apps.

3.3.2 Interoperability and Standardization of Data Format

One significant technical challenge in the combination of both traditional infrastructure and modern sensing is interoperability, or compatibility, the ability to use various systems and tools to communicate with each other. This requires:

- **Standard APIs:** APIs that enable the new social sensing systems to push the data into existing systems without significant redesigns.

- **Common Data Formats:** JSON or XML, or other emerging standards, such as SensorML (Sensor Model Language), can be used to make other systems receive and parse information that is in data streams.
- **Middleware Solutions:** Different software is used as a translational overlay between the old systems and the new systems. These serve as translators, which can even out data structure, frequency of updates, or protection requirements.

Visualization of smart city platforms as central locations for all data sources (legacy sensors and/or mobile applications) is also on the rise. On these platforms, cloud computing and edge processing would help standardize inputs and provide insights to sanitation, traffic, and emergency response, among other departments.

4. MOTIVATING PEOPLE TO CONTRIBUTE TO THE COMMUNITY

Social sensing has become an important means of collecting real-time, location-specific, and varied information created by ordinary citizens in the highly dynamic digital world. This collaborative data collection framework allows individuals to become sensors through the exchange of insights, experiences, and observations through mobile phones and the web. To realize the transformative potential of social sensing in various sectors like smart cities, disaster management, healthcare, traffic systems, and environmental monitoring, there is a pressing need to secure strong public involvement. The goal is to uncover the multifaceted nature of the process of promoting meaningful civic participation in data contribution, specifically in terms of accessibility, incentive, education, trust, and community development (Xie et al., 2023; Yang et al., 2012; Zeng et al., 2023).

4.1 The role of the Public in Social Sensing

Social sensing is based on voluntary participation. Compared to more conventional approaches to sensing, where people rely on installed equipment or government infrastructure, social sensing leverages individual devices, actions, and reports to learn about phenomena on a real-time basis. During disasters, such as eyewitness accounts of blocked roads or injured victims, can be more timely and useful than satellite images. Similarly, traffic applications such as Waze use user information to identify congestion and road hazards effectively.

However, one of the main assumptions in this model is that the individuals should be ready to reveal information repeatedly and consistently. A lack of long-term engagement among people can reduce the quality, coverage, and reliability of information, which might result in partial or biased information.

4.2 Designing Accessible and User-Friendly Interfaces

One of the main steps towards an increased number of user' participations is the creation of easily understandable and accessible user interfaces. Applications that are complex or require some technical skills may cause the users irritation, especially among less tech-savvy groups such as the elderly or people with low educational levels.

4.2.1 Simplicity and Clarity

The main goal of apps and web services should be simplicity and clarity in the interaction. For example, the whole process of reporting a case should be so easy that one can complete it in no more than three steps: choosing a category (like flood, traffic jam), taking a picture, and submitting the form with one tap.

4.2.2 Multilingual Support

One of the greatest advantages of multilingual societies is the ability to include regional language support, thus making more people able to get involved and be part of a community that is more tolerant and diverse. When users realize that platforms are speaking their language both literally and figuratively, they are more likely to get involved.

4.2.3 Accessibility for Differently-Abled Users

The main accessibility features are voice command, screen reader, and haptic feedback, which can help persons with disabilities to take part in the civic life of the community, thus ensuring equity in the contribution of civic data.

4.3 Incentivizing Participation: Gamification and Recognition

In our technologically advanced times, the role of motivation through incentives becomes very important. Most of the platforms resort to gamification use of gaming principles in a non-gaming environment to enhance the engagement of their users.

4.3.1 Points and Leaderboards

By creating an environment where points, badges, or levels can be achieved, the platform may become more attractive to regular visitors. Leaderboards of "Top Contributors of the Week" can be a good way to arouse the spirits of competition, thus allowing both pride and motivation to appear in the ball of active users.

4.3.2 Tangible Rewards

Some online platforms may propose material incentives, e.g., discount coupons, free access to their premium services, or gift vouchers in exchange for loyalty of users, i.e., continuous participation. For instance, in urban community reporting apps, it is possible to facilitate the use of a metro card and grocery vouchers to create enthusiasm for an app by providing recharge points or vouchers, respectively.

4.3.3 Social Recognition

One of the ways to give users a feeling of pride and uniqueness is by exposing them in newsletters, social media, or at community meetups. Just an appearance on the list of the top users, even if only symbolic, generates psychological gratification, thus raising the user's loyalty to the platform.

4.4. Building Awareness and Educating the Public

Apart from the technical background, the social digital literacy is the main foundation of the whole public initiative. A large minority of people may not understand what social sensing is or how their inputs can result in tangible benefits in their community(Restuccia et al., 2016).

4.4.1 Public Information Campaigns

It is advised that governments, NGOs, and local institutions launch educational campaigns with the use of digital platforms, community radio, posters, and workshops. Explaining that citizen reports can influence policy, making or improving emergency response can help transform passive users into active contributors.

4.4.2 Integration with education

The introduction of sensing in school curricula will enable students to be acquainted with the subjects at an extremely tender age. They may even participate in local sense-making projects as part of science, civics, or computer education classes. Digital training Community-based digital training, especially in rural or marginalized communities, can build the confidence and skills needed to competently use sensing platforms through education.

4.5. Community-Based Engagement and Local Leadership

The sense of being part of a community and collective responsibility is what community-level engagement, beyond individual motivation, can raise.

4.5.1 Engaging Local Leaders and Volunteers

Community leaders, social workers, teachers, and student groups can take the role of emissaries of social sensing projects. Community members' participation is higher when programs are supported by the trusted local figures.

4.5.2 Participatory Events and Workshops

Communities will be revived by organizing hackathons, data walks, or "citizen science" days. Interactive sessions allow people to see the impact of their inputs directly.

4.5.3 Feedback Mechanisms

One of the common causes for disengagement is the feeling that contributions are going unnoticed. Actions like updates on how reports have resulted in specific actions (e.g., pothole fixed) can attract participation by establishing trust and providing motivation for continued engagement.

4.6 Case Studies and Real-world Applications

- **Safetipin (India):** Safetipin customers can file reports on street safety elements such as lighting, accessibility, and crowd density. They make the reporting process fun and provide dashboards for cities. This platform has been a game-changer in the city's urban safety policies, especially for women.
- **Waze (Global):** Waze, owned by Google, is an area covered by real-time user traffic data worldwide. The drivers give out the information about car crashes, speed traps, and traffic jams. Real-time benefits, better navigation, and badges are used to motivate.

- **Ushahidi (Kenya):** At first, it was created to visualize violence reports during Kenya's post-election crisis in 2008. Ushahidi has then made the transition to a global crisis mapping platform. It monitors elections, disasters, and health emergencies mainly with crowd-sourced data.
- **MyGOV (India):** A platform that gathers citizen feedback on governance and that actively integrates social sensing into policymaking, MyGOV. Quizzes, feedback forums, and contests are some of the means by which it engages citizen opinion.

5. TECHNOLOGICAL ADVANCES AND NEW TRENDS IN SOCIAL SENSING FOR PUBLIC UTILITIES

Social sensing refers to the merging of individual data that is created through a variety of digital tools. This has become a major factor in the handling and preparation of public utilities. The community-generated real-time information has revolutionized the manner in which services are planned and delivered in sectors like water, energy, transportation, waste management, and emergency response (Z. Xu et al., 2018).

The future of social sensing continues to be informed by several technology trends as this ecosystem continues to develop. These are the growing applications of AI and predictive analytics, the integration of blockchain to ensure the security of the information, the spread of standardization to enable the interoperability of the systems, and the tendency toward inclusive and accessible design. These aspects not only enable the social sensing to do more but also address those issues that have existed long enough in terms of social trust, usability, and policy compatibility (A. Xu et al., 2024).

5.1 Examining AI and Predictive Analytics

AI, with its capability to derive insights from complicated datasets, has, basically, extended the horizon for the social sensing technology to be implemented in the field of public utilities. The data generated by social sensing platforms, citizen reporting apps, GPS-based transit data, social media posts, and crowdfunded environmental alerts is vast and consists mostly of unstructured or semi-structured data. AI, particularly through ML and deep learning models, is quite competent to perform the tasks of data filtering, categorization, and interpretation, which eventually yields a link between these data and the discovery of new patterns (Tang et al., 2023).

5.1.1 Data Processing and Interpretation

NLP enables AI systems to get the meaning of user-generated text data, such as complaints, tweets, or posts about local service conditions. To illustrate, an NLP model can group complaints of "low water pressure" or "frequent power cuts" to identify which areas are experiencing these issues repeatedly. In the same manner, image recognition models comprehend the pictures of potholes, damaged infrastructures, or garbage bins that are overflowing, and thus, by doing it automatically, they confirm the problem reports.

5.1.2 Predictive Analytics and Future Planning

The scope of predictive analytics is extended to planning future activities by utility services, apart from just understanding. The use of energy in hot weather situations in the past, combined with the corresponding data, will be an example of historical data that can be used to estimate future energy demands. The grid load balancing and energy buying decisions can be guided in this way. The data on traffic congestion that comes from the use of navigation apps can be used by authorities in the transportation sector to locate areas that will be crowded and where accidents are more likely to happen in the future.

5.1.3 Decision Support for Public Officials

Utility managers get a real-time illustration of potential hazards and suggestions for actions that may be of help via the dashboard and alert systems that employ AI technology. For instance, the decision support mechanism can perform the function of rerouting the refuse trucks in the event where there is a blockage on the normal route that is to be taken. Or it can send a signal to the maintenance staff when the sensors indicate that a water pipe blockage is near. Public officials can, therefore, make use of AI not only for diagnosis but also as a support system to their strategic plans, which, in turn, help them to respond quickly and efficiently to community demands.

5.2 Using Blockchain to Safeguard Data

Without a doubt, the benefits of the analysis of social sensing data are huge; however, the trustworthiness of data sharing remains a basic concern. Citizens are often suspicious of the manner in which their information, especially geolocation or identification data, will be used, stored, and maybe even misused. Blockchain technology offers a strong alternative to those worries (Rathod et al., 2023; Sangeetha et al., 2023).

5.2.1 Immutable and Transparent Records

Technically, blockchain serves as a decentralized ledger that logs data records in an immutable and chronological order. No modification can be made to a record except if all the participants of the network agree on it. Such a degree of data honesty is very important, especially in public utilities where it is things like accountability and traceability are demanded.

5.2.2 Decentralization and Public Trust

In contrast to the centralized databases, blockchains are not authorized by a particular entity to control data; rather, they rely on the entire network collectively as a whole. Therefore, it is not only immune to any form of tampering, it also builds trust among users. The presence of a smart city is supported by the mutual trust among government departments, utility providers, and civic tech start-ups who have access to the same data pool, which is secure and free from the danger of monopoly or data manipulation.

5.2.3 Smart Contracts and Automated Enforcement

Moreover, another blockchain breakthrough for utilities is smart contracts, which refer to the autonomous agreements whose execution is done by the blockchain ledger itself. For instance, it could be just a smart contract that automatically releases a refund or triggers a maintenance request if a power outage lasts more than a specific duration. Such contracts lessen the workload of the administration and increase the accountability of the service providers. In addition, identity management being supported by blockchain is also one of the ways. This method allows people to share their information, which is still personalized; however, it is anonymous.

5.3 Standardization and the Ability to Communicate

There is a technological challenge amidst all the promises of tech, and that is the problem with the fragmented nature of platforms, devices, and datasets, which in turn results in silos in public utilities. The differences between cities or departments also include the use of incompatible formats, which is a great obstacle in collaborative endeavors. In this way, standardization is the base for social sensing ecosystem scaling (Daousis et al., 2024).

5.3.1 Data Format and API Uniformity

For the public to be able to use the data obtained from social sensing in an efficient manner, the data should be in a commonly agreed schema and format. The standards as JSON, LD for semantic data representation, or GTFS (General Transit Feed Specification) for public transport data give the ability to a system to comprehend and process other systems' data accurately. The definition of the Application Programming Interface (API) should be done in a way that will make it possible for different services, such as a weather application, municipal service portals, or citizen reporting tools, to have an easy and straightforward interaction.

5.3.2 Cross, Jurisdictional Comparisons and Collaboration

Through standardization, cities will have access to data that will help them evaluate policies, their impact on them, and find the best approaches. Environment planners from Mumbai may learn from the achievement of leakage water control in Hyderabad if both the performance indicators and the reporting structures of these two services are similar. In addition, this possibility of joint action also exists in emergencies. Standardized APIs, for example, during a natural disaster, allow emergency services, hospitals, and civic authorities that are in different districts or even different states to coordinate better. This close work between the agencies is very important to the efficient and timely response.

5.3.3 Policy and Regulatory Alignment

The development of national and international regulatory frameworks is the first step towards bringing permanent standardization at the policy level. They can assign the duty of developing the guidelines to the Bureau of Indian Standards (BIS), the NITI Aayog, ISO, or any other organization of this category. The engagement of the industry, academic researchers, and civil society would result in standard setting being not only technically sound but inclusive socially as well.

5.4 Making Designs Available to All

The contributions of social sensing may be maximized only when all population groups can equally share the benefits and take part. This implies the need for designing digital services that are accessible, inclusive, and adaptable in terms of language and culture.

5.4.1 Bridging the Digital Divide

The lack of internet access and digital literacy is a major problem in many developing areas. To include such people in social sensing, the designed solutions should be suitable for low-end devices, offline or low-bandwidth usage, and the user's need for technical literacy should not be assumed while offering user interfaces. SMS, based reporting tools and IVR (Interactive Voice Response) systems can be as powerful as mobile apps, yet much more accessible when they are customized according to local constraints. Government campaigns for the promotion of digital inclusion, e.g., Digital India, can be used for expanding training, access, and participation subsidies in civic tech initiatives.

5.4.2 Multilingual and Inclusive Interfaces

Social sensing tools must be designed to take into account the linguistic diversity of their users. There are officially 22 languages and hundreds of dialects in India alone. According to this point, developers will have to incorporate multilingual support and local idioms in applications so that more people will have a chance to participate, and data quality will be improved as well. The addition of features for people with disabilities, such as screen readers, haptic feedback, or voice commands, also contributes to the expansion of social inclusiveness. Thus, it not only serves the social aspect but also makes social diversity evident in the collected data.

5.4.3 Cultural Sensitivity and Trust

Mobile application development should take into consideration the cultural context and local power structures. For instance, women belonging to conservative areas and being shy may not report problems unless they are given secure, anonymous channels. Young people might prefer chatbots or gamified reporting mechanisms, whereas elderly users may be devoted to neighborhood kiosks or community mediators. Communities being involved in solution and co-creation, through participatory design workshops, pilot programs, and iterative feedback loops, make the system familiar to the user base. It has made it possible for the public to enjoy utilities at transformative levels thanks to innovation, based solutions. The technological advances that are leading to this futuristic scenario are nothing but AI's interpretive power, blockchains, data integrity, standardization, interoperability, and inclusive design. Together, these trends redefine the entire process of data capturing, trusting, sharing, and using for decision-making. The integration of these technologies brings to the table not just intelligent cities but also more just and accountable governance.

6. CONCLUSION

Summing up, social sensing is redefining how it governs better and functions in the smart zones. Since cities continue to grow and create uncertainty in the provision of the masses, it is not only logical but also a prudent move to base their focus on the collective voice of the citizens. Cities can be made to be more effective in their response to infrastructure-related, transportation-related, energy- and water-related, waste management, and public safety challenges by using real-time and human-generated data that can be accessed via Smartphones, mobile applications, and social media. This fusion changes the dynamics between city planners and the populace, thereby making citizens not only recipients of the services but also proactive participants in the ecosystem.

The triumph of social sensing, still, is dependent on several essential factors. The most important thing is trust. Since the data being contributed is non-standardized and prone to errors or even deliberate misuse, strong trust models are paramount. The models assess the trustworthiness of contributors via factors like behavioral history, the accuracy of the location, cross-validation, and social verification. Also, the efforts to ensure the privacy of the data and the ethical use of data are on equal footing. This involves compliance with international data protection laws such as GDPR and the development of privacy-friendly and clear privacy policies. Many urban systems continue to be very dependent on old technologies with minimal capacity to support high amounts of unstructured real-time human information. The issue is resolvable through standardization, APIs, shared data formats, and shared reporting protocols, and may also enable the smooth integration of systems. Moreover, the ability to compare, partner in the management, and deliver the same level of service to cities and their tools is ensured by the fact that they can communicate with one another.

Another fact that is equally important is the social aspect: active citizen participation. Social sensing is highly unlikely to harness its full potential if engagement is not widespread. To keep participation alive, the platforms should not be difficult and must be accessible to people of all walks of life and with varying degrees of technological skills. Using techniques such as gamification can help participation grow. One of these techniques is community leaderboards. People are motivated by competition and can increase participation just by this way. Besides, multilingualism, which is one part of inclusiveness, supports people to take part regardless of the language they speak, while disabilities, which is another part, helps the presence of physically challenged individuals in the dataset. Lastly, features can suggest the socioeconomically disadvantaged rather than the rich, which means the dataset reflects the needs of the worthy group.

Future advancements in smart cities have the potential to elevate social sensing to unprecedented levels. For example, AI and predictive analytics are highly effective resources that can be very helpful in predicting trends, locating oddities, and planning the use of a service before any inconveniences turn into crises. By doing this, it will not only prolong the sustainability of city systems but also, at the same time, make them change with the times, as well as be self-sustaining.

Also, data shared from the use of blockchain technology is more traceable and trustworthy, thereby making citizens assured that their inputs are secure and cannot be tampered with. This additional security guarantees trust, which is one of the most important aspects of public interaction. Besides, uniform communication models between different sectors and their partnerships are also essential aspects. In order to develop solutions that are scalable and interoperable, governments, enterprises, the civic community, and research institutions need to put their heads together.

In the end, smart zone development largely depends on tripartite cooperation among technology, governance, and the people. Technology breakthroughs have limited impact without accompanying governance held accountable and responsible, and without citizens who are active and participative. Social sensing, in essence, is moving away from command and control planning to a participatory, data-driven governance model that acknowledges input from each resident. This is a step closer to an urban future that will be more transparent, fair, and environmentally friendly. Since the digital and physical worlds are gradually integrating, smart zones are the ones that will be using the social sensing technology to have better control of urban problems, be they routine maintenance issues or emergencies caused by natural disasters. These areas will receive the advantages of quickened feedback loops, better decision-making, and more resourceful and effective utilisation. Smart areas can set new standards for intelligent, citizen-focused urban living that includes mutual trust building, equitable access, and adoption of new ideas and solutions.

Social sensing is not simply one more technological innovation but rather a philosophical change that favors' democratized urban governance. The metropolis of the future will not be solely characterised by sensors and software, but rather one whose population is people whose voices, through technology, are policy makers and the directors of their surroundings. Through ethical and technological frameworks, the citizens gain not only the intellectual tools but also the moral framework; thus, they become the best sensors of the urban setting. The insights of the citizens, which are collected responsibly, can transform cities to become intelligent, clean, safe, and humane ones.

REFERENCES

Alvarez, H. I. G., Peña, H. H. Z., del Consuelo Cortes Velázquez, M., & Ypiña, C. P. F. (2023). Multimedia Technology and Smart Cities; an Inevitable Combination of Urban Progress. *Communications in Computer and Information Science, 1906 CCIS*, 86–97. DOI: 10.1007/978-3-031-45316-8_8

Castrellon, J. P., Sanchez-Diaz, I., & Gil, J. (2024). Smart loading zones. A data analytics approach for loading zones network design. *Transportation Research Interdisciplinary Perspectives*, *24*, 101034. Advance online publication. DOI: 10.1016/j.trip.2024.101034

Daousis, S., Peladarinos, N., Cheimaras, V., Papageorgas, P., Piromalis, D. D., & Munteanu, R. A. (2024). Overview of Protocols and Standards for Wireless Sensor Networks in Critical Infrastructures. In *Future Internet* (Vol. 16, Issue 1). DOI: 10.3390/fi16010033

Debnath, B., Roychoudhuri, R., & Ghosh, S. K. (2016). E-Waste Management – A Potential Route to Green Computing. *Procedia Environmental Sciences*, *35*, 669–675. Advance online publication. DOI: 10.1016/j.proenv.2016.07.063

Gubbi, J., Buyya, R., Marusic, S., & Palaniswami, M. (2013). Internet of Things (IoT): A vision, architectural elements, and future directions. *Future Generation Computer Systems*, *29*(7), 1645–1660. Advance online publication. DOI: 10.1016/j.future.2013.01.010

Jan, F., Min-Allah, N., & Düştegör, D. (2021). Iot based smart water quality monitoring: Recent techniques, trends and challenges for domestic applications. In *Water (Switzerland)* (Vol. 13, Issue 13). DOI: 10.3390/w13131729

Okoli, N. J., & Kabaso, B. (2024). Building a Smart Water City: IoT Smart Water Technologies, Applications, and Future Directions. In *Water (Switzerland)* (Vol. 16, Issue 4). DOI: 10.3390/w16040557

Ouallane, A. A., Bahnasse, A., Bakali, A., & Talea, M. (2021). Overview of Road Traffic Management Solutions based on IoT and AI. *Procedia Computer Science*, *198*, 518–523. DOI: 10.1016/j.procs.2021.12.279

Rashid, M. T., Wei, N., & Wang, D. (2023). A survey on social-physical sensing: An emerging sensing paradigm that explores the collective intelligence of humans and machines. *Collective Intelligence*, *2*(2), 263391372311708. DOI: 10.1177/26339137231170825

Rathod, T., Jadav, N. K., Tanwar, S., Polkowski, Z., Yamsani, N., Sharma, R., Alqahtani, F., & Gafar, A. (2023). AI and Blockchain-Based Secure Data Dissemination Architecture for IoT-Enabled Critical Infrastructure. *Sensors (Basel)*, *23*(21), 8928. Advance online publication. DOI: 10.3390/s23218928 PMID: 37960626

Restuccia, F., Das, S. K., & Payton, J. (2016). Incentive mechanisms for participatory sensing: Survey and research challenges. *ACM Transactions on Sensor Networks*, *12*(2), 1–40. Advance online publication. DOI: 10.1145/2888398

Sangeetha, S., Kumari, K. A., Shrinika, M., Sujaybharath, P., Varsini, S. M., & Kumar, K. A. (2023). Ensuring Location Privacy in Crowdsensing System Using Blockchain. *Lecture Notes in Electrical Engineering, 995 LNEE*, 35–47. DOI: 10.1007/978-981-19-9748-8_4

Schmidtke, N., Behrendt, F., Gerpott, F. T., & Wagner, M. (2022). Integration of New Business Models in Smart Logistics Zones. *International Journal of Supply and Operations Management*, *9*(1), 15–33. DOI: 10.22034/ijsom.2021.109063.2160

Taher, T., Ahmed, R. U., Haider, M. A., Das, S., Yasmin, M. N., & Mamun, N. (2018). Accident prevention smart zone sensing system. *5th IEEE Region 10 Humanitarian Technology Conference 2017, R10-HTC 2017, 2018-Janua*, 67–70. DOI: 10.1109/R10-HTC.2017.8288908

Tang, C., Qiu, C., & Xiao, X. (2023). Machine-learning and water energy harvesting based wireless water consumption sensing system in buildings. *E-Prime - Advances in Electrical Engineering. Electronics and Energy*, *5*, 100260. Advance online publication. DOI: 10.1016/j.prime.2023.100260

Ullo, S. L., & Sinha, G. R. (2020). Advances in smart environment monitoring systems using iot and sensors. In *Sensors (Switzerland)* (Vol. 20, Issue 11). DOI: 10.3390/s20113113

Wang, D., Abdelzaher, T., & Kaplan, L. (2015). Social Sensing: Building Reliable Systems on Unreliable Data. In *Social Sensing: Building Reliable Systems on Unreliable Data*. DOI: 10.1016/C2013-0-18808-3

Wang, D., Szymanski, B. K., Abdelzaher, T., Ji, H., & Kaplan, L. (2019). The age of social sensing. *Computer*, *52*(1), 36–45. DOI: 10.1109/MC.2018.2890173

Wijayaningsih, N., Arifah, M. N., & Izzati, M. N. (2020). Smart zone: Application improving zero waste lifestyle based on community development. *Jurnal Ilmiah Penalaran Dan Penelitian Mahasiswa*, *4*(2), 102–117.

Xie, L., Su, Z., Chen, N., & Li, R. (2023). Differential Privacy-Based Incentive Scheme for App-Assisted Mobile Edge Crowdsensing. *Proceedings - IEEE Global Communications Conference, GLOBECOM*, 5518–5523. DOI: 10.1109/GLOBECOM54140.2023.10437797

Xu, A., Biehl, J., & Lee, A. (2024). Getting it Just Right: Towards Balanced Utility, Privacy, and Equity in Shared Space Sensing. *ACM Transactions on the Internet of Things*, *5*(2), 1–26. Advance online publication. DOI: 10.1145/3648479

Xu, Z., Mei, L., Choo, K. K. R., Lv, Z., Hu, C., Luo, X., & Liu, Y. (2018). Mobile crowd sensing of human-like intelligence using social sensors: A survey. *Neurocomputing*, *279*, 3–10. DOI: 10.1016/j.neucom.2017.01.127

Yang, D., Xue, G., Fang, X., & Tang, J. (2012). Crowdsourcing to smartphones: Incentive mechanism design for mobile phone sensing. *Proceedings of the Annual International Conference on Mobile Computing and Networking, MOBICOM*, 173–184. DOI: 10.1145/2348543.2348567

Zeng, F., Pang, C., & Tang, H. (2023). Sensors on the Internet of Things Systems for Urban Disaster Management: A Systematic Literature Review. In *Sensors* (Vol. 23, Issue 17). DOI: 10.3390/s23177475

Chapter 11
Social Sensing for Effective Public Utilities in Smart Zones

Binastya Anggara Sekti
http://orcid.org/0000-0001-5489-4888
Universitas Esa Unggul, Indonesia

ABSTRACT

The integration of social sensing into smart zone infrastructures presents a transformative approach to public utility management, where citizens actively contribute to the monitoring and enhancement of urban services. By leveraging both participatory data—such as mobile app reports and social media posts—and opportunistic data from IoT devices and smartphones, cities can detect issues related to water, electricity, sanitation, and waste in near real-time. This chapter explores the architecture, technologies, and case studies that demonstrate how social sensing enhances situational awareness, reduces response times, and fosters citizen engagement. We also address the challenges related to data privacy, inclusivity, and digital equity. The chapter concludes with a forward-looking perspective on how combining AI, federated learning, and decentralized systems such as blockchain can ensure resilient, ethical, and adaptive smart zone ecosystems for the future.

1. INTRODUCTION

Urbanization is a phenomenon experienced in all major cities in the world, where residents from surrounding areas come to seek good education or get more decent jobs and others that are not available in their home areas. The impact of urbanization is on increasing the need for water supply, electricity, as well as a very rapid and

DOI: 10.4018/979-8-3373-4962-6.ch011

significant increase in waste. This very rapid urbanization growth is in line with UN data released in 2024, which by 2050, around 68% of the world's pollution will live in urban areas. While there are still other factors that can hinder population growth, such as declining fertility rates in the most populous cities, there are still other factors that can hinder population growth. (UN-DESA, 2024).

The rapid increase in population encourages the creation of a smart city concept to overcome social and economic impacts that can cause negative problems for its citizens. Various urban city problems that appear on the surface are traffic congestion, energy crisis, food crisis, flood disasters and others that are increasingly difficult to handle. The rapid development of information technology and the expansion of internet networks have encouraged the establishment of smart zones in smart cities. Smart zones represent certain areas in the city that have digital infrastructure, good internet connections, IoT devices, and AI systems to provide better public services than ever before. In this smart zone environment, it is very possible to detect and deploy in real-time about the use of various utilities of public services, anomaly detection, prediction of demand level growth so that it can be used by city managers to make accurate decisions based on data. The increasing number of Internet of Drones (IoD) has been used by the military and civilians to keep their citizens safe in real-time (Ali, Chaudhry, Ramzan, & Al-Turjman, 2020).

Social sensing - The utilization of data from individuals and the devices they use has evolved into a new paradigm in the improvement of public utility services in smart zones. This approach enables real-time data collection and analysis from a variety of sources, such as mobile phones, social media, Internet of Things (IoT) devices and other digital applications. The insights generated from this data are critical to understanding utility consumption patterns, identifying inefficiencies, and enabling predictive and proactive service delivery. The identification process is done by tracing behavioral patterns from data acquired through mobile devices, and analyzing the correlation between micro-scale social functions and mobile-based sensing features. One of the study results showed that this approach can be used to predict social functioning of outpatients in the context of healthcare services (Wang et al., 2020). Figure 1 shows the role of social sensing to improve the operation of public utility services in smart zones.

Figure 1. Social sensing network monitoring to improve the operation of public utility services in smart zones.

Social sensing technology can improve the quality of public utility management in smart zones by utilizing technology to monitor real-world activities, with a good social sensing architecture, all public utilities can continue to innovate in shaping urban services in a sustainable manner.

2. UNDERSTANDING SOCIAL SENSING IN SMART ZONES

Smart zones are micro-environments in cities that digitally integrate various technologies such as CCTV, smartphones, social media, and various other IoT devices, to AI-based systems that have been integrated through internet networks that aim to improve the quality and convenience of public services. At the heart of these smart zones lies social sensing, a dynamic method that collects and processes explicit and implicit human-generated data for use in interpreting and predicting to improve urban living conditions. Social activities that occur in urban areas are dynamic and intense human activities. Observations of these social activities can be taken from street views using crowdsourced smart devices to identify critical needs for further development (Zhang et al., 2025).

Social sensing differs from traditional environmental sensing in its ability to quickly capture the subjective experiences, contextual insights, and collective behavioral trends of citizens living in cities. With the utilization of technology, such information can be used to enhance the value of a more human-centered urban infrastructure, where the system will be adaptable to people's actual needs, preferences, and anomalies. Social sensing dengan analisis spasial aktivitas manusia dalam satu wilayah perkotaan dapat dilakukan digunakan sebagai strategi agar terjadi integrasi spasial atar kota sekitarnya (He, Zhai, & Zhao, 2025).

Cities are complex environments inhabited by diverse communities with varying needs, habits, and expectations. Urban systems must continuously develop innovative solutions using technology to improve services in complex situations. In smart city programs with high population density and severe air pollution due to high energy consumption, innovations are implemented through a multi-level system starting from local communities. If not addressed, urban emissions will become extremely problematic. Some measures that can be taken include promoting the development of self-produced energy generation systems (solar power plants, biogas, etc.) within their communities and implementing car-free zones. (Sofeska & Sofeski, 2022).

2.1 Conceptualizing Social Sensing

Social sensing can be broadly categorized into two types:

- Participatory sensing: Citizens voluntarily share data through mobile applications, surveys, or dedicated platforms (e.g., reporting water leaks or garbage pileups).
- Opportunistic sensing: Data is passively collected from sensors embedded in smartphones, wearables, vehicles, or household IoT devices without requiring direct user action.

This dual-mode data collection enables continuous monitoring of urban systems from multiple perspectives—technical, social, behavioral, and environmental.

Social sensing is a very fast approach to collecting and analyzing data from the community to understand the dynamics of a city in real time. This approach makes the citizens of a city not only recipients of services; but also active and passive sensors involved in observing the real conditions of the environment, infrastructure, and public services. By leveraging the various digital devices that residents use daily (smartphones, vehicles, stationary devices, IoT-based household appliances, etc.), social sensing can become a comprehensive, detailed, and dynamic collective observation ecosystemSmartphone applications such as voice calls, text messages, and other applications can be used to collect data on a person's social habits. In this way, assessments of real-world social interactions can be carried out using a pre-prepared and planned daily recording method. (Haucke, Heinzel, & Liu, 2024).

Social sensing can be measured using two approaches: participatory sensing and opportunistic sensing. Participatory sensing relies on the active participation of citizens, who consciously and voluntarily report certain conditions through mobile applications, online surveys, or public reporting platforms. For example, someone can report flooding, damaged roads, or piles of trash using the reporting feature in a city services app. This model can foster a sense of community ownership of the environment and encourage greater involvement in improving public services (Schulte et al., 2025). Meanwhile, opportunistic sensing collects passive data from devices used daily by the public without requiring direct intervention. For example, GPS sensors in mobile phones can be used to monitor residents' movement patterns, accelerometers in wearables can be used to detect physical activity, or temperature and humidity sensors in smart home devices can help map micro environmental conditions. This model offers high scalability and efficiency in data acquisition, but also poses major challenges related to privacy and data regulation (Hara & Abe, 2025).

The main strength of social sensing lies in its ability to combine data from various perspectives—technical, social, behavioral, and environmental—into one comprehensive framework. When combined with technologies such as AI and big data analytics, social sensing can transform raw data into valuable insights for urban decision-making. For example, analyzing patterns in citizen reports and water pressure sensors can be used together to detect hidden water leaks, while combining social and environmental data can help anticipate potential sanitation crises in densely populated areas (Li, Zhang, & Lu, 2015). Thus, the concept of social sensing not only presents technological innovations, but also redefines the relationship between government and society in building a more responsive and humane city. This is a step towards future governance that is data-driven, collaborative, and inclusive.

2.2 Sources of Social Sensing Data

The source of social sensing data comes from every city resident, for example from the smartphone they own. The data can be obtained from smartphones so that the physical activity of residents can also be known. After a trial, a good approach is with effectiveness and usability (Li et al., 2015). Table 1 is an explanation of a social sensing data.

Table 1. Social Sensing Data

Source	Description	Example Use Case
Mobile Devices	GPS, camera, accelerometer data	Detect foot traffic in public spaces
Social Media	Geotagged posts, sentiment analysis	Identify citizen dissatisfaction with public transport
Smart Meters & Appliances	Real-time consumption data	Monitor electricity/water usage trends
Wearable Devices	Biometric and location data	Monitor public health metrics (e.g., during a pandemic)
Crowdsourcing Platforms	User-submitted reports or ratings	Highlight sanitation or waste collection delays

In the era of smart cities, the flow of information is no longer one-way from the government to citizens, but rather an interactive ecosystem that involves active community participation. Social sensing changes the way we understand and manage public services—by making citizens not only recipients of services, but also part of the city's sensing system itself. To understand how this works, we need to look at the functional flow that forms the framework for social sensing in smart zones.

The social sensing process in smart zones involves:

- Sensing and Data Collection: Smart devices and citizen contributions generate raw data from daily activities, both directly and indirectly.
- Data Integration and Aggregation: Data from various sources is then consolidated through edge or cloud platforms to form a complete and contextual picture.
- Intelligent Processing: Artificial intelligence (AI) and machine learning (ML) algorithms are used to recognize patterns, detect anomalies, and predict trends based on existing data.
- Service Optimization: The insights generated are used by city managers to adjust and improve utility services in near real-time, such as water distribution, electricity, and waste management.
- Citizen Feedback and Iteration: Every improvement and change is communicated back to the community, while new data collected continues to refine the system, creating a continuous learning process.

Social sensing has the potential to act as a "sensory nervous system throughout the city" that supports real-time collective intelligence-based decision-making.

2.4 Advantages in Smart Zone

Smart zone is a term for the success of development in an area. The success of the establishment of a smart zone is influenced by transformation planning, adequate physical infrastructure, availability of technological infrastructure, high human resource competence, and partnership cooperation and commitment to regional development. Where the availability of technological infrastructure is the dominant determining factor (Hardjomuljadi & Anondho, 2025). Smart zones can start from a specific domain, such as the urban traffic domain. In Sweden with a project called "Smart Urban Traffic Zones" which started from September 2020 to September 2022, it has been successfully implemented in 2 major cities in improving public spaces, pedestrian comfort and urban transportation efficiency (Salmi & Hökars, 2023).

Amidst the increasingly complex dynamics of urban life, a city's ability to listen and respond quickly to the needs of its citizens is key. Social sensing technology opens a new chapter for smart cities—by making citizens not just users of services, but active partners in monitoring and improving their environment. By collecting data from everyday devices like smartphones and social media, cities can gain a more dynamic and contextual understanding of real-world conditions on the ground. The result is public services that are not only more efficient but also more humane—because they are designed based on the realities and voices of the residents themselves.

Some of the key advantages of this approach include:

- Contextual Awareness – The system can capture when and where an issue occurs, enabling solutions that are more targeted and aligned with local needs.
- Scalability – By leveraging devices already owned by the community, such as smartphones and digital platforms, cities can reach vast areas without the need for expensive new infrastructure.
- Timeliness – Social sensing enables early detection of disruptions or anomalies, allowing for faster responses compared to traditional systems.
- Cost Efficiency – Since it does not require the installation of large-scale sensors, this approach is much more cost-effective. Cities can function more intelligently without straining budgets, while still improving service quality.

Figure 2 shows domain of integrated public utility services and controlling.

Figure 2. Domain of Integrated Public Utility Services and Controlling

3. APPLICATIONS IN PUBLIC UTILITY MANAGEMENT

The development of smart cities is a rapidly growing trend in today's urban environment. A smart city is a city that uses advanced technology, data analytics, and the Internet of Things (IoT) to optimize operational efficiency and city services. The goal of a smart city is to improve the quality of life for residents, attract businesses and investors, and create a sustainable and resilient environment. Smart cities typically have various interconnected systems and devices that can be managed remotely, including transportation systems, energy systems, public safety, healthcare, and waste management. They also rely on data and analytics to make informed decisions and enhance overall city performance. (Tekinerdogan, Köksal, & Çelik, 2023). Figure 3 shows IoT based Smart City.

Figure 3. IoT Based Smart City

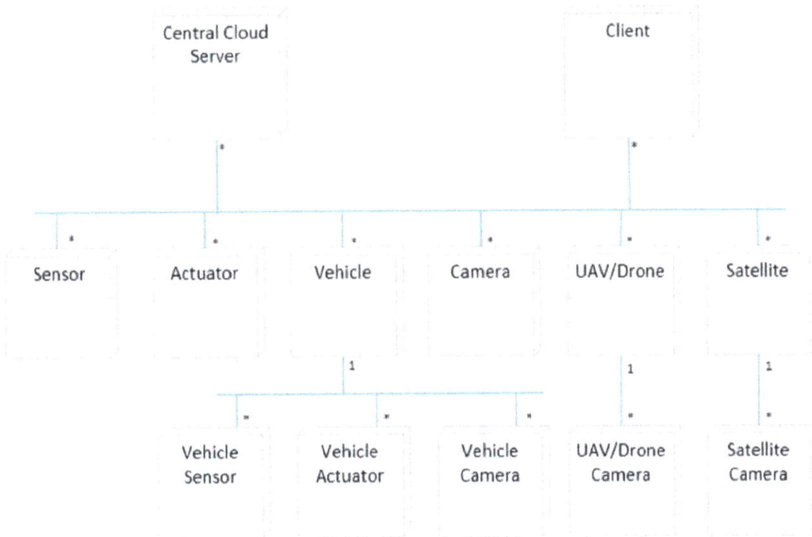

Collaboration between social sensing, AI, and IoT technologies is changing the face of public service management such as water, electricity, waste, and sanitation in smart cities. These technologies not only work behind the scenes, but also deliver systems that can respond to the needs of the community quickly and accurately. When sensors and smart devices are connected with artificial intelligence, public services become more responsive, adaptive, and able to predict problems before they occur-for example, detecting potential water leaks or power interruptions. Security in smart cities is also an important part because it is vulnerable to threats from cybercriminals. Sustainability needs to be applied to smart transportation systems, smart health, smart environment, and smart living (Toh, 2020).

More than just technical efficiency, this combination opens up space for active citizen participation in city life. Through a simple smartphone or app, people can report problems, share data, or simply provide direct feedback. This marks a major shift from a slow, bureaucratic system to an open, collaborative and humanized service ecosystem-where citizens are no longer just served, but also shape the services they use.

3.1 Water Supply and Management

Situasi pada kota urban dimana jumlah kendaraan semakin banyak, jumlah sampah semakin banyak, kebutuhan air semakin besar, perluasan dan kompleksitas infrastruktur penerangan umum, dan lain-lain. Untuk menjaga dan memastikan pembangunan ang berkelanjutan, maka hal-hal tersebut perlu ditangani, termasuk kebutuhan vital, seperti air (Sánchez, Elicegui, Cuesta, Muñoz, & Lanza, 2013).

Effective water management requires both granular data and community feedback. Social sensing contributes to this by integrating:

- Smart water meters that record usage patterns and detect anomalies (e.g., unexpected surges).
- Mobile app reports from citizens on water quality issues or supply outages.
- Geotagged images or social media posts indicating flood risks or leaks.

For example, in Tokyo, a combination of AI-driven analytics and social sensing enabled early leak detection, reducing non-revenue water losses by 17%. Generative AI has attracted everyone's interest in petrifying their work as it has a wide impact on various aspects of society (Miyazaki, Murayama, Uchiba, An, & Kwak, 2024). Figure 4 shows the smart water solution.

Figure 4. Smart Water Solution

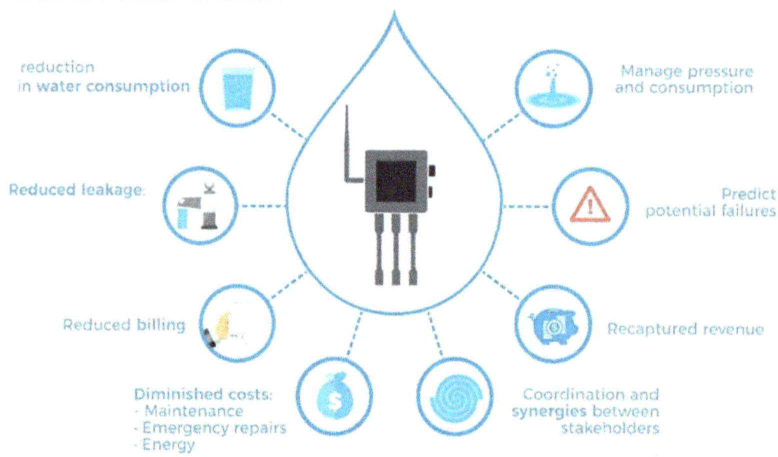

AI Integration: Machine learning algorithms analyze historical and real-time data to detect irregular consumption, optimize pressure zones, and predict seasonal demand changes.

IoT Role: Smart valves and connected sensors allow automated shutdowns in leak zones, improving response time and reducing water loss.

3.2 Electricity Distribution and Management

Electricity grids in smart zones benefit immensely from both user feedback and IoT data. Social sensing helps in:

- Crowdsourcing blackout reports to identify and localize power outages faster than utility sensors alone.
- Analyzing social sentiment to assess satisfaction and demand patterns.
- Detecting energy theft through anomalous patterns reported by smart meters and citizens.

An AI-enhanced energy management system in Amsterdam combined meter data and citizen feedback to cut peak loads by 22% during summer months (GreenTech Journal, 2022).

AI Integration: Reinforcement learning models predict peak loads and recommend load-shedding strategies or energy redistribution.

IoT Role: Smart home devices communicate directly with utility operators for demand-response actions.

3.3 Waste Management and Recycling

Waste management in urban cities is a problem that should not be taken lightly because waste growth is very fast. The community's commitment to cope with typhoid in urban areas is very necessary because it affects the quality of the environment. Population growth and urbanization have led to an increase in the amount of solid waste that is difficult to recycle in the traditional way. However, with the support of AI technology, recycling can be done more quickly by fostering cross-sector partnerships (Dawar, Srivastava, Singal, Dhyani, & Rastogi, 2025).

Inefficient waste collection leads to environmental hazards and community dissatisfaction. Social sensing addresses this through:

- Crowdsourced photos and location data on overflowing bins or illegal dumping.
- Smart bins that report fill levels and trigger route optimization.
- Citizen-reported satisfaction ratings on collection timeliness and cleanliness.

In Seoul, AI-powered waste analytics combined with mobile citizen feedback improved collection route efficiency by 35% and reduced missed pickups by half. The waste management crisis in South Korea can cause inconvenience and considerable environmental impact, this requires policies and public concern during overcoming the solid waste crisis (Ko, Kim, Shin, & Shin, 2020). Figure 5 shows the smart waste management. In the image, first, with IoT, all waste infrastructure is connected to the smuggling control center. Second, sensors on garbage trucks can facilitate the garbage hauler fleet. Third, sensors in each garbage collection tub. Fourth, the policy center can work with the cement processing industry to be recyclable.

Figure 5. Smart Waste Management

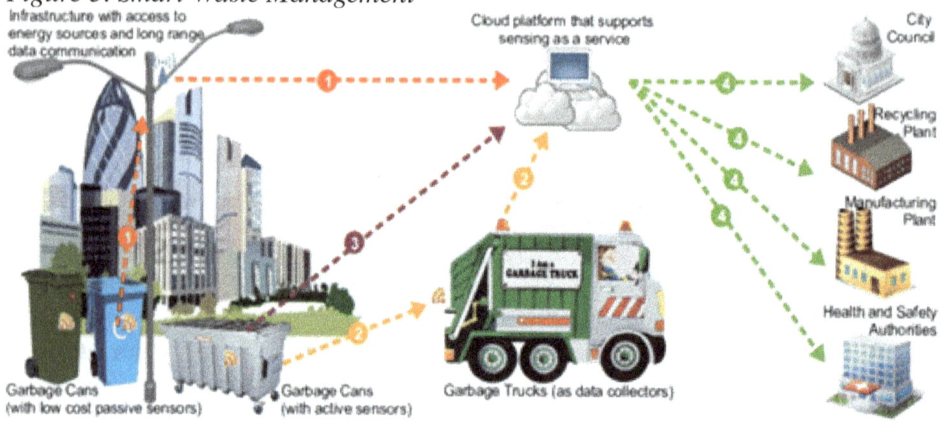

AI Integration: Clustering algorithms group high-waste zones for dynamic collection scheduling.

IoT Role: RFID-tagged bins track waste types and collection frequencies to support smart recycling incentives.

3.4 Sanitation and Public Hygiene

Public sanitation is an important aspect of maintaining public health in every city. By utilizing the power of social sensing, smart cities can now significantly improve the cleanliness of public facilities. Residents can upload photos of dirty areas in real time, rate the cleanliness of public toilets, and submit complaints via social media. This information is then analyzed by an AI system to identify previously unreported cleanliness issues. For example, the CleanLoo app in Mumbai allows residents to report the condition of public restrooms directly. Data from residents is then integrated with the city government's AI system, reducing the time required to address cleanliness complaints by up to 40%. A comprehensive approach involving stakeholders, technology, data, and alignment with sustainable goals is crucial for water management in India. Water scarcity could occur, and India is preparing to address the potential scarcity in 12 river basins by 2050 (Kumar & Goyal, 2025).

The integration of artificial intelligence in such systems can utilize Natural Language Processing (NLP) to analyze citizen input and complaints, as well as image classification to detect hygiene violations from submitted photos. Additionally, the role of the Internet of Things (IoT) is also highly crucial. Public toilets equipped with sensors can monitor the number of users and automatically send notifications to cleaning staff when they become dirty. With this approach, cities can maintain and ensure a healthier and more comfortable public environment for everyone.

3.5 Comparative Table of Social Sensing in Public Utilities

Some factors of public utilities can be a measure of social needs, such as water, electricity, garbage, sanitation, and others. Table 2 shows a comparative table of social sensing in public utilities.

Table 2. Comparative Table of Social Sensing

Utility Type	Social Sensing Input	AI Functionality	IoT Contribution	Key Outcome
Water	Leak reports, smart meter data	Anomaly detection, demand forecasting	Smart valves, pressure sensors	Reduced leakage, optimized supply
Electricity	Blackout alerts, usage feedback	Load prediction, fault detection	Smart meters, smart plugs	Stable grid, reduced outages
Waste	Geotagged bin photos	Route optimization	Smart bins, RFID tracking	Efficient collection, cleaner environment
Sanitation	Hygiene complaints, photo uploads	Sentiment/image analysis	Restroom sensors, footfall counters	Improved cleanliness, faster response

4. SYSTEM ARCHITECTURE FOR SOCIAL SENSING IN PUBLIC UTILITIES

An effective social sensing system in smart zones depends on a layered architecture that supports scalable data collection, intelligent processing, actionable analytics, and real-time citizen interaction. The architecture must integrate diverse data sources, ensure interoperability across platforms, and deliver insights that empower public utility managers to make informed decisions. Previous research was by modeling the requirements of smart cities, followed by reference architectural designs for IoT-based smart cities. This architecture is modeled using the selected architectural look by involving stakeholders (Han & Kim, 2025).

4.1 Layered Architecture Overview

The proposed system architecture comprises the following five key layers:

a. Data Collection Layer

This foundational layer gathers data from both human-centric and device-centric sources:

- Human-centric (social sensing): Crowdsourced reports, mobile app feedback, social media posts, image uploads, voice inputs.
- Device-centric (IoT): Smart meters, waste bin sensors, water flow monitors, energy usage devices, sanitation facility sensors.

Example: In smart water grids, smart flow meters continuously send telemetry data, while users can report issues like low water pressure through a city app.

b. Data Aggregation and Preprocessing Layer

Raw data from various sources is cleaned, validated, and fused for analysis:

- Edge computing performs initial processing close to data sources to reduce latency.
- Cloud platforms aggregate city-wide data and standardize formats for analysis.
- Data fusion techniques combine structured sensor data with unstructured human reports.

For example, image metadata from users' sanitation reports is combined with foot traffic sensor data to prioritize cleaning schedules.

c. Decision Support and Action Layer

This layer enables human operators and city officials to make informed interventions:

- Interactive dashboards for utility monitoring and visualization.
- Automated alerts and action triggers (e.g., dispatching a maintenance crew).
- Rule-based and AI-driven recommendations for resource optimization.

For exapmle, in a surge in water use and temperature forecast indicate upcoming peak demand, the system can recommend preemptive load balancing.

d. Citizen Interaction and Feedback Layer

Engaging users is crucial for both data input and accountability:

- Mobile apps for issue reporting and status updates.
- Chatbots and voice assistants for citizen queries.
- Gamification elements (e.g., reward points for valid reports).
- Feedback loops to show how citizen input led to action.

Residents can see the response to their utility complaints via real-time app notifications, enhancing transparency and engagement.

4.2 System Architecture Diagram

Below is an illustrative diagram of the social sensing architecture for public utilities shows ini figure 6.

Figure 6. System Architecture for Social Sensing in Smart Zone Utilities

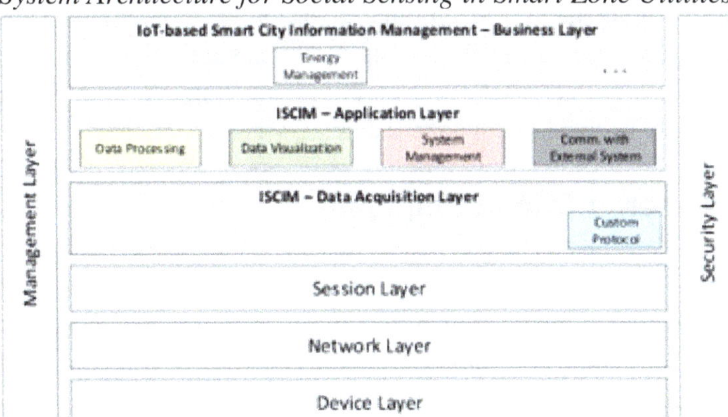

5. CASE STUDIES

To truly understand how social sensing can have a real impact on people's lives, we can look at best practices that have been implemented in various cities around the world. Through a combination of citizen participation, artificial intelligence, and IoT devices, these cities have succeeded in creating smarter, more efficient, and more inclusive public utility management systems. Here are four inspiring examples from Barcelona and Singapore, Jakarta and Seoul:

5.1 Barcelona Smart Water Initiative

Barcelona has long been known as a pioneer in the implementation of smart city technology, and their smart water management initiative is one of the most notable. The city has developed an integrated water management system that combines water quality sensors, weather data, and direct reporting from residents via the mobile app "BCN Water Watch." With this app, residents can easily report any incidents in their surroundings. The AI system processes data from flow and pressure sensors installed in the city's pipeline network, enabling faster and more accurate identification of issues. The increase in air pollution in Barcelona has also prompted the city government to adopt a smart and sustainable approach to address this challenge (Soriano-Gonzalez, Martin, Perez-Bernabeu, & Carracedo, 2024).

The positive impact is that the leak detection rate has increased by 40%, water consumption has decreased by up to 25% over a two-year period (2019–2021), and the response time of field technicians has been drastically reduced from an average of three days to less than 24 hours. This success demonstrates that the combination of community engagement and automatic sensor technology makes the system more responsive in addressing issues and provides significant efficiency in water management. This initiative serves as an important example of how social sensing can be effectively utilized to enhance public services.

5.2 Smart Energy Grid in Singapore

As a "Smart Nation" proclaimed by Singapore, with the Intelligent Energy System (IES) project, the Singapore energy authority has built a smart power grid that integrates user data, smart household devices, and artificial intelligence technologies. The aim is to create a more efficient, adaptive and data-driven electricity distribution system to support Singapore's growing population and energy needs. Singapore is currently ranked 5th in the Smart City category in the World (StraitsTimes, 2024).

The social sensing approach is carried out by directly involving the community in monitoring their energy use. Each home is equipped with a smart meter capable of recording electricity usage directly. In addition, the public can also access the national energy portal, where they can submit feedback on unusual electricity bill increases or blackouts occurring in their area. The combination of automated data from the tool and voluntary reports from the community results in a comprehensive monitoring ecosystem.

The results of the implementation of this system are very influential. In a short period of time, electricity use in households was reduced by 15% on average, thanks to initiatives that take advantage of feedback from the community to foster energy awareness. The system is also able to automatically manage power allocation to prevent overloading, while utilizing the collected data to predict demand and optimize energy storage through battery storage technology.

This experience in Singapore conveys valuable information: direct community involvement through social sensing not only improves energy efficiency, but also creates a culture of sustainable energy saving. When people are empowered to observe and be aware of their own energy use, they are taking part in solutions towards smarter, greener cities..

5.3 Jakarta WasteWatch Program

In Indonesia, the DKI Jakarta Provincial Government is collaborating with local startups to develop the WasteWatch platform, a smart waste management initiative that combines citizen reporting and AI-based visual detection. Through the WasteWatch app, citizens can upload photos of waste accumulation sites complete with GPS coordinates. The AI system then analyzes the images to verify and classify the type of waste, and integrates the citizen report data into the waste collection scheduling system. As a result, waste truck routes can be dynamically adjusted to prioritize high-priority areas. Global climate change affects all cities, and the Jakarta municipal government must manage unprecedented threats to the livelihood of its residents. (Reshetova & Hartley, 2025).

Within the first six months, collection route efficiency increased by 30%, the number of public complaints related to waste accumulation decreased by 40%, and the response time of the Sanitation Department was drastically reduced from 48 hours to just 12 hours. This program demonstrates how mobile-based crowdsourcing and visual analytics can enhance a city's monitoring capabilities without requiring significant increases in operational costs. The success of WasteWatch serves as a concrete example of how citizen engagement can accelerate the transformation of public services into more efficient and responsive systems.

5.4 Seoul Smart Sanitation Monitoring

The Seoul Metropolitan Government has launched the “Clean Seoul” platform as a smart solution for monitoring public sanitation conditions in real time. Through this system, residents can report the condition of public toilets simply by scanning the

QR code available at each facility. Residents can also rate cleanliness, record usage times, and add comments. All this information is then analyzed by an AI system to quickly and accurately identify maintenance needs, replacing the less responsive fixed-schedule maintenance approach. Research in Seoul is being conducted to create a healthy smart environment with three dimensions: urban cognition, interactive experience, and resident well-being. (Han & Kim, 2025).

The results of the implementation show a significant increase in public satisfaction with sanitation facilities, from 68% to 92%. Cleaning schedules are now based on demand and actual conditions, rather than simply routine timetables. More than just improving services, this approach builds a sense of ownership and shared responsibility for maintaining the cleanliness of public spaces—creating a culture of collaboration between the government and citizens in keeping the city clean and healthy..

6. CHALLENGES AND LIMITATIONS

Although social sensing offers many advantages in the management of public utilities in smart zones, its implementation cannot be separated from various challenges. These challenges include technical, ethical, legal, and socio-cultural aspects that can affect the effectiveness of the system as a whole.

6.1 Data Privacy and Security

One of the main challenges is how to protect citizens' personal data collected through devices such as smartphones, smart meters, and social media.

- Risk: The collected location, consumption behavior, and public activity data can lead to privacy breaches if not encrypted and accessed by unauthorized parties.
- Example: In 2021, there was a data leak in the energy monitoring system of a city in Europe due to weak API authentication from smart meters.

Potential solutions:

- Implementation of security standards such as TLS 1.3 and end-to-end encryption.
- The use of a federated learning approach to analyze data without the need to upload sensitive information to the cloud.

6.2 Data Quality and Noise

Data obtained from citizens through social media or reporting applications is often unstructured, incomplete, or even contains hoaxes.

- Challenge: The system must be able to filter out information that is irrelevant, redundant, or misleading.
- Case in point: In a trial of a water leak reporting system in India, 32% of citizen reports proved to be duplicate or unverifiable due to a lack of metadata.

Potential solutions:

- The application of natural language processing (NLP) and image verification AI for content validation.
- Gamification and user reputation schemes to encourage accurate reporting.

6.3 Digital Divide

Not all citizens have the same access to digital technology which is the main source of social sensing data.

- Risk: Data will be biased against certain socio-economic groups, resulting in asymmetric sensing that harms vulnerable populations.
- Fact: Studies show that reports from upper-middle-urban areas dominate 60% of public reporting system inputs, while marginalized areas are almost underrepresented.

Potential solutions:

- Provision of offline reporting terminals such as smart kiosks in public places.
- Community-based digital literacy training so that residents who are less technologically literate can still participate.

6.4 Integration Complexity and Interoperability

Combining different types of sensors, applications, and protocols from different vendors into one unified system is a significant technical challenge.

- Common issues: Data formats are inconsistent, APIs are closed, and not all platforms follow international IoT standards such as MQTT or CoAP.

- Impact: Integration delays lead to non-real-time decision-making.

Potential solutions:

- The use of open data standards and IoT middleware to unify the system.
- Implementation of semantic data models so that data from various domains can understand each other (interoperable).

6.5 Trust and User Participation

The effectiveness of social sensing is highly dependent on the active participation of the community. However, a lack of trust in the government or technology system can reduce engagement.

- Contributing factors: Lack of transparency, fear of data misuse, and lack of feedback from submitted reports.
- Effect: Decreased reporting or false reporting as a form of protest.

Potential solutions:

- A blockchain-based transparency system that records reports and public responses openly.
- Provide incentives such as points, badges, or public recognition for active citizen participation.

Table 3 shows a summary of challenges and implications

Table 3. Summary of Challenges and Implications

Challenge	Impact	Potential Solutions
Privacy & Security	Loss of trust, violation of the law	Encryption, federated learning
Data Quality	Misinterpretation, inefficiency	AI-based validation, whistleblower reputation
The Digital Divide	Service inequality	Reporting kiosks, digital literacy
Integration Complexity	Out of sync system	Open standard, middleware IoT
Low Trust	Citizen participation decreases	Transparency, social incentives

7. FUTURE DIRECTIONS

As the smart city ecosystem develops, social sensing is expected to undergo a significant transformation in the next few years. Innovations in artificial intelligence, edge processing, 6G networks, and collaborative models will strengthen its ability to manage public utilities efficiently, inclusively, and sustainably.

7.1 Federated Learning for Privacy-Preserving Intelligence

Federated learning (FL) allows AI modeling to be done in a decentralized manner without the need to upload raw data to a central server. This becomes an important key in the processing of sensitive social data such as water consumption, energy, or public sanitation reports.

FL Advantages:

1. Maintaining the confidentiality of personal data
2. Reduce the risk of data leaks
3. Increasing technology adoption in privacy-conscious regions

Supporting study before; shows that federated learning improves the accuracy of power consumption anomaly detection by 18% compared to centralized methods, while maintaining user privacy.

7.2 Digital Twin Integration

Digital twins are real-time virtual representations of physical systems that allow for simulation and predictive analysis.

Application: Digital twin integration with social sensing data allows:

1. Prediction of water demand during the dry season based on residents' reports
2. Simulation of electricity load based on community behavior data
3. Evaluate the impact of sanitation policies before implementation

Illustration: The following image shows the integration of citizen data with digital twins for water pressure simulation in densely populated environments.

7.3 AI-Powered Chatbots and Voice Interfaces

To increase citizen engagement, AI-based reporting systems such as chatbots and voice assistants will play an important role.

Feature:

- Reporting utility conditions by voice only
- Automatic explanation of the status of the complaint
- Personalize utility information according to the user's location

For example, a "Smart-City AI Assistant" that allows reporting the condition of damaged water lines with just a voice command via a smartphone or smart speaker.

7.4 Distributed Ledger and Blockchain for Transparency

In a digital age that demands openness, building trust between the government and the public is a major challenge. One way to overcome this is by introducing a system that is not only sophisticated, but also transparent and fair. Blockchain technology offers new opportunities to create a public reporting system that is free from manipulation, open, and verifiable by anyone. As a result, citizens no longer feel like passive reporters but become part of a system that truly listens to and acts on their voices.

Key benefits of implementing blockchain in public service reporting:

- Building public trust through immutable and transparent record-keeping
- Providing a clear audit trail for every report and improvement step
- Enabling public validation of reports through crowdsourcing, strengthening accuracy and community participation

Example technology works in practice shows all public interactions related to water, sanitation, and electricity services are recorded in a publicly accessible digital ledger, creating a more honest, responsive, and accountable service ecosystem.

7.5 Ethics and Policy-Oriented Frameworks

In order for social sensing to develop responsibly, policies are needed that support innovation while ensuring safety, ethics, and inclusivity.

Main components:

- Regulation of the use of social data by governments and technology vendors

- Algorithm bias removal guide
- Standardization of interoperability between sensing systems and utilities

Policy recommendations is Develop a national framework based on the principles of AI for the Public Good, which promotes transparency, accountability, and public participation.

7.6 Next-Gen Sensing Networks and Edge AI

The advancement of 5G/6G networks and edge computing will enable real-time sensing systems with ultra-low latency, improving decision-making efficiency.

Future developments:

1. Energy-efficient micro-sensors that can be integrated in public infrastructure
2. Edge AI for direct inference in devices such as smart meters
3. Data interconnection between smart zones in one city

Projection, by 2030, it is estimated that 70% of smart zones in developed countries will have an edge AI-based social sensing network with real-time feedback and auto-correction capabilities.

7.7 Toward Resilient, Inclusive, and Adaptive Smart Zones

Building a truly smart zone is not just about sensors, data, or algorithms. Its essence lies in how technology can bring services closer to people, without leaving anyone behind. The future of social sensing is about creating urban spaces that can hear all voices—including those that have often been marginalized. This means that inclusivity and diversity must be core values in every design and implementation of smart city technology. To achieve this, the following strategic steps are needed:

- Co-design with local communities – Develop systems together with residents, not for them, so that solutions truly reflect real needs.
- Local language-based reporting platforms and inclusive interfaces – Ensure there are no language or technological barriers hindering participation.
- Digital volunteer networks – Organizing tech-savvy residents to assist those without digital access or skills, ensuring everyone can participate and benefit.

With this approach, social sensing becomes not just a data collection tool but a bridge to social solidarity and urban transformation toward a more resilient, fair, and humane future.

8. CONCLUSION

Social sensing has become an important pillar in urban areas in the development revolution in the management of public utilities in the ever-changing and increasingly complex urban environment. With IoT technology, data from urban residents from social media, smartphones, CCTV, and various sensors on shorthouses can be used in real-time to answer public needs. In the context of smart zones, the social sensing approach will increase efficiency and can build good relationships between residents and local governments.

As discussed in this chapter, social sensing plays an important role in four key sectors: water distribution, energy, waste management, and sanitation. It enables early detection of issues such as water leaks, power outages, waste accumulation, or poor sanitation conditions. It is like an "extra set of senses" for the city-where citizens are not only the recipients of services, but also play an active role as problem detectors and part of the solution. This approach reflects a shift from a completely top-down management system to a collaborative, data-driven model, where government and citizens work together to create a more responsive and sustainable environment.

While social sensing offers great potential, its true impact can only be achieved if challenges such as privacy protection, data quality, and digital access gaps are seriously and thoroughly addressed. The future of resilient smart zones depends on several important things:

- Ethical system development that adapts to the diversity of users,
- Strong policies to ensure data security and accountability,
- Implementation of cutting-edge technologies such as federated learning, digital twin, and blockchain.

However, more than just technology, active community engagement is key. A truly smart city is not just about digital infrastructure, but also about inclusivity and equity. If social sensing is implemented with a transparent and participatory approach, citizens will not only be recipients of services, but also partners in shaping the system that serves them. The integration of AI, IoT, and social sensing in public services is not just a step forward in terms of technology, but also a paradigm shift-towards a more responsive, efficient, and human-centered model of urban governance. The success of this approach will be an important foundation for smart

REFERENCES

Ali, Z., Chaudhry, S. A., Ramzan, M. S., & Al-Turjman, F. (2020). Securing Smart City Surveillance: A Lightweight Authentication Mechanism for Unmanned Vehicles. *IEEE Access : Practical Innovations, Open Solutions*, *8*, 43711–43724. DOI: 10.1109/ACCESS.2020.2977817

Dawar, I., Srivastava, A., Singal, M., Dhyani, N., & Rastogi, S. (2025). A systematic literature review on municipal solid waste management using machine learning and deep learning. *Artificial Intelligence Review*, *58*(6), 183. Advance online publication. DOI: 10.1007/s10462-025-11196-9

Han, M. J. N., & Kim, M. J. (2025). A smart social sustainability model for smart city to enhance human context experience. *Cities (London, England)*, *159*, 105788. Advance online publication. DOI: 10.1016/j.cities.2025.105788

Hara, S., & Abe, M. (2025). Case study: OtologMap: a case study on the construction of an environmental sound map recorded by smart devices at Okayama and Kurashiki. *Noise Control Engineering Journal*, *73*(1), 57–68. DOI: 10.3397/1/37731

Hardjomuljadi, S., & Anondho, B. (2025). A model for transforming conventional industrial areas into eco-smart zones: The role of development planning, infrastructure, competence, and partnerships on success of the transformation process. *International Journal of Innovative Research and Scientific Studies*, *8*(4), 32–44. DOI: 10.53894/ijirss.v8i3.7737

Haucke, M., Heinzel, S., & Liu, S. (2024). Social mobile sensing and problematic alcohol consumption: Insights from smartphone metadata. *International Journal of Medical Informatics*, *188*, 105486. Advance online publication. DOI: 10.1016/j.ijmedinf.2024.105486 PMID: 38754285

He, Z., Zhai, W., & Zhao, C. (2025). Using social sensing to evaluate the effect of spatial planning: A case study of Yangtze River Delta Urban Agglomeration in China. *Frontiers of Architectural Research*, *14*(4), 1048–1062. DOI: 10.1016/j.foar.2024.12.004

Ko, S., Kim, W., Shin, S.-C., & Shin, J. (2020). The economic value of sustainable recycling and waste management policies: The case of a waste management crisis in South Korea. *Waste Management*, *104*, 220–227. https://doi.org/https://doi.org/10.1016/j.wasman.2020.01.020

Kumar, S., & Goyal, M. K. (2025). Water policy review: Ensuring sustainable water management for India. *Journal of Environmental Management*, *388*, 125823. Advance online publication. DOI: 10.1016/j.jenvman.2025.125823 PMID: 40440942

Li, H., Zhang, Q., & Lu, K. (2015). Integrating mobile sensing and social network for personalized health-care application. In *Proceedings of the ACM Symposium on Applied Computing* (Vol. 13-17-Apri, pp. 527–534). DOI: 10.1145/2695664.2695767

Miyazaki, K., Murayama, T., Uchiba, T., An, J., & Kwak, H. (2024). Public perception of generative AI on Twitter: An empirical study based on occupation and usage. *EPJ Data Science, 13*(1), 2. Advance online publication. DOI: 10.1140/epjds/s13688-023-00445-y

Reshetova, E., & Hartley, K. (2025). Strengthening governance accountability in smart cities: The case of urban energy transition in Jakarta, Indonesia. In *Smart Cities to Smart Societies: Moving beyond Technology* (pp. 99–130). https://doi.org/DOI: 10.4324/9781003439325_9

Salmi, A.-K., & Hökars, F. (2023). Smart urban traffic zones - demonstration of three smart zones in Sweden based on geofencing and sensor data. In *Transportation Research Procedia* (Vol. 72, pp. 1459–1466). DOI: 10.1016/j.trpro.2023.11.611

Sánchez, L., Elicegui, I., Cuesta, J., Muñoz, L., & Lanza, J. (2013). Integration of utilities infrastructures in a future internet enabled smart city framework. *Sensors (Basel), 13*(11), 14438–14465. DOI: 10.3390/s131114438 PMID: 24233072

Schulte, K., Grieve, A., Barratt, B., Baker, T., Coonjobeeharry, H., & Mead, M. (2025). Advancing participatory sensing and knowledge production methods for city air quality governance: Applying the Breathe London Community Programme model. *Environmental Science & Policy, 170*, 104092. Advance online publication. DOI: 10.1016/j.envsci.2025.104092

Sofeska, E., & Sofeski, E. (2022). Developing Projects for Realizing of the Program "Skopje 2020 Smart Strategy" by Enhancing Citizen Approach, Engineering, Digitalization, and Sensing of the City District Toward Smarter Sustainability Urban Potential in the Small Ring of Skopje. In *Advances in Science, Technology and Innovation* (pp. 69–84). DOI: 10.1007/978-3-030-86499-6_7

Soriano-Gonzalez, R., Martin, X. A., Perez-Bernabeu, E., & Carracedo, P. (2024). Modeling and Optimization of NO2 Stations in the Smart City of Barcelona. *Applied Sciences (Basel, Switzerland), 14*(22), 10355. Advance online publication. DOI: 10.3390/app142210355

StraitsTimes. (2024). Singapore is world's 5th smartest city. Retrieved April 3, 2025, from https://www.straitstimes.com/singapore/singapore-is-5th-smartest-city-in-the-world-top-in-asia-global-index

Tekinerdogan, B., Köksal, Ö., & Çelik, T. (2023). System Architecture Design of IoT-Based Smart Cities. *Applied Sciences (Basel, Switzerland), 13*(7), 4173. Advance online publication. DOI: 10.3390/app13074173

Toh, C. K. (2020). Security for smart cities. *IET Smart Cities*, *2*(2), 95–104. https://doi.org/https://doi.org/10.1049/iet-smc.2020.0001. DOI: 10.1049/iet-smc.2020.0001

UN-DESA. (2024). World Population Prospects 2024. Retrieved May 27, 2025, from https://www.un.org/development/desa/pd/world-population-prospects-2024

Wang, W., Mirjafari, S., Harari, G., Ben-Zeev, D., Brian, R., Choudhury, T., & Campbell, A. (2020). Social Sensing: Assessing Social Functioning of Patients Living with Schizophrenia using Mobile Phone Sensing. In *Conference on Human Factors in Computing Systems - Proceedings*. DOI: 10.1145/3313831.3376855

Zhang, Y., Kwan, M.-P., Yu, B., Liu, Y., Song, L., & Chen, N. (2025). Quantitative Identification of Mixed Urban Functions: A Probabilistic Approach Based on Physical and Social Sensing Data. *Transactions in GIS*, *29*(1), e13272. Advance online publication. DOI: 10.1111/tgis.13272

Chapter 12

Towards Smart and Sustainable Waste Management Through IoT Integration for Automated Sorting

Mariem Smaoui
National Engineering School of Sfax, Tunisia

Mouna Rekik
http://orcid.org/0000-0003-1869-2394
National Engineering School of Sfax, Tunisia

Lotfi Krichen
National Engineering School of Sfax, Tunisia

ABSTRACT

Waste management is a major environmental challenge, driven by urban growth and modern consumption, which surpasses the capacity of traditional sorting methods. To address this challenge, integrating Artificial Intelligence (AI) and the Internet of Things (IoT) into waste management offers an innovative solution. This chapter presents an autonomous and intelligent system capable of automatically identifying and sorting waste (plastic, metal, glass, cardboard, organic) using smart cameras and specialized sensors. A central unit equipped with a camera and AI models detects plastic and cardboard in real time, while other sensors identify the characteristics of glass, metals, and organic waste. Mechanical actuators then direct the waste to the appropriate bins. Level sensors prevent overflows and optimize collection, with

DOI: 10.4018/979-8-3373-4962-6.ch012

data accessible via a mobile application. A functional prototype was tested, confirming technical feasibility, effective integration, and reliable automated sorting, thus laying the groundwork for industrial or large-scale deployment.

INTRODUCTION

Municipal Solid Waste: A Rising Global Concern

In recent years, the management of municipal solid waste (MSW) has become a major issue due to the rapid increase in the volume of waste generated worldwide. This growth is mainly driven by accelerated urbanization, industrialization, and often inadequate management practices. In 2020, approximately 2.1 billion tons of MSW were generated globally each year, and this figure could reach nearly 3.8 billion tons by 2050 if urgent measures are not taken. This sharp increase puts significant pressure on authorities to implement waste management strategies that are both sustainable and effective, capable of reducing environmental impacts while considering social and economic dimensions (UN Environment Programme, 2024; Kaza et al., 2018; Maalouf et al., 2022).

The evolution of lifestyles and consumption patterns in developing countries, linked to urbanization and industrialization, leads to a notable rise in waste produced per capita. This trend is often related to a country's wealth level, with a higher GDP generally associated with greater waste generation per person. For example, the United States, although accounting for less than 5% of the world's population, generates about 12% of global solid waste, illustrating the impact of lifestyle, consumption, and urbanization on waste volumes (UN Environment Programme, 2024).

When examining waste generation per capita across regions, it is noticeable that areas such as North America and Central and South Asia produce similar amounts, even though their economic and demographic dynamics differ. In middle-income countries where urbanization is rapidly progressing, waste management becomes an increasing challenge, especially where collection and disposal infrastructures are insufficient. For instance, in India, issues such as overloaded landfills, inefficient waste sorting, and weak enforcement of regulations are major obstacles. These challenges highlight the urgent need to integrate technological innovations and strengthen collaboration with the private sector to improve waste management, processing, and recycling (Singh, 2019; Worldbank, 2018).

Transitioning to Sustainable Waste Management

Sustainable waste management has become a key issue in addressing the environmental, social, and economic challenges facing the modern world. Efficient waste management not only helps limit negative impacts on ecosystems, such as soil, air, and water pollution, but also promotes social justice by reducing inequalities, particularly for vulnerable populations who are often the most affected by poor waste practices (Adeniyi, 2023; Ferronato et al., 2019).

At the heart of this dynamic, waste sorting plays a crucial role. It is the essential first step toward the effective recovery of materials and largely determines the success of strategies aimed at reduction, reuse, and recycling. Rigorous sorting maximizes the recovery of recyclable materials, reduces the volume of waste sent to landfills, and facilitates the integration of organic waste into composting or anaerobic digestion systems (Almtireen et al., 2025; Haitao, 2021).

In this context, the circular economy emerges as a major strategic approach. It aims to extend the life cycle of products and materials while promoting the responsible management of natural resources. By reducing waste production at the source and enhancing residual materials through efficient sorting, the circular economy helps preserve resources, strengthen industrial competitiveness, and actively contribute to the fight against climate change particularly in high-emission sectors such as steel, plastics, aluminum, and cement (MacArthur, 2013; Kirchherr et al., 2017).

For example, a report by Material Economics indicates that the European Union could reduce its CO_2 emissions by up to 296 million tonnes annually by 2050 by optimizing the use of existing materials, notably through better source sorting and improved recovery pathways (Material Economics 2018).

The success of this transition relies on coordinated efforts between governments, businesses, and civil society. This requires sustained investment in sorting infrastructure, public awareness campaigns, and technological innovation.

The digital revolution, in particular, offers new opportunities: artificial intelligence, the Internet of Things, and automation now make it possible to significantly improve sorting processes, optimize waste logistics, and enhance decision-making through real-time data. These tools make waste management smarter, more efficient, and better adapted to the evolving needs of territories (Abdullah et al., 2020; Arindam et al., 2022).

In summary, sustainable waste management depends on an integrated approach that combines environmental principles, social equity, and innovation. Waste sorting, as a key lever, is the cornerstone of this approach, paving the way for optimal resource recovery and the construction of a more circular and resilient society.

Analysis of Existing Approaches to Waste Sorting and Classification

Traditional automated sorting systems, widely used in industrial chains, are mainly based on Programmable Logic Controllers (PLCs) combined with simple sensors (proximity sensors, limit switches, push buttons, etc.). These sensors can detect the presence or position of an object but are unable to accurately identify the material type. This presents a challenge when sorting mixed waste known as heterogeneous waste streams. In such cases, it is essential to recognize the materials (paper, plastic, metal, etc.) in order to recycle them properly. However, traditional sensors cannot distinguish between these materials, making the sorting process imprecise, inefficient, and sometimes entirely ineffective. Furthermore, irregular shapes, varied dimensions, and random positions of objects on the conveyor belt further exacerbate the limitations of these systems. For example, a sensor may detect the presence of an object but cannot determine whether it is a plastic bottle or a metal can. This inability significantly hinders the effective separation of recyclable materials (Almtireen et al., 2025, Sathish et al., 2017).

According to (Bozma et al., 2002), traditional sensor-based systems are ineffective at classifying objects with random shapes or orientations, which is common in waste treatment facilities. These technical limitations highlight the need for smarter systems capable of recognizing materials not only by their physical shape but also by visual characteristics such as color, texture, or surface structure.

To overcome these limitations, optical sorters have been developed. These machines use cameras and image processing technologies to analyze the visual aspects of waste such as color, shape, or texture under proper lighting. Based on these characteristics, the machine identifies the type of waste (plastic, paper, metal, etc.) and triggers a mechanism, typically pneumatic, to direct the waste to the appropriate bin. This system is particularly effective for light and homogeneous waste such as paper, thin cardboard, or soft plastics. However, these pneumatic sorters reach their limits when handling heavier waste, such as rigid plastics or metal objects. Indeed, a simple jet of compressed air is not sufficient to effectively move these items. Moreover, when waste items are of very diverse shapes, stacked, or poorly positioned, the optical system alone may lack the necessary precision (Tianchen et al., 2023; Xiaoyu et al., 2020).

To overcome these obstacles, robotic solutions have been introduced. These advanced machines, equipped with smart cameras and robotic arms, can visually identify waste items and handle them one by one. This enables the system to manage more complex waste, increase sorting accuracy, and adapt the process to a wider variety of materials (Maria et al., 2021; Zeli et al., 2020).

Among these innovations, delta robots equipped with suction cups (vacuum grippers) have proven particularly effective. Using artificial intelligence algorithms such as Mask R-CNN, these robots can automatically detect, grasp, and sort waste. Suction cups are often preferred over mechanical grippers because they are sufficient to move recyclable objects without requiring a very precise grip. However, they perform best on flat or smooth surfaces; if the object is too irregular or rough, suction becomes less reliable (Laski et al., 2015; Konstantia et al., 2022; Hassan et al., 2022; Xinxing et al., 2022).

The integration of Artificial Intelligence (AI) into sorting systems has led to significant improvements in recycling efficiency and material quality. For example, Wilts et al., (2021) demonstrated that AI not only increases sorting speed but also enhances the purity of recovered materials. The "ZRR for municipal waste" project clearly illustrates these benefits by showing the positive impact of AI on operational performance and safety, thus contributing to the goal of a sustainable circular economy.

Numerous studies confirm the relevance of AI and machine learning in improving and automating waste sorting. For instance, Lubongo et al., (2024) explored the combination of AI with spectroscopic methods and computer vision systems for plastic sorting, achieving impressive results: up to 99.99% accuracy and a sorting capacity of up to 10 tons per hour. Despite some persistent challenges with complex plastics, these systems show great potential.

The intelligent robot developed by (Cheng et al., 2024) can recognize containers such as PET bottles or colored glass using computer vision and specialized grippers. Meanwhile, Feng et al., (2022) designed a smart bin integrating an AI model, GECM-EfficientNet, which enables automatic waste classification with an accuracy exceeding 94%.

However, despite these advances, certain limitations remain. Current systems, even those equipped with AI, are not always able to reliably distinguish more than three to five categories of waste under real-world conditions. Moreover, their implementation cost remains high, and adapting them to diverse environments (urban, industrial, rural) can be challenging. Training AI models also requires rich, representative, and annotated datasets, which limits the generalization of existing solutions. Therefore, it is necessary to develop hybrid approaches that combine visual recognition, physical sensors, and connected platforms to deliver intelligent, cost-effective, accurate, and scalable sorting solutions.

Another common issue affecting the environment and human health concerns waste management in cities. In many municipalities, trash bins are not monitored in real time. As a result, cleaning services are not always aware of when a bin is full, which sometimes leads to overflows and the spread of waste in the streets.

This situation harms public hygiene and degrades the quality of life for residents (Arindam et al., 2022).

To address this, smart cities must implement a monitoring system for trash bins to ensure they are emptied in a timely manner. Currently, some authorities already use route optimization techniques for waste collection, such as those proposed by Shan-Huen et al., (2015), but these methods do not take into account real-time data provided by connected smart bins.

The use of modern technologies based on the Internet of Things (IoT) opens the door to more effective waste monitoring and control than traditional approaches. For example, Kellow et al., (2020) presented an integrated hardware and software solution to improve waste management.

Many researchers have investigated this area, each from a specific angle. Vera et al., (2013) proposed a method combining bin allocation and garbage truck routing, although these two aspects were handled separately and without the use of smart bins. Other studies developed containers capable of estimating their fill level, but without incorporating recent technologies such as Internet-connected systems (Vicentini et al., 2009).

Later, Yann et al., (2015) introduced the concept of a smart bin to contribute to urban cleanliness. Soumyabrata et al., (2023) designed a complete IoT-compatible waste management system specifically intended for smart cities. Ramson et al., (2017) also proposed a system utilizing wireless sensors to monitor the fill level of bins in real-time.

However, few researchers have truly explored the full integration of IoT into a comprehensive waste management system. To address this gap, Patric et al., (2019) developed a smart city infrastructure integrating IoT into waste management. Furthermore, Nidhya et al., (2020) implemented an intelligent algorithm, called ERS, to optimize waste collection in this context.

To address the limitations identified in the literature review and to develop a reliable and efficient waste management system, this paper proposes an integrated approach combining Artificial Intelligence (AI) and the Internet of Things (IoT). The goal is to automate waste sorting, enhance real-time monitoring, and optimize the management of collection bins.

The proposed system is autonomous, intelligent, and connected. It automatically identifies and classifies different types of waste (plastic, metal, glass, cardboard, organic) using high-resolution cameras and AI models trained on large annotated datasets. This visual recognition is supported by specialized sensors: electromagnetic sensors for metals, optical sensors for glass, and gas sensors for organic waste. Once the material is identified, mechanical actuators guide the waste to the appropriate bin.

In parallel, level sensors installed in the bins monitor their fill status to prevent overflow. All of this data is transmitted in real time via the IoT network to a platform accessible to waste management personnel, including through a mobile application. This facilitates supervision, route planning for waste collection, and overall improvement of the waste management process.

Compared to the solutions previously described in the literature, our proposed system introduces several significant advancements:

- Comprehensive Integration of AI, IoT, and sensors: Unlike many existing works that focus on a single component such as; optical sorting, smart bins, or sensor-based detection, our solution combines intelligent classification, multi-sensor detection, automated physical sorting, and connected bin management into one coherent and fully integrated platform.
- Real-Time Multi-Material Recognition: Our system moves beyond the basic categories (plastic, paper, metal) by incorporating a variety of sensors capable of recognizing multiple material types under realistic conditions, including irregularly shaped and moving objects. This improves classification accuracy and robustness in complex waste streams.
- Dynamic Optimization of Collection Routes: By leveraging real-time fill-level data transmitted via IoT, our approach enables dynamic, intelligent scheduling of waste collection. Field agents are equipped with a dedicated mobile application for real-time monitoring and decision-making—an aspect not addressed in prior works such as those of Shan-Huen et al., (2015) or Vera et al., (2013).

The remainder of this paper is organized as follows. Section 2 provides an overview of the intelligent waste sorting system. Section 3 outlines the design of the proposed solution. Section 4 describes the implementation of the prototype, while Section 5 presents and discusses the experimental results. Finally, Section 6 concludes the paper.

DESCRIPTION OF THE INTELLIGENT WASTE SORTING SYSTEM

The intelligent waste sorting and management system is based on the integration of Artificial Intelligence (AI) and Internet of Things (IoT) technologies. It is designed to automate all stages of the sorting process, ensure optimized monitoring, and enable real-time management of waste flows. This innovative approach aims to enhance processing efficiency while minimizing human intervention (Figure 1).

At the core of the system architecture, a motorized conveyor belt automatically transports waste from the deposit point to the appropriate collection bins. During this transfer, a vision module captures images of the objects in motion. These images are analyzed using AI algorithms trained on databases containing representative examples of common materials such as plastic or cardboard. Thanks to this machine learning, the system can accurately identify the nature of the waste and instantly determine the appropriate sorting action.

In addition to visual recognition, the system incorporates various specialized sensors capable of detecting other types of materials. Some sensors are designed to identify specific characteristics of organic waste, while others analyze the optical properties of surfaces, such as the distinctive light reflection of glass. Electromagnetic sensors detect the presence of metals, and additional sensors ensure reliable detection of objects entering the sorting area.

Once the type of material is identified, motorized mechanisms guide each piece of waste to its designated compartment, in accordance with the decision made by the control system. All operations are carried out smoothly and in a synchronized manner, based on real-time updated data.

Bin management is also automated. Level sensors continuously monitor the fill status of each compartment, allowing for timely interventions and preventing overflow. This information also helps optimize collection routes.

Thanks to its IoT connectivity, the system transmits all collected data to a remote platform. An intuitive mobile application allows operators to monitor bin status, access-sorting statistics, and receive alerts in case of anomalies or bin saturation.

Figure 1. Block diagram of the intelligent waste sorting system

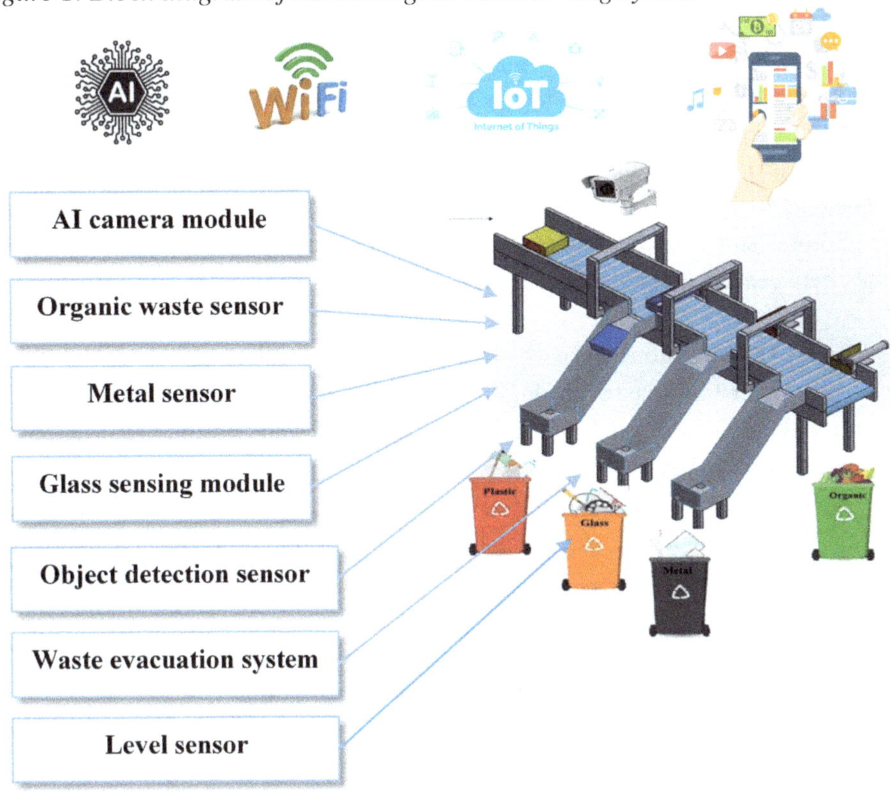

Cardboard And Plastic Detection System

In the waste sorting system, an intelligent camera, specifically the ESP32-CAM equipped with the OV2640 sensor, plays a central role in material identification. Positioned above the conveyor belt, this camera captures real-time images of the waste deposited on the moving belt. These images are then analyzed locally using embedded artificial intelligence algorithms that have been specifically trained on large annotated datasets containing various types of waste, particularly cardboard and plastic. The AI is capable of extracting relevant visual features (such as shapes, colors, and textures) to classify the observed objects with a high level of accuracy.

This recognition process enables the sorting decision to be automated without human intervention. As soon as a piece of waste is identified as plastic or cardboard, a signal is sent to the actuators responsible for directing it to the appropriate collection bin.

Glass Detection System

Glass detection within the sorting system relies on a simple yet effective opto-electronic method that combines a laser module with a photoresistor (LDR – Light Dependent Resistor). This setup is designed to differentiate transparent materials, such as glass, from opaque materials like plastic, metal, or cardboard.

The operating principle is based on whether or not the laser beam is interrupted. The photoresistor is positioned to directly receive the laser light. When an object passes between the laser and the LDR, two scenarios can occur:

- Beam Transmission: If the object is transparent (such as glass), the laser light partially or fully passes through the material and continues to reach the photoresistor. The detected light level remains high, indicating the presence of a transparent material.
- Beam Interruption: If the object is opaque (plastic, metal, cardboard, etc.), the beam is blocked, and the LDR receives little or no light. This drop in luminosity is interpreted as the presence of a non-transparent material.

This detection method offers several advantages: it is fast, cost-effective, easy to integrate, and does not require complex data processing.

Metal Detection System

Metal detection in the sorting system relies on the use of inductive proximity sensors, which allow for the identification of metallic objects without any physical contact. These sensors operate by generating an electromagnetic field around their sensing head. When a metallic material, such as steel, iron, or other conductive alloys, enters this field, it disrupts the electromagnetic lines. This disturbance is immediately detected by the sensor, which then emits a signal indicating the presence of metal. This operating principle, based on the conductive properties of metals, enables fast, reliable, and precise detection, even in industrial environments or in the presence of vibrations. Thanks to this technology, metallic waste can be automatically identified and directed to the appropriate collection bin, without requiring visual analysis or complex processing.

Organic Matter Detection System

Organic waste, such as food scraps or decomposing plants, naturally releases gases like carbon dioxide, methane, and ammonia. This characteristic gas emission serves as a reliable indicator of the presence of organic matter. That is why integrating an MQ135 gas sensor into our automated sorting system allows for the effective detection of these substances. Thanks to this detection, the system can identify organic waste in real time and automatically separate it from other types of waste, thereby improving the accuracy and efficiency of sorting while contributing to better waste management.

Object Detection Sensor

The object detection sensor plays a key role in this sorting system. It detects the presence of a waste item as soon as it arrives on the conveyor belt, allowing the system to stop the conveyor in order to analyze and classify the waste properly. The sensor is also used after the waste has been identified, when it reaches the appropriate sorting bin. In this system, an active infrared sensor has been selected for object detection. Its ability to detect nearby objects in real time ensures a reliable, accurate, and efficient operation of the automated sorting process.

Waste Evacuation System

In this system, servomotors are used to direct each piece of waste from the conveyor belt to the appropriate bin, based on the decision made by the sorting system. Each type of waste is associated with a specific servomotor and object detection sensor. Once the material is identified, the corresponding servomotor moves into position. The conveyor then restarts to advance the object. When a sensor placed near the corresponding bin detects the presence of the waste, the conveyor briefly stops, and the servomotor activates to direct the object into the appropriate bin.

Fill Level Detection System

Level sensors installed in the bins enable real-time monitoring of each container's fill level, preventing any overflow or saturation. These sensors play a crucial role in optimizing waste management by ensuring collections are efficient and calibrated to actual needs. The data they collect is transmitted automatically via the Internet of Things (IoT) to the management team, who can access it at any time through a dedicated mobile application. This connected solution streamlines remote supervision, enhances the planning of collection routes, and helps reduce operational costs while improving the overall efficiency of the sorting system.

PROPOSED SYSTEM DESIGN

The intelligent sorting system relies on a harmonious combination of carefully selected hardware and software components to ensure optimal performance. In this section, the technical specifications of all the electronic components used are presented, along with the various programming, modeling, and communication tools employed to develop processing algorithms, manage user interfaces, and guarantee smooth and secure data transmission.

ESP32-WROOM-32

The microcontroller forms the core of the system, responsible for analyzing and managing the various stages of automated waste sorting. To determine the most suitable solution for our connected project, a comparison was conducted between several boards: the ESP32, NodeMCU-ESP8266, and Raspberry Pi Pico W. Following this analysis, the ESP32-WROOM-32 (Figure 2) emerged as the optimal choice due to its excellent balance between performance, features, and cost. This board features a dual-core processor running up to 240 MHz, 520 KB of SRAM, and typically 4 MB of flash memory, enabling it to handle complex real-time applications. It also integrates Wi-Fi and Bluetooth, offering reliable and versatile wireless connectivity suitable for a wide range of IoT applications, from home automation to industrial monitoring. Moreover, the ESP32 supports multiple communication interfaces (UART, SPI, I2C, ADC, PWM), making it easy to integrate with various sensors and actuators. Thanks to its low-power modes and affordable price, it is a powerful and cost-effective platform for developing smart devices (Hussein et al., 2024).

In this project, two ESP32 boards are used. The first one is connected to the ESP32-CAM and various sensors (glass, metal, organic, etc.) to automate the sorting process and control the servomotors that direct the waste into the appropriate bins. The second ESP32 board is dedicated to measuring the fill level of the bins using sensors, and it sends the collected data in real time to the managers via a mobile application, thereby facilitating supervision and collection management.

Figure 2. ESP32-WROOM-32

ESP32-CAM

Object detection is a computer vision technique that enables the recognition of objects in various environments to classify and analyze them. Thanks to advances in machine learning and artificial intelligence, this technology has significantly expanded its range of applications. Within the framework of the waste sorting system, the integration of AI-enabled cameras was explored to automatically detect and classify plastic and cardboard waste. Several smart cameras, such as the Raspberry Pi with camera, the NVIDIA Jetson Nano camera, and the ESP32-CAM, were compared based on several criteria: cost, ease of integration, power consumption, processing speed, and compatibility with the development environment.

Following this comparison, the AI-Thinker ESP32-CAM camera (Figure 3) proved to be the ideal choice for this cost-effective project aimed at detecting plastic and cardboard waste. It integrates an OV2640 camera with flash, capable of capturing images up to 2 megapixels (1600 × 1200 pixels). This module features 32 Mbit of built-in flash memory, 512 KB of internal RAM, 4 MB of external PSRAM, and a powerful ESP32-D0WD processor. With its Wi-Fi and Bluetooth capabilities, the ESP32-CAM is particularly suitable for a wide range of IoT applications such as home automation devices, wireless industrial control, remote monitoring, and wireless identification and positioning systems. It thus constitutes a complete and economical solution for most connected projects (Shraddha et al., 2025).

Figure 3. AI-Thinker ESP32-CAM camera

KY-008 Laser Module And Photoresistor

The KY-008 laser module and the photoresistor (Figure 4) are used together to detect the presence of glass in the sorting system. The laser emits a light beam directed at the object, while the photoresistor measures the reflected light. Thanks to the specific properties of glass, which transmits light differently from other materials, the system can effectively identify its presence. These components are easy to integrate and operate with low power consumption (3.3 V to 5 V), offering a simple, cost-effective solution suited for low-cost waste sorting applications (Maria et al., 2022).

Figure 4. KY-008 laser module and photoresistor

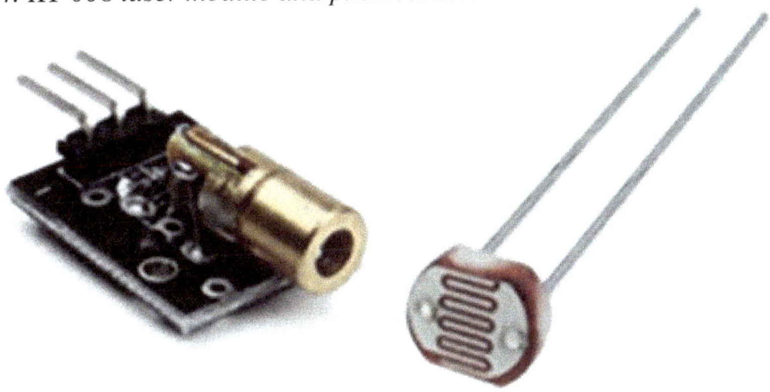

Metal Sensor

A comparison between the inductive proximity sensors CJM18M-8A1 and LJ12A3-4-Z/BY was conducted. However, these sensors proved to be either ineffective for long-range detection or expensive for our application. To overcome these limitations, we designed and built our own inductive proximity sensor, capable of detecting metallic objects at a greater distance while remaining cost-effective (Rui et al., 2022).

This sensor operates on the same principle as conventional inductive sensors: it uses an electromagnetic field generated by a copper coil to detect the presence of metal. Four main components are used in its construction: a copper coil, a diode, a 10 nF capacitor, and a 220-ohm resistor. These components are connected as shown in Figure 5, forming an RC (resistor-capacitor) circuit integrated into the sensor's electronics.

When a metallic object approaches the coil, it disturbs the emitted electromagnetic field. This disturbance is detected by the circuit, which then sends a signal to the ESP32. The ESP32 processes the signal, compares it to a reference threshold, and activates or deactivates a digital output accordingly. This simple yet effective system enables reliable long-range metal detection while keeping manufacturing costs low (Texas Instruments Incorporated 2021).

Figure 5. Inductive proximity sensor

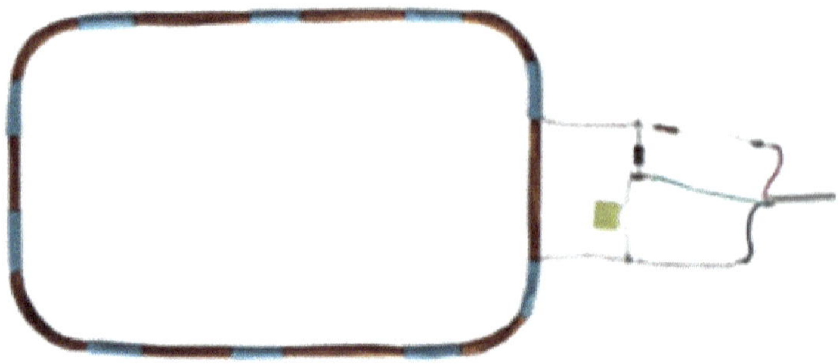

MQ135 Gas Sensor

As part of this project, several gas sensors were evaluated to determine the most suitable one for detecting gases emitted by organic waste. The MQ7 sensor is designed specifically for detecting carbon monoxide, while the MQ-3 is sensitive to alcohol, ethanol, and smoke. Although these sensors are useful in certain situations, they are not sufficient for accurately identifying the typical gases produced by organic decomposition. In contrast, the MQ135 sensor (Figure 6) proved to be the most appropriate for this application, thanks to its ability to detect a broad range of gases such as ammonia, benzene, and various chemical vapors, all of which are commonly released by decaying organic matter. It is also fully compatible with the ESP32, making integration into our system straightforward. From a technical perspective, the MQ135 operates at 5V (also functional at 3.3V with reduced stability), consumes approximately 60 mA, and functions within a temperature range of –20 °C to +50 °C, with an average response time of about 30 seconds. Its high sensitivity, broad gas detection range, and ease of integration make it an ideal choice for detecting organic waste in smart sorting applications (Astuti et al., 2021).

Figure 6. MQ135 gas sensor

LM393 fc-51 IR Sensor

The LM393 FC-51 infrared sensor (Figure 7) is used in our system for proximity object detection. It operates based on the principle of infrared light reflection: an IR LED emits a light beam, and a photoreceiver detects the light reflected by an object within its detection range. This sensor is characterized by an adjustable detection range from 2 cm to 30 cm, which can be set using an integrated potentiometer, making it adaptable to various sorting scenarios. It operates at a supply voltage between 3.0 V and 6.0 V, with a detection angle of approximately 35°. Thanks to its low power consumption, ease of use, and reliability, the LM393 sensor is perfectly suited for object detection in automated sorting systems (Yelbek et al., 2024).

Figure 7. LM393 FC-51 infrared sensor

SG90 Digital Servo 180°

Servomotors are activated to direct each waste item toward the appropriate bin. For our waste sorting system, the SG-90 mini servomotor (Figure 8) was chosen due to its speed and ease of programming, making it more efficient for evacuation tasks than a robotic hand. Weighing only 9 grams, this compact motor measures 23 × 12.2 × 29 mm, making it ideal for embedded systems with limited space. It operates at a voltage of 4.8 V and provides a stall torque of 1.8 kg/cm, with a rotation speed of 0.1 seconds per 60 degrees, ensuring fast and precise action. It can operate within a temperature range of 0 °C to 55 °C, making it perfectly suited for the precise and consistent movements required in automated sorting applications (Shraddha et al., 2025).

Figure 8. SG90 digital servo

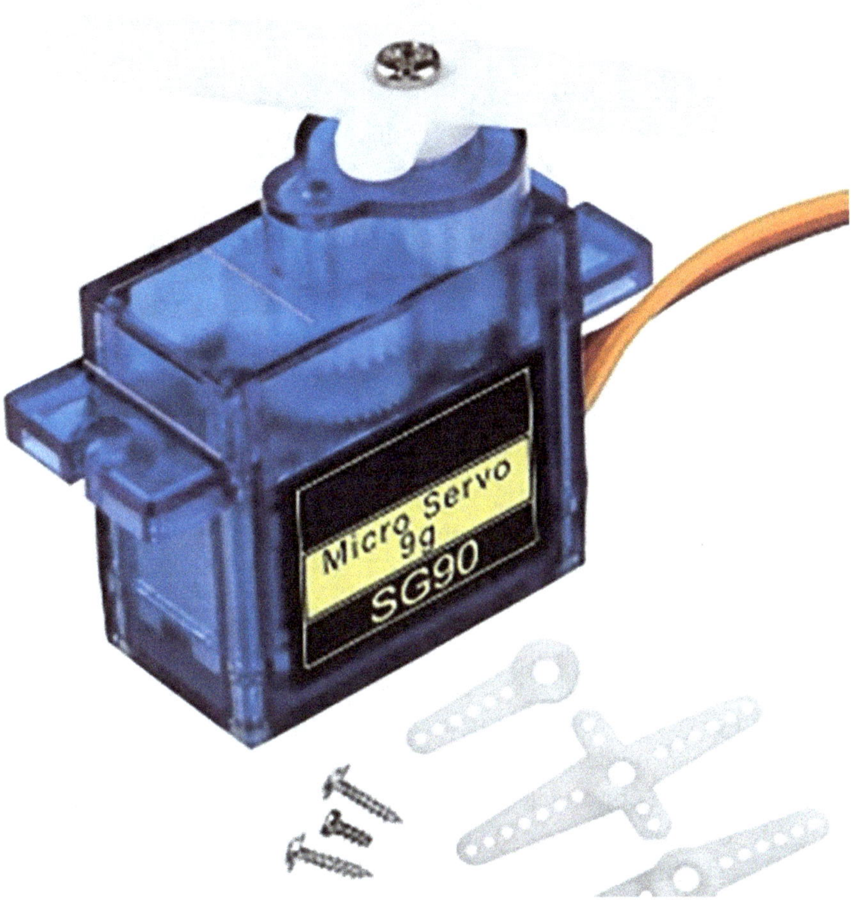

HC-SR04 Sensor

The HC-SR04 (Figure 9) ultrasonic sensor was selected for our smart sorting system due to its excellent cost-performance ratio, low power consumption, and ease of integration. It operates with a 3.3V or 5V DC power supply, consumes approximately 15 mA, and uses an ultrasonic frequency of 40 kHz. It offers a detection range from 2 cm to 4 meters, with an accuracy of ±3 mm, making it perfectly suited for measuring the fill level of waste bags or bins. Compatible with microcontrollers such as the ESP32, this sensor is a reliable, simple, and economical solution for real-time monitoring in automated sorting systems (Arindam et al., 2022).

Figure 9. HC-SR04 sensor

Software Development Tools

In the software section, the intelligent waste detection (plastic, cardboard, etc.) relies on a machine learning model trained using the Edge Impulse platform, an advanced solution dedicated to data collection, model training, and deployment on embedded devices. This platform easily integrates into edge MLOps workflows, enabling efficient model training on the ESP32-CAM AI Thinker camera and seamless generation of object detection code (Shraddha et al., 2025). Additionally, the ESP32 board is programmed using the Arduino IDE, a development environment known for its simplicity and broad compatibility with various microcontrollers (Arindam et al., 2022). For the user interface, the Smart Waste mobile application was developed using the MIT App Inventor platform, facilitating Android app creation even for users with limited programming experience (Anuradha et al., 2023). Finally, communication between the mobile application and the embedded system is handled via Firebase, a cloud server chosen for its ease of integration, full compatibility with App Inventor, and advanced features such as real-time database, authentication, hosting, and cloud functions, ensuring smooth and secure data synchronization (Wen-Tsai et al., 2023).

IMPLEMENTATION OF THE SYSTEM

This section presents the key stages of the project in detail, beginning with the training of artificial intelligence models, followed by the development of the control programs. The precise wiring of the electronic components necessary for the proper functioning of the system is then addressed. The fabrication and assembly of the conveyor are also described. Finally, the complete assembly of the system is presented, highlighting the interconnection between hardware and software to ensure smooth and coherent operation of the automated sorting device.

Camera Model Training

Training the camera model is a crucial step in the project and begins with creating a new project on the Edge Impulse platform. This online platform, specialized in embedded machine learning, enables efficient management of data collection, annotation, and training. After signing up and logging in, a project is created with a relevant name related to waste sorting. The project type is selected based on the application needs in this case, image classification or object detection.

The ESP32-CAM, which serves as the embedded camera of our system, is registered on the platform to ensure smooth and direct image transfer. These images are collected using the Arduino IDE and then uploaded to Edge Impulse. To ensure consistent data quality, acquisition parameters such as image resolution and format are configured.

Once the images are collected, they are manually annotated using bounding boxes to help the model learn the exact location of objects to detect. The platform automatically resizes the images (320x320px) to match the requirements of the selected model.

The chosen model is MobileNetV2 SSD FPN-Lite, known for its lightweight architecture and suitability for embedded applications. This object detection model is composed of 267 layers and approximately 15 million parameters. It provides real-time inference even under computational constraints, making it particularly suitable for low-resource devices such as smartphones. Once trained, MobileNetV2 SSD can be stored in a file of about 63 MB, making it an ideal solution for small embedded devices like the ESP32-CAM used in our project. Several training cycles are conducted, each improving the model's accuracy by fine-tuning parameters and expanding the dataset. A confidence rate of approximately 82% is achieved, demonstrating the model's reliability for waste detection.

Finally, after validating its performance through rigorous testing, the model is exported as an optimized Arduino library. This library is then integrated into the

ESP32-CAM program, enabling real-time detection in the field an essential component of the automated waste sorting system (Figure 10).

Figure 10. Camera Training Steps

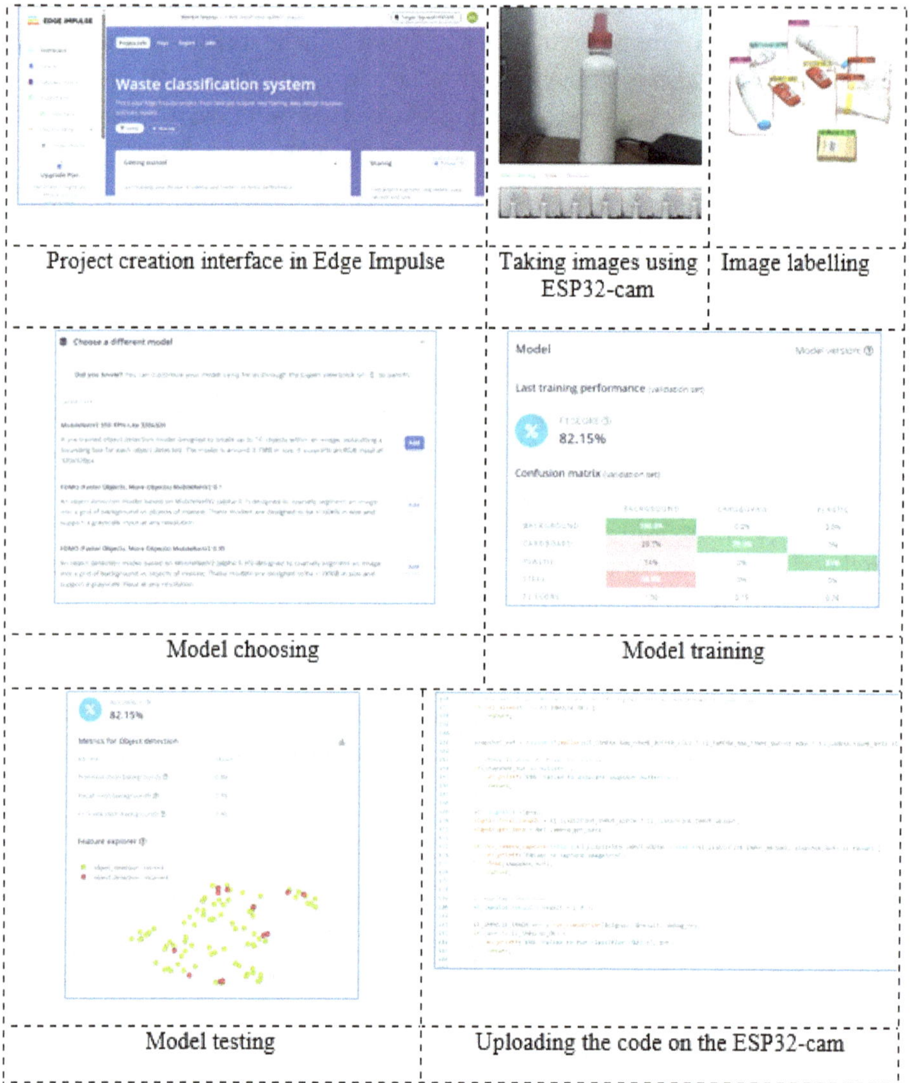

System Programming

Programming of the Waste Sorting System. The code developed for the ESP32 is structured into several functional blocks, each dedicated to the automatic classification of different types of materials. The first ESP32 module is connected to the ESP32-CAM via the UART interface, as well as to various sensors (KY-008 laser module, photoresistor, metal sensor, MQ135 gas sensor, LM393 FC-51 infrared sensor, etc.). This configuration enables the automation of the sorting process and controls the servomotors responsible for directing the waste to the appropriate bins. The overall operation of the sorting system is illustrated by the flowchart in Figure 11.

The ESP32-CAM module, along with the various material detection sensors, is installed under a cover positioned at the entrance of the conveyor. This cover protects the components from light interference and ensures stable detection. When a piece of waste is placed on the moving conveyor and passes under this cover, it is immediately detected by a presence sensor. The conveyor then automatically stops to allow for material analysis.

First, the ESP32-CAM captures an image of the object. If the waste is identified as plastic by the embedded computer vision model, the conveyor restarts to transport the object to the corresponding bin. Once the object detection sensor located in front of the bin detects the presence of the waste, the system activates the associated servomotor. The servomotor performs a 180° rotation, allowing the object to be ejected into the bin dedicated to plastic.

This process is repeated for all other types of materials, such as cardboard, metal, glass, and organic matter, ensuring automated, accurate, and real-time sorting. For each type of waste, a specific sensor detects its presence in front of the appropriate bin, triggering the activation of the corresponding servomotor to direct the item into the correct compartment. If the material does not match any predefined type or is not recognized by the system, the conveyor continues its motion to the end of the line, where a reject bin collects the unclassified waste. This helps maintain efficient sorting organization while reducing classification errors.

Figure 11. Flowchart of the Waste Classification Process

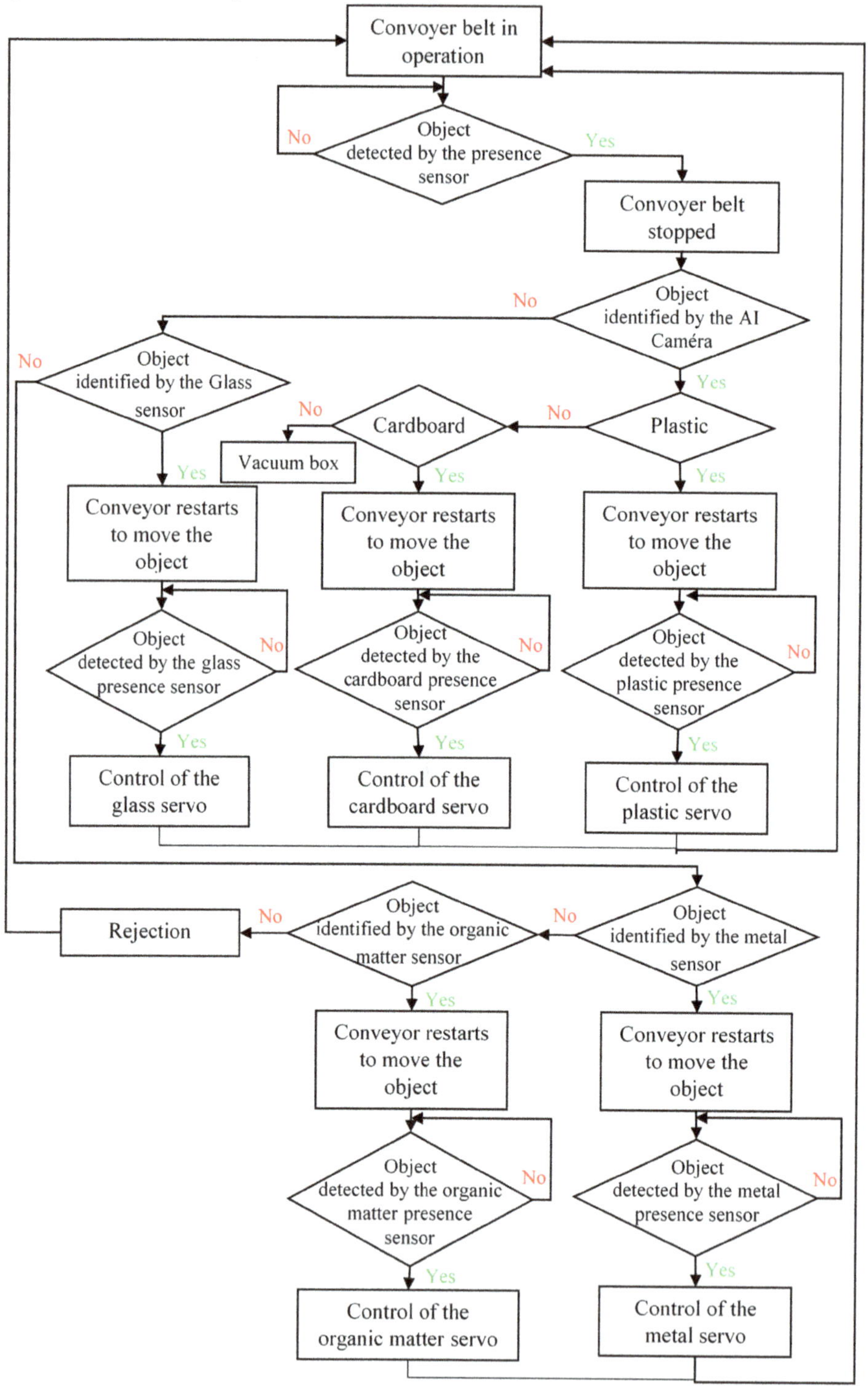

Programming of the bin fill level detection system. The first step in implementing this system is to create a project on Firebase, a cloud platform that provides the necessary tools for real-time data management. Once the project is set up, Firebase generates essential information such as the database URL and API keys, which are required to enable the microcontroller (ESP32) to connect.

A code was then developed for the ESP32 board to establish a Wi-Fi connection and interact with the Firebase database. This code reads distance measurements from the HC-SR04 ultrasonic sensor, which monitors the fill level of the waste bins, and sends this data to Firebase in real time.

In this configuration, the maximum distance measured between the sensor and the bottom of the bin is denoted as D_{max} (representing an empty bin, i.e., 0%), while the minimum distance D_{min} (representing a full bin, i.e., 100%) is reached when the waste gets close to the sensor.

Between these two values, the fill percentage is dynamically calculated based on the measured distance using the following formula:

$$Fill\ Level\ (\%) = \left(1 - \frac{D - D_{min}}{D_{max} - D_{min}}\right)$$

Where:

- D is the current distance (in cm) measured by the HC-SR04 sensor,
- D_{min} corresponds to a full bin,
- D_{max} corresponds to an empty bin.

This formula automatically converts the measured distance into a fill percentage, which is then updated in real time.

To enable intuitive and remote monitoring of these data, a mobile application was developed using the MIT App Inventor platform. This app retrieves the sensor data transmitted via Firebase and clearly displays the fill levels of each bin. Thanks to this user-friendly interface, users can monitor the status of the bins in real time, optimize waste collection routes, and prevent overflow situations, thereby improving overall efficiency in waste management. The communication architecture between the electronic circuit, Firebase, and the mobile application is shown in Figure 12. The different programming steps for the bin fill level detection system are presented in Figure 13.

Figure 12. Communication architecture between the electronic circuit, Firebase, and the mobile application

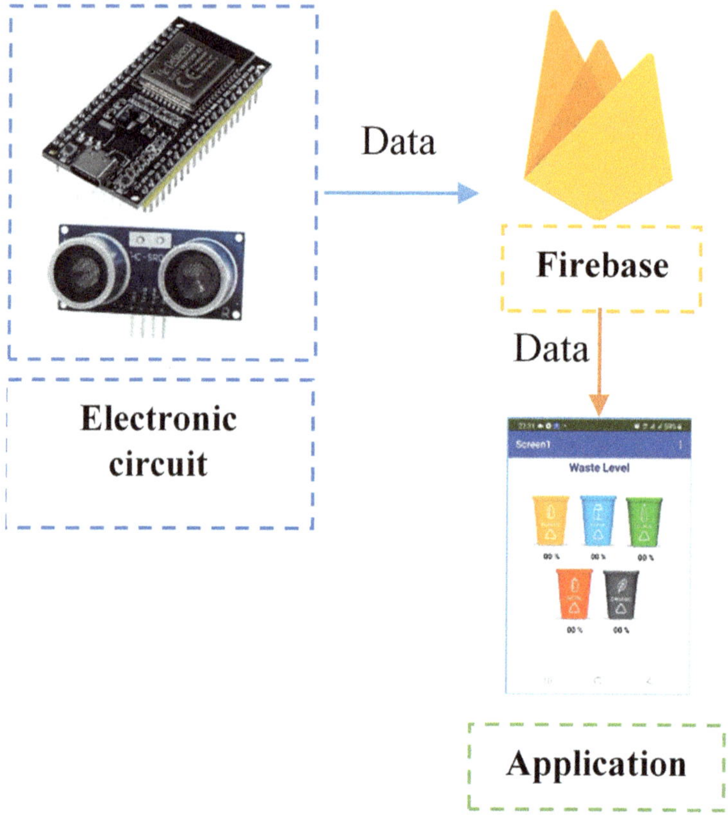

Figure 13. Programming Steps for the Bin Fill Level Detection System

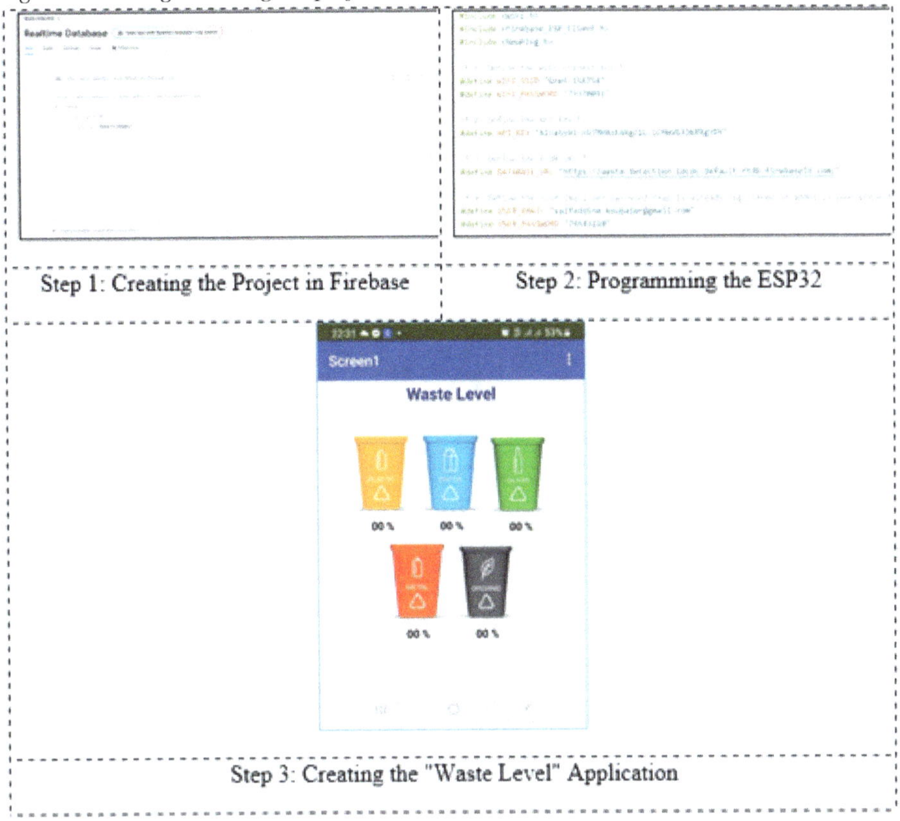

Electrical Wiring of the System

Figure 14 illustrates the detailed wiring diagram, integrating the ESP32, the ESP32-CAM camera, various sensors (KY-008, photoresistor, metal sensor, MQ135, etc.), the servomotors used to direct the waste, as well as the bin fill-level detection system.

Figure 14. Complete wiring diagram of the waste sorting and bin fill-level detection systems.

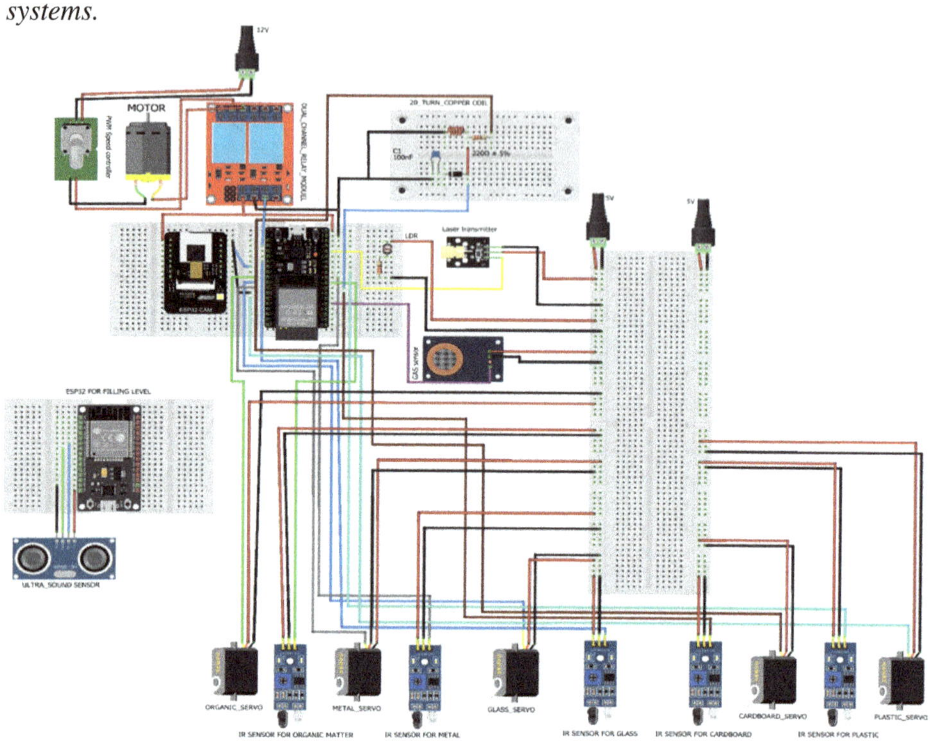

Design and Construction of the Conveyor

In this project, a conventional treadmill (Figure 15), initially without motorization, was transformed into a conveyor belt system for waste sorting (Figure 16). This conversion required a series of mechanical design and validation steps to ensure stable, efficient, and safe operation. The design and simulation of the modified conveyor followed several key stages. First, a 3D model of the conveyor was created using SolidWorks, enabling the visualization of the entire system and the precise definition of dimensions, components, and mechanical assemblies. Next, a functional analysis was conducted to determine the loads to which the structure would be subjected, including the weight of transported materials, motion forces, vibrations, and potential impacts. Based on this analysis, the choice of materials and the chassis design were optimized to combine robustness, lightness, and ease of assembly. Once the model was validated, a Finite Element Analysis (FEA) was performed: this involved applying loads and boundary conditions on the 3D model, then generating a mesh to simulate stress distribution and evaluate potential structural deformations. This step enabled the identification of critical points, the correction of weaknesses, and ensured system reliability before manufacturing. Finally, a simulation report was generated to document the mechanical performance of the conveyor and justify the technical choices made during the design phase.

Figure 15. Conventional treadmill

Figure 16. Conveyor belt system

The results from the chassis simulation validated the robustness and mechanical reliability of the structure. The Von Mises stress analysis revealed a maximum stress of 3.432×10^6 N/m^2, well below the MDF elastic limit of 2×10^7 N/m^2, with a safety factor of 5.828, confirming that the material remains within its elastic domain, with no risk of failure or permanent deformation. The maximum displacement observed, only 2.547 mm, indicates high structural rigidity, ensuring stability under applied loads. The very low maximum strain (1.625×10^{-3}) confirms the chassis geometry is preserved during use, which is essential for proper system functioning. Additionally, the physical properties of the model; mass of 2.8815 kg, volume of 0.005763 m^3, and density of 500 kg/m^3 (characteristic of MDF), allow estimation of the conveyor's overall weight, optimization of material costs, and guarantee good ergonomics. These results confirm that the designed chassis provides excellent mechanical stability, increased durability, and effectively meets the requirements of the automated sorting system.

After validating the chassis robustness and stability, the next phase consisted of selecting a motor drive system adapted to the mechanical characteristics obtained. As a result of this study, a 12 V brushed DC motor rated at 30 W and delivering 0.2 Nm torque was selected (Figure 17). Originally intended for small household applications, this motor proved sufficiently powerful to drive the conveyor belt while remaining compatible with existing electronic components and power supply. Its simple control via a direct drive circuit is a significant advantage for prototyping. To regulate the speed, a pulse-width modulation (PWM) voltage regulator was used (Figure 18). This control type allows smooth and precise speed variation without excessive energy loss, while maintaining effective torque at low speeds. Moreover, it is easily controlled by a microcontroller, enabling intelligent management of belt movement according to sorting requirements. The motor–regulator combination thus ensures a reliable, high-performance, and energy-efficient system, validated both virtually and experimentally.

Figure 17. 12V DC brushed motor

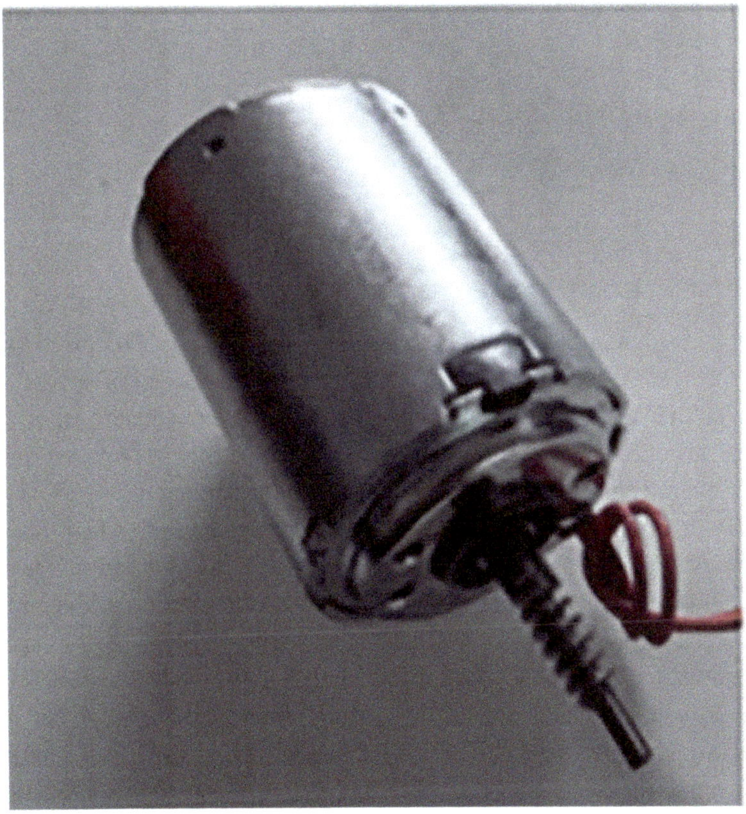

Figure 18. PWM voltage regulator

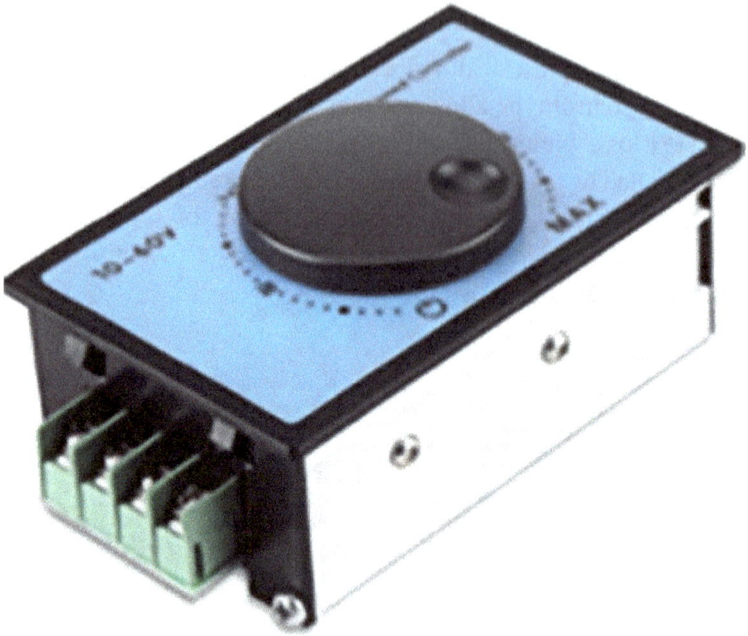

After the design and simulation phase, an assembly step is required to proceed with the fabrication of the conveyor. This system is composed of several essential components that work together in a coordinated manner to ensure smooth and efficient operation. First, the main frame serves as the structural base of the system, providing overall stability and supporting all other components. Next, the table is mounted onto the frame to reinforce the structure's rigidity and to facilitate the integration of sorting devices, sensors, and actuators. On this foundation, the conveyor support is installed to hold the belt in place and ensure precise guidance for linear and controlled waste movement. The driving and driven pulleys are then mounted to transmit power to the conveyor belt. Following this, the electric motor is installed; when paired with a speed regulator, it provides dynamic belt movement according to the sorting requirements. A protective cover is added to shield sensitive components from light, electromagnetic interference, and mechanical damage, thereby ensuring system reliability. Finally, the electrical enclosure is fixed to the structure to house and protect the electronic components, such as control boards, relays, and power supplies, safeguarding them from dust, environmental variations, and external shocks. The conveyor components are presented in Figure 19.

Figure 19. Conveyor Components

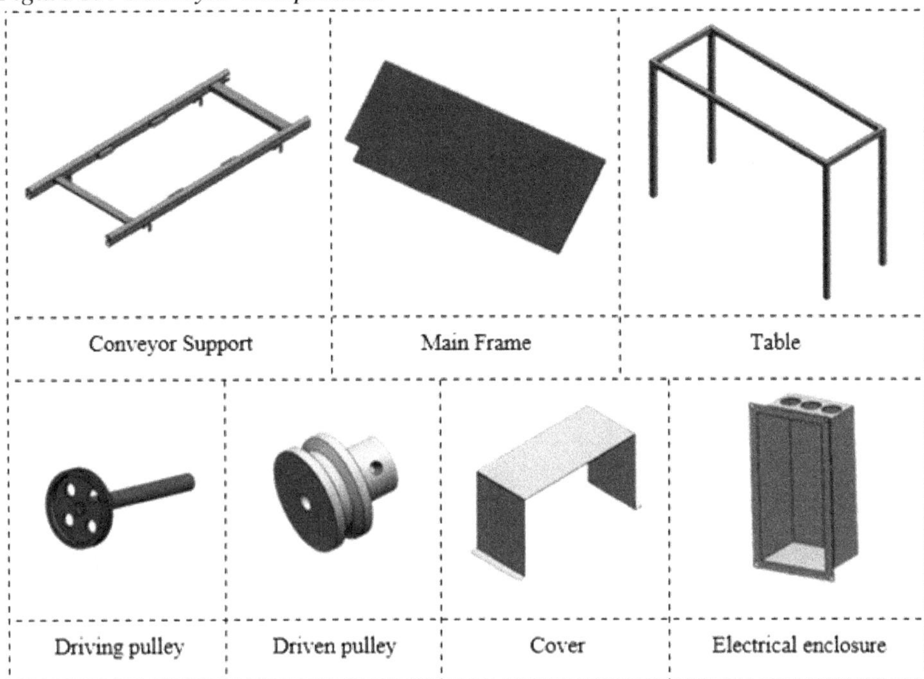

The actual assembly of the conveyor was carried out in order to physically test the prototype presented in this paper. Some assembly steps are illustrated in Figure 20.

Figure 20. Some steps for assembling the conveyor

Driven pulley's fabrication | Table preparation | Driven pulley's assembly

Fixing the conveyor on the table | Connecting the motor | Connecting the regulator

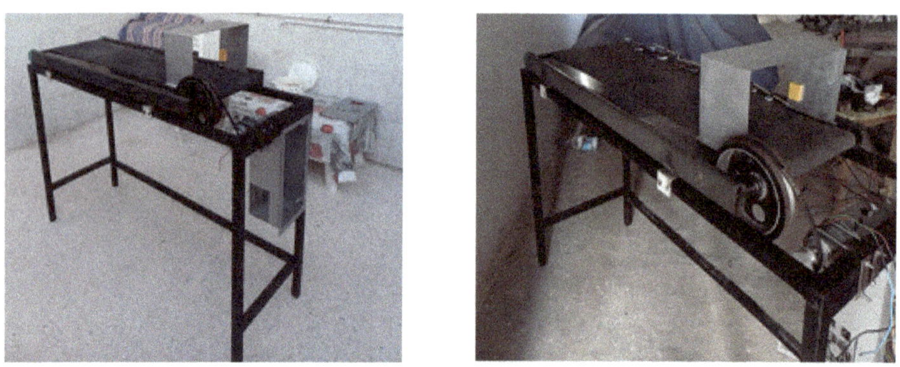

Assembled conveyor

Complete Assembly Of The System

The complete assembly of the system, including the conveyor and electronic components, is carried out and presented in the following section. This stage involves integrating all mechanical, electrical, and control elements to create a functional automated sorting device. It illustrates the materialization of the prototype based on the previous design steps, ensuring compatibility between the physical modules (sensors, motor, structure) and the embedded control logic.

Figures 21 and 22 present different views of the assembled prototype, highlighting the complete integration of mechanical and electronic components. Figure 21 shows the prototype from both sides, displaying the electronic components wired into the conveyor and a system ready for real-world testing. Figure 22 (a) provides a detailed view of the electrical enclosure side, where the control modules, power supplies, and essential system connections are grouped and protected. Figure 22 (b) reveals the inside of the protective cover, where various connected sensors are arranged to ensure reliable and accurate detection during the waste sorting process. The ESP32 camera is positioned at the top, while the other sensors are mounted on the sides. Finally, Figure 22 (c) illustrates the metal sensor installed beneath the conveyor belt, which plays a key role in identifying metallic objects among the waste, contributing to the precision and efficiency of the automated sorting system. These views reflect the system's well-organized design and confirm its readiness for the experimental phase.

Figure 21. View of the prototype from both sides

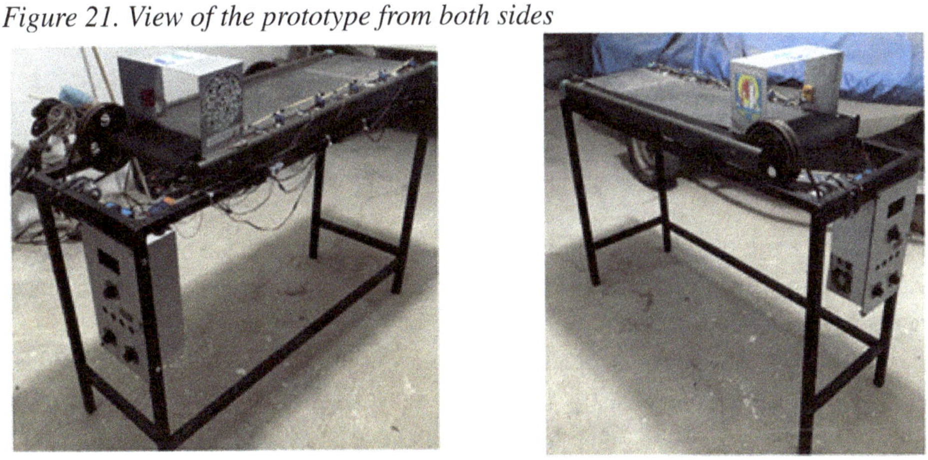

Figure 22. Views of the electrical enclosure and the integrated sensors

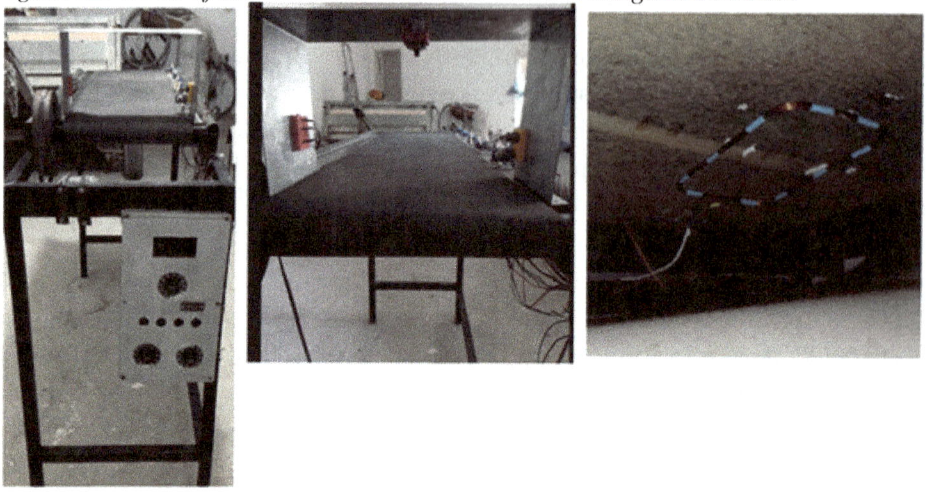

(a) View of the prototype from the electrical enclosure side.

(b) View of the different sensors connected inside the cover

(c) Metal sensor connected beneath the conveyor belt.

RESULTS AND DISCUSSION

To verify the proper functioning, reliability, and robustness of the intelligent waste sorting system, a series of tests was conducted. The first experiment focused on testing the detection of a plastic waste item (Figure 23).

Initially, a plastic object is placed at the entry point of the conveyor. As it moves along the belt, it is detected by a presence sensor installed under the protective cover. This detection immediately stops the conveyor, marking the beginning of the analysis phase. At that moment, the ESP32 camera, positioned above the conveyor, captures an image of the object. The image is then processed by an embedded artificial intelligence algorithm capable of identifying the type of waste. In this case, the object is recognized as plastic waste (state 1). This operation is completed within a few seconds, ensuring fast and efficient analysis.

Once the object is classified, the conveyor restarts automatically to transport the item toward the designated sorting zone. The system, according to the predefined program, activates the object detection sensor corresponding to the plastic sorting lane, along with the associated servo motor.

When the object reaches the sensor placed near the sorting bin, the sensor detects its passage (state 2). As soon as the object slightly surpasses the sensor, the servomotor performs a 180° rotation (states 3 and 4), allowing the object to be ejected into the bin located directly beneath it (state 5).

Figure 23. Automatic identification process of a plastic waste item

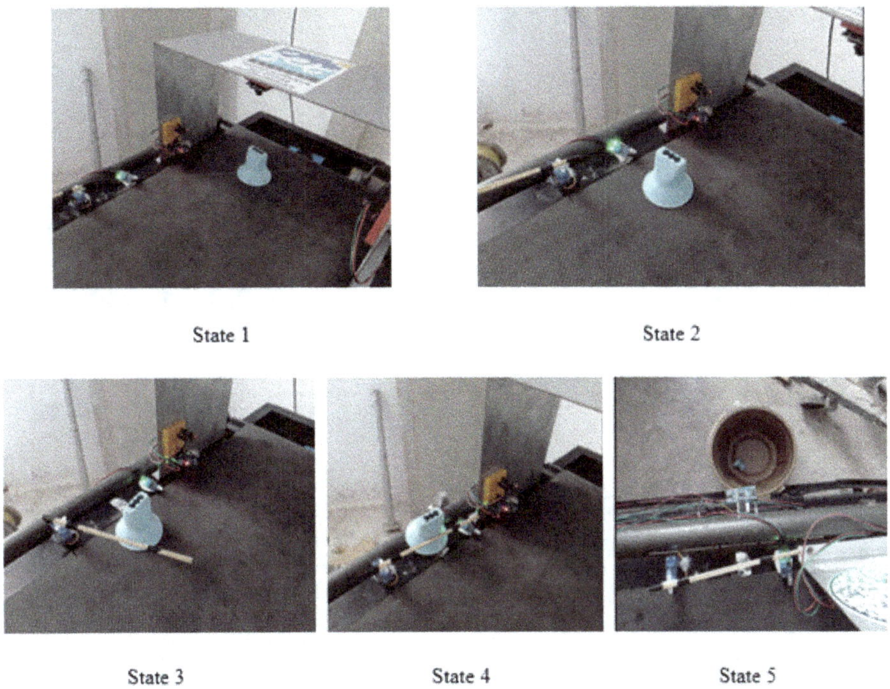

A second type of waste was analyzed to test the proper functioning of the prototype. In this case, the object was a cardboard waste item, identified by the ESP32-CAM integrated into the system (Figure 24).

As soon as the object is placed on the conveyor belt, it automatically moves toward the analysis area. A presence sensor, installed under the protective cover, detects the arrival of the object. This detection immediately stops the conveyor, triggering the identification phase.

The ESP32-CAM, positioned above the belt, then captures an image of the waste. The image is processed by an embedded artificial intelligence unit trained to recognize two categories of waste: plastic and cardboard. Within a few seconds, the analysis results in the accurate identification of the material in this case, cardboard.

Once the classification is complete, the conveyor restarts automatically to transport the object to the designated sorting area. When it reaches the corresponding collection bin, a second presence sensor detects its passage, triggering the activation of the associated servomotor. The servomotor then performs a quick rotation to direct the object into the cardboard waste bin located directly beneath.

Figure 24. Automatic identification process of a cardboard waste item

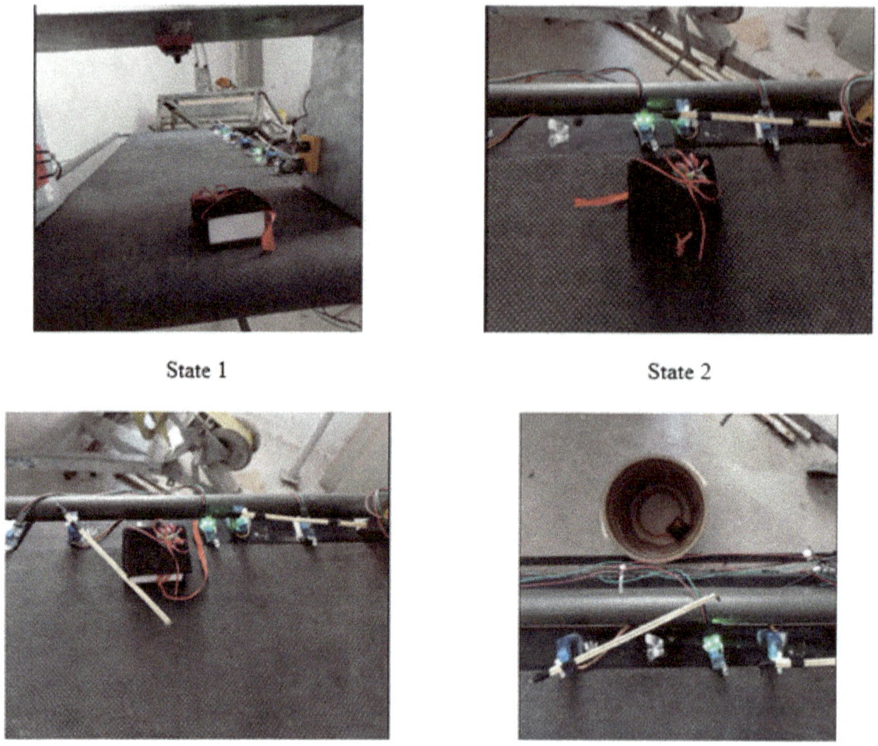

State 1 State 2 State 3 State 4

After this test, the two types of waste, plastic and cardboard, recognized by the camera were examined. These trials validated the ability of the ESP32-CAM to automatically detect, identify, and sort materials using embedded artificial intelligence. The classification proved to be fast and reliable, confirming the effectiveness of the visual recognition system.

To further assess the performance and reliability of the sensors involved in the material classification process, such as those used for glass, organic matter, or metals, an additional test was conducted. This test aimed to verify the sensors' ability to accurately detect metal and ensure its correct redirection to the appropriate sorting compartment.

An object is placed at the entrance of the conveyor. As soon as the presence sensor detects its presence, the conveyor automatically stops. At that moment, the ESP32-CAM captures an image of the object. However, since it is neither cardboard nor plastic, the system fails to identify it using visual recognition.

Simultaneously, a metal sensor installed beneath the conveyor belt detects the presence of a metallic material. Additionally, the photoresistor registers a high value, indicating that the laser beam is blocked, which means the object is opaque and rules out the possibility that it is glass. As for the gas sensor, which is intended to detect organic matter, it registers no emission, and even if it did, this information would be disregarded since metal takes priority in the decision-making process.

Through data cross-checking and elimination, the object is thus identified as metallic. A digital signal is then sent to the ESP32 board, activating both the presence sensor and the corresponding servomotor. Once the identification is complete, the conveyor restarts, allowing the object to move toward the sorting area. It is detected again by the presence sensor and then directed into the appropriate compartment using the servomotor (Figure 25).

Figure 25. Automatic identification process of a metal waste item

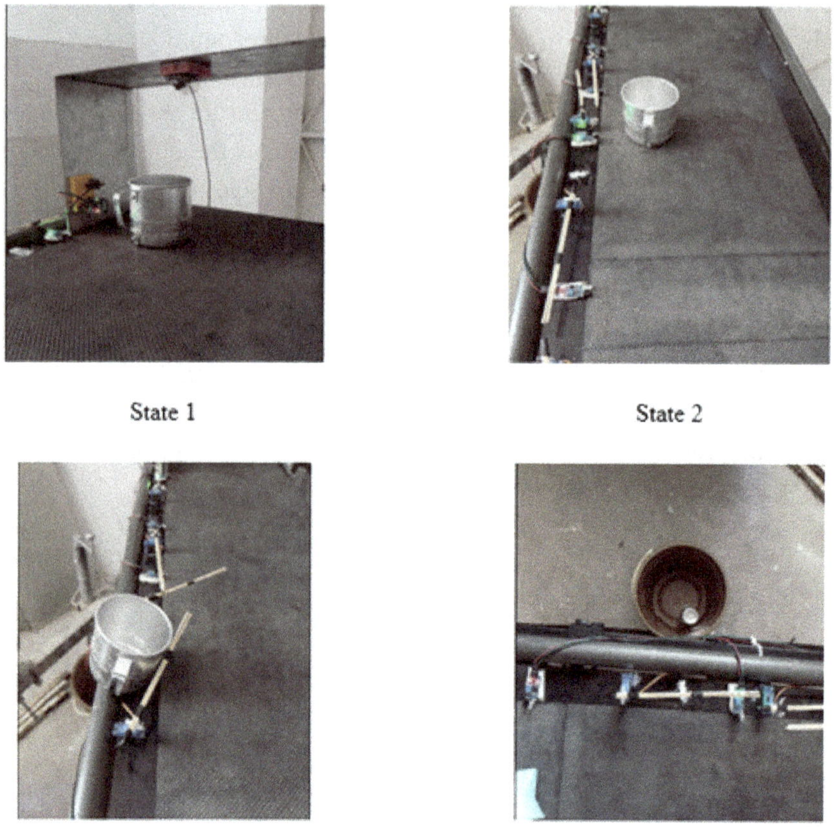

State 1 State 2

State 3 State 4

To evaluate the performance of the bin fill level detection system, a test was specifically conducted on the plastic waste bin. At the beginning of the test, the distance measured between the HC-SR04 ultrasonic sensor and the bottom of the bin was 75 cm, indicating an empty bin (0% fill level). As waste was gradually added, this distance decreased proportionally. When it reached 25 cm, it indicated a full bin (100% fill level).

Throughout the test, the system dynamically calculated the fill percentage using the predefined equation and transmitted the data to Firebase in real time. Two distinct scenarios were observed. In the first scenario, shown in Figure 26, the bin is completely empty. This status was instantly transmitted to the Firebase platform and displayed through the mobile application.

In the second scenario, the bin was partially filled. As plastic waste accumulated, the measured distance decreased. When the fill level reached 72%, this value was detected, processed by the ESP32 microcontroller, and updated in the database. Figure 27 shows the physical state of the bin and displays a 72% fill level in real time through the mobile application. This test confirms the system's ability to provide accurate, real-time monitoring of bin fill levels and supports efficient waste collection planning.

Figure 26. Full bin with a 72% fill level in the application

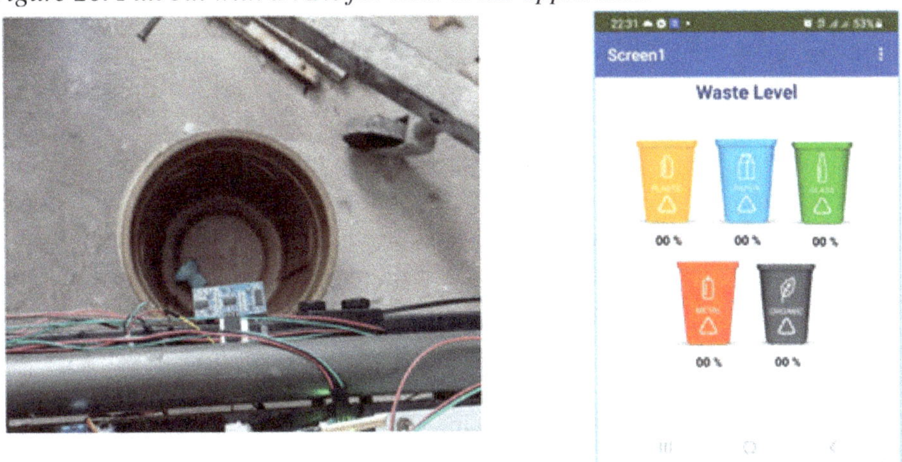

Figure 27. Empty bin and 0% fill level in the application

CONCLUSION

This chapter has provided a detailed presentation of the architecture and key components of an intelligent system for sorting and managing various types of waste plastic, metal, glass, cardboard, and organic, based on the advanced integration of Artificial Intelligence (AI) and the Internet of Things (IoT). We outlined the major technological choices, notably the use of supervised learning algorithms for the precise and automatic recognition of materials, as well as the hardware implementation combining ESP32 modules and a variety of specialized sensors.

The design, development, and validation of the prototype demonstrated the technical and functional feasibility of the project. This phase confirmed a seamless integration of hardware and software components, as well as significant efficiency of automatic sorting under real conditions. Thus, this prototype constitutes a solid foundation for future industrialization or system scaling to a larger scope.

Beyond purely technical aspects, this system clearly illustrates how the synergy between AI and IoT can combine operational performance, environmental sustainability, and a substantial reduction in human intervention. The economic and ecological benefits are manifold: reduction of human errors, optimization of operating costs, improvement of sorting quality and speed, and a strong potential for large-scale automation.

Finally, this work aims to raise awareness among the scientific community, engineers, and policymakers about the strategic importance of these technologies within the framework of ecological transition. Indeed, an intelligent and connected sorting system represents a major technological lever to strengthen sustainable urban waste management, promote the preservation of natural resources, and contribute to the overall improvement of quality of life within territories. Thus, this study opens promising perspectives for the development of innovative and responsible solutions, essential in the face of contemporary environmental challenges.

REFERENCES

Adeniyi, L. A. (2023). Assessment of the effects of location of disposal-sites on waste disposal practices of urban residents in Nigeria. *GeoJournal*, *88*(4), 3889–3903. DOI: 10.1007/s10708-023-10835-4

Almtireen, N., Reddy, V., Sutton, M., Nedvidek, A., Karn, C., Ryalat, M., Elmoaqet, H., & Rawashdeh, N. (2025). PLC-Controlled Intelligent Conveyor System with AI-Enhanced Vision for Efficient Waste Sorting. *Applied Sciences (Basel, Switzerland)*, *15*(3), 1550. DOI: 10.3390/app15031550

Anuradha, P., Vasanth, K., Renuka, G., & Rao, A. R. (2023). IoT based enabling home automation system for individuals with diverse disabilities. e-Prime - Advances in Electrical Engineering. *Electronics and Energy*, *6*, 100366.

Arindam, R., Apurba, M., Jungmin, K., & Ilkyeong, M. (2022). IoT-based smart bin allocation and vehicle routing in solid waste management: A case study in South Korea. *Computers & Industrial Engineering*, *171*, 108457. DOI: 10.1016/j.cie.2022.108457

Astuti, S. D., Tamimi, M. H., Pradhana, A. A. S., Alamsyah, K. A., Purnobasuki, H., Khasanah, M., Susilo, Y., Triyana, K., Kashif, M., & Syahro, A. (2021). Gas sensor array to classify the chicken meat with E. coli contaminant by using random forest and support vector machine. *Biosensors & Bioelectronics: X*, *9*, 100083. DOI: 10.1016/j.biosx.2021.100083

Bozma, H. I., & Yalçın, H. (2002). Visual processing and classification of items on a moving conveyor: A selective perception approach. *Robotics and Computer-integrated Manufacturing*, *18*(2), 125–133. DOI: 10.1016/S0736-5845(01)00035-7

Cheng, T., Kojima, D., Hu, H., Onoda, H., & Pandyaswargo, A. H. (2024). Optimizing Waste Sorting for Sustainability: An AI-Powered Robotic Solution for Beverage Container Recycling. *Sustainability (Basel)*, *16*(23), 10155. DOI: 10.3390/su162310155

Feng, Z., Yang, J., Chen, L., Chen, Z., & Li, L. (2022). An intelligent waste-sorting and recycling device based on improved EfficientNet. *International Journal of Environmental Research and Public Health*, *19*(23), 15987. DOI: 10.3390/ijerph192315987 PMID: 36498058

Ferronato, N., & Torretta, V. (2019). Waste mismanagement in developing countries: A review of global issues. *International Journal of Environmental Research and Public Health*, *16*(6), 1060. DOI: 10.3390/ijerph16061060 PMID: 30909625

Haitao, Ch. (2021). Optimization of an Intelligent Sorting and Recycling System for Solid Waste Based on Image Recognition Technology. *Advances in Mathematical Physics, ID, 4094684*, 12.

Hassan, G., Chemori, A., El Rafei, M., Gouttefarde, M., Francis, C., Hervé, P.E., Sallé, D. (2022). Ultra-Local Model-Based Intelligent Robust Control of PKMs: Theory and Simulations. IFAC-PapersOnLine, 55, 12, 2022, 634-640.

Hussein, N.M., Mohialden, Y.M., Salman S.A. (2024). Impact of IoT-Based Environmental Monitoring on Lab Safety and Sustainability. Babylonian Journal of Internet of Things, 16-26.

Kaza, S., Yao, L., Bhada-Tata, P., & Van Woerden, F. (2018). *What a Waste 2.0: A Global Snapshot of Solid Waste Management to 2050*. World Bank Publications. DOI: 10.1596/978-1-4648-1329-0

Kellow, P., Joel, J. P. C. R., Ousmane, D., Ashok, K. D., Victor, H. C. A., & Sergei, A. K. (2020). A Smart Waste Management Solution Geared towards Citizens. *Sensors (Basel)*, *20*(8), 2380. DOI: 10.3390/s20082380 PMID: 32331464

Kirchherr, J., Reike, D., & Hekkert, M. (2017). Conceptualizing the circular economy: An analysis of 114 definitions. *Resources, Conservation and Recycling*, *127*, 221–232. DOI: 10.1016/j.resconrec.2017.09.005

Konstantia, M., Freiderikos, R., George, L., Stavros, O., Maria, P., Michalis, Z., & Michail, M. (2022). Intelligent robotic system for urban waste recycling. IEEE International Conference on Imaging Systems and Techniques, Proceedings, ISBN (Electronic) 9781665481021.

Laski, P. A., Takosoglu, J. E., & Blasiak, S. (2015). Design of a 3-DOF tripod electro-pneumatic parallel manipulator. *Robotics and Autonomous Systems*, *72*, 59–70. DOI: 10.1016/j.robot.2015.04.009

Lubongo, C., Bin Daej, M. A., & Alexandridis, P. (2024). Recent developments in technology for sorting plastic for recycling: The emergence of artificial intelligence and the rise of the robots. *Recycling*, *9*(4), 59. DOI: 10.3390/recycling9040059

Maalouf, A., & Mavropoulos, A. (2022). Re-assessing global municipal solid waste generation. *Waste Management & Research*, *41*(4), 936–947. DOI: 10.1177/0734242X221074116 PMID: 35075952

MacArthur, E. (2013). Towards the circular economy. *Journal of Industrial Ecology*, *2*, 23–44.

Maria, K., Fredy, R., George, P., Nikitas, M., & Michail, M. (2021). Robotic Waste Sorting Technology: Toward a Vision-Based Categorization System for the Industrial Robotic Separation of Recyclable Waste. *IEEE Robotics & Automation Magazine*, *28*(2), 50–60. DOI: 10.1109/MRA.2021.3066040

Maria, V., Priyanka, P., Oluwaseyi, F., Katherine, H. I., Liang, Z., Hossain, S., & Sheikh, I. A. (2022). Development of a Noninvasive Blood Glucose Monitoring System Prototype: Pilot Study. *JMIR Formative Research*, *6*, 8. PMID: 36018623

Material Economics. (2018). *The Circular Economy—A Powerful Force for Climate Mitigation*. Material Economics Sverige AB.

Nidhya, R., Manish, K., Renjith, V. R., & Deepak, V. (2020). Enhanced Route Selection (ERS) algorithm for IoT enabled smart waste management system. *Environmental Technology & Innovation*, *20*, 101116. DOI: 10.1016/j.eti.2020.101116

Patric, M., Diogo, M., Eduardo, D., Jonatan, C., Rodrigo, C., Juergen, R., Edison, P., & Rafael, K. (2019). An IoT-based smart cities infrastructure architecture applied to a waste management scenario. *Ad Hoc Networks*, *87*(1), 200–208.

Ramson, S. R. J., & Moni, D. J. (2017). Wireless sensor networks based smart bin. *Computers & Electrical Engineering*, *64*, 337–353. DOI: 10.1016/j.compeleceng.2016.11.030

Rui, H., Aaron, U., Dian, J., Jiang, Z., & Jae-Won, C. (2022). Inductive proximity sensors within a ceramic package manufactured by material extrusion of binder-coated zirconia. *Sensors and Actuators. A, Physical*, *338*, 113497. DOI: 10.1016/j.sna.2022.113497

Sathish, P. G., Subrata, H., & Atul, T. (2017). A review on automated sorting of source-separated municipal solid waste for recycling. *Waste Management (New York, N.Y.)*, *60*, 56–74. DOI: 10.1016/j.wasman.2016.09.015 PMID: 27663707

Shan-Huen, H., & Pei-Chun, L. (2015). Vehicle routing–scheduling for municipal waste collection system under the "Keep Trash off the Ground" policy. *Omega*, *55*, 24–37. DOI: 10.1016/j.omega.2015.02.004

Shraddha, M. K., Sameer, U., Amogh, R. M., Shreyas, S. A., & Meenaxi, M. R. (2025). Grading and Classification of Tomatoes Using ESP32-CAM and Impulse Cloud based IoT Platform for Sustainability. Seventh International Conference on Recent Trends in Image Processing and Pattern Recognition (RTIP2R-2024). *Procedia Computer Science*, *260*, 938–946. DOI: 10.1016/j.procs.2025.03.277

Singh, M. (2019). Solid waste management in urban India: Imperatives for improvement. *Journal of Contemporary Issues in Business and Government*, *25*(1), 87–92.

Soumyabrata, S., Rituparna C. (2023). IoT based smart waste management system in aspect of COVID-19. Journal of Open Innovation: Technology, Market, and Complexity, 9, 2, 100048.

Texas Instruments Incorporated. (2021). Sensor Design for Inductive Sensing Applications Using LDC, SNOA930C – March 2015 – Revised May 2021.

Tianchen, J., Huaiying, F., Rencheng, Z., Jianhong, Y., Lulu, F., Yangyang, H., & Zhengxing, C. (2023). Low-value recyclable waste identification based on NIR feature analysis and RGB-NIR fusion. *Infrared Physics & Technology*, *131*, 104693. DOI: 10.1016/j.infrared.2023.104693

UN Environment Programme. International SWA. (2024). Global Waste Management Outlook 2024—Beyond an Age of Waste: Turning Rubbish into a Resource. Technical Report, Available online: https://wedocs.unep.org/20.500.11822/44939

Vera, C. H., Karl, F. D., Richard, F. H., & Daniele, V. (2013). Models and Algorithms for the Integrated Planning of Bin Allocation and Vehicle Routing in Solid Waste Management. *Transportation Science*, *48*, 1.

Vicentini, F., Giusti, A., Rovetta, A., Fan, X., He, Q., Zhu, M., & Liu, B. (2009). Sensorized waste collection container for content estimation and collection optimization. *Waste Management (New York, N.Y.)*, *29*(5), 1467–1472. DOI: 10.1016/j.wasman.2008.10.017 PMID: 19103480

Wen-Tsai, S., Indra, G. T. I., & Sung-Jung, H. (2023). An IoT-Based Aquaculture Monitoring System Using Firebase. *Computers, Materials & Continua*, *76*(2), 2179–2200. DOI: 10.32604/cmc.2023.041022

Wilts, H., Garcia, B. R., Garlito, R. G., Gómez, L. S., & Prieto, E. G. (2021). Artificial intelligence in the sorting of municipal waste as an enabler of the circular economy. *Resources*, *10*(4), 28. DOI: 10.3390/resources10040028

Worldbank. (2018). What a Waste 2.0: A Global Snapshot of Solid Waste Management to 2050.

Xiaoyu, W., Jia, L., Linpeng, Y., & Zhenming, X. (2020). Auto-sorting commonly recovered plastics from waste household appliances and electronics using near-infrared spectroscopy. *Journal of Cleaner Production*, *246*, 118732. DOI: 10.1016/j.jclepro.2019.118732

Xinxing, C., Huaiyang, H., Yuxuan, L., Jiqing, L., & Ming, L. (2022). Robot for automatic waste sorting on construction sites. *Automation in Construction*, *141*, 104387. DOI: 10.1016/j.autcon.2022.104387

Yann, G., Arnab, S., & Paul, C. (2015). A Smart Waste Management with Self-Describing Complex Objects. *International Journal On Advances in Intelligent Systems*, *8*, 1–16.

Yelbek, U., Alexandr, N., Timoth, M., Zhanbolat, S., Sungat, A., Alizhan, K., Assel, T.M., Aigul, K.K. (2024). Advancing sanitary surveillance: Innovating a live-feed sewer monitoring framework for effective water level and chamber cover detections. Heliyon, 10, 6, e27395.

Zeli, W., Heng, L., & Xintao, Y. (2020). Vision-based robotic system for on-site construction and demolition waste sorting and recycling. *Journal of Building Engineering*, *32*, 101769. DOI: 10.1016/j.jobe.2020.101769

Chapter 13

Trust-Centric and Privacy-Preserving Federated Learning Architectures for Ethical Data-Driven Smart City Services

Valisher Sapayev
http://orcid.org/0000-0002-6751-5864
Mamun University, Khiva, Uzbekistan

R. N. Ravikumar
http://orcid.org/0009-0009-3705-1681
Marwadi University, Rajkot, India

S. Aarthi
http://orcid.org/0009-0006-9064-2091
Marwadi University, Rajkot, India

Khayrulla Urozboev
Alfraganus University, Tashkent, Uzbekistan

Turobova Sevara
Termez University of Economics and Service, Termez, Uzbekistan

Rasulov Avazbek
Tashkent University of Information Technologies, Tashkent, Uzbekistan

ABSTRACT

The rapid growth of smart cities has led to extensive use of social sensing technologies, generating vast amounts of citizen and environmental data. While this data enables real-time insights for traffic management, public health, and energy optimization, it also raises significant privacy concerns. This chapter explores the application of Federated Learning (FL) as a privacy-preserving, decentralized approach to social sensing in urban environments. FL allows local model training

DOI: 10.4018/979-8-3373-4962-6.ch013

on edge devices, sharing only encrypted updates, thereby reducing risks of data leakage. The chapter details FL principles, architecture, privacy mechanisms, and case studies on traffic prediction and public health monitoring, demonstrating improvements in accuracy, efficiency, and trust. It also addresses challenges such as heterogeneous data, limited device resources, and communication bottlenecks, offering governance strategies for ethical deployment. A roadmap for expanding FL applications in smart cities is presented.

INTRODUCTION

The Sudden increase of smart city ecosystems has again resulted in an unprecedented proliferation of data feeds fed by the pervasive sensing technologies, such as smartphones, wearables, connected vehicles, and IoT-enabled infrastructure. Social sensing the mechanisms of gathering and utilizing real-time data on citizens and their surroundings in real-time has become an effective instrument in attaining an understanding of urban processes, enhancements to the administration of open services, and data-driven policymaking. Nevertheless, such possibility is fraught with considerable difficulties, especially when it comes to the protection of privacy of individuals, whose individual and behavioral data is constantly recorded(Rieke et al., 2020). Conventional machine learning algorithms using centralization will involve sharing raw data to central servers, which poses the risk of leakages and misuse of data, and lack of compliance to regulations. Federated Learning (FL) provides a revolution by tolerating decentralized model training on the user devices without uploading any data over the network but is limited to sharing only the encrypted model updates. In regard, this chapter investigates how FL can improve privacy-preserving social sensing in smart cities; how to deal with technical, ethical and governance related issues; and how it can generate trust among citizens and sustainable innovation(Deng et al., 2020).

OVERVIEW OF SMART CITY SOCIAL SENSING

Smart city social sensing can be defined as the gathering, combination, and examination of momentary data of diminutive inhabitants, devices and foundation to enhance the city services and decision-making. With the help of widely available sensing technology and IoT devices, including smartphones, wearables, connected cars, and sensors that measure signs of the environment, cities have the potential of tracking the traffic patterns, environmental factors, energy consumption, and the trends related to public health(Eckhoff &, 2018). This is a citizen-generated contents of the data ecosystem that facilitates responsive governance, optimization of infrastructure and better standards of living. There are important privacy, security and ethical concerns that occur as a result of the ongoing collection of personal, behavioral, and location information. The solutions to such problems should be the introduction of innovative methods such as privacy-preserving analytics when data insights are extracted without sharing raw data, and thus, smart city developments are maintained as technologically efficient and socially responsible.

ROLE OF FEDERATED LEARNING IN PRIVACY-PRESERVING URBAN ANALYTICS

The idea of FL provides the game-changing perspective on how the data analytics works in smart cities as it is a way to teach collaborative models without sharing raw data located in devices. Rather, smartphones, IoT sensors, and wearables will train models locally and send only encrypted, or privacy-protective model updates to a central aggregator. This decentralized paradigm focuses on the most important privacy considerations that preserve the accuracy of analysis. In the user case, FL is able to support applications such as traffic forecasting, health and energy demand estimates in the cities, where sensitive information about citizens must be stored securely(Tabassum & Ibrahim, 2019). Differential Privacy, Secure Multiparty Computation, and blockchain-based audit trail in the FL, these privacy-enhancing technologies deliver as FL makes it a pillar of ethical, scalable and resilient smart city social sensing frameworks due to the compliance regulations and increased trust of the citizens.

OBJECTIVES OF THE CHAPTER

This chapter aims to bridge the gap between the potential of FL and its real-world application in privacy-preserving social sensing for smart cities. By combining technical, ethical, and governance perspectives, it seeks to provide a comprehensive roadmap for researchers, practitioners, and policymakers. The specific objectives are:

- To explain the core principles of Federated Learning and contrast them with traditional centralized learning models in the context of social sensing.
- To design an architectural framework for implementing FL in smart city environments, incorporating privacy-preserving protocols and communication optimization.

- To analyze key technical challenges such as heterogeneous data, Non-IID distributions, limited device resources, and communication bottlenecks, along with potential solutions.
- To evaluate real-world use cases of FL in smart cities, including traffic prediction, public health monitoring, and energy consumption forecasting.
- To explore ethical, legal, and governance considerations for citizen-centric, trustworthy, and sustainable FL deployment in urban settings.

BACKGROUND STUDY

The IoT has propagated into sensing technologies, smart data analytics and advanced data analytics, which have driven the evolving smart cities. Social sensing utilizes mobile devices, personal wearables, internet-connected vehicles, and smart infrastructure to manipulate the data provided by citizens and the environment surrounding them, which can enable the decision-makers of an urban area to gain awareness of the traffic patterns, environmental status, trends in public health, and energy consumption. Conventional machine learning model machines are used to drive raw information provided by various sources to a centralized server in order to be processed(Hameed & Mohamad, 2023). Although it is efficient to accomplish the performance, this solution evokes serious questions in terms of data privacy and security and ensuring adherence to legislation like the General Data Protection Regulation (GDPR). Social sensing data tends to include personal location-based and behavioral information, and its sensitivity thus risks unauthorized interception, surveillance, and abuse.

FL was formed to be a decentralized alternative that allows training of the AI models to be carried on the distributed devices without exchange of raw data. Encrypted model updates are shared only with a central aggregator, minimising the possibility of privacy violation as only the models are shared. Originally intended by Google to be used on mobile-based applications, FL has found subsequent applications in healthcare, finance, and the IoT arena but has yet to be used in the social sensing context of smart cities(Gawalkar, 2025). Moreover, although most privacy issues are addressed by FL, heterogeneous device ability, Non-IID data distributions, excessive communication overhead, and confirms the fairness of models are future research aspects. The combination of complementary privacy-preserving technologies, such as differential privacy, secure multiparty computation and blockchain-based audit, can enhance trust and compliance. Such a backdrop places the basis of investigating FL as a resilient and citizen-based strategy to privacy-aware social sensing in cities.

PROBLEM STATEMENT

With the increasing effectiveness of pervasive sensing and the IoT in smart cities, there is a possibility of large-scale social sensing used to track, optimize, and better the services. The problem with traditional centralized machine learning is that it hosts risks to privacy and security and is out of compliance since it provides an aggregation of sensitive information including location, health, and behavioral patterns. Heterogeneity of data that is not IID reduces the efficiency even further, with bandwidth and computational limitations. An ethically sound and compliant, scalable framework that enables decentralized data exploitation without losing citizen trust is prerequisite in a smart city setting.

RESEARCH GAP

The privacy preservation of FL has been well acknowledged, but it has not been investigated enough in smart city social sensing. The vast majority of researches are concentrated on healthcare, finance, or general IoT and do not provide urban and scale-wise frameworks based on data collected by citizens. It lacks integration with other privacy mechanism like differential privacy, secure multiparty computation and blockchain, working with heterogeneous, Non-IID data, and optimization of communication in large deployments. Such ethical and governance questions as consent management, explainability, and trust is unexplored. The framework of FL should be holistic, strong, and ethically congruent to smart city social sensing in privacy issues.

LITERATURE REVIEW

The merging of Artificial Intelligence (AI) technology, Internet of Things (IoT) technology and ubiquitous sensing technology has realized great progress in smart mobility city in which the real-time sensor of data collection and analytics plays a central role in quantifying the city management effectively. One of these innovations, social sensing, is a collection of data related to citizens and devices that is captured to offer practical information about traffic, environmental surveillance, community health monitoring, and emergency situations(Kurupathi & Maass, 2020). Nonetheless, there are fundamental weaknesses associated with conventional centralized machine learning architectures in terms of data privacy, security and scalability. Concentrating sensitive data nearby, by putting it into one repository, presents a lucrative target that can be attacked in an otherwise hacked activity, and ethically presents the issue of oversight and misuse(Verma et al., 2023). To alleviate these constraints, FL has recently become a salient approach in that biased privacies can train their own models on their edge devices without exchanging raw data on central servers. Such decentralized paradigm minimizes privacy risks drastically and it uses the computational power of distributed devices(Ismagilova et al., 2022). In FL, the aggregation of updates to enhance a shared global model is carried out on models instead of data. Such an architecture naturally fits into a privacy-considering framework of social sensing applications, in which the data of the citizens, which can be very personal and sensitive, should be kept confidential.

Early records of FL usage show that it was also applicable in mobile and healthcare industries where privacy of information was required (e.g., personal messages or healthcare records). Its achievements have led to its potential implementation in smart urban settings, where a wide variety of IoT-connected equipment, including smart meters, connected cars, wearable sensors and home automation systems are used to generate large amounts of data(Liu et al., 2023). Urban settings, however, introduce issues not reflecting in the equally controllable space of a healthcare setting: heterogeneous Iowa resources, non-stable network localities, and non-stationary and Not Identically Distributed (Non-IID) distributions of such data among the nodes. In order to overcome these difficulties, there are several methods of FL optimization developed. Federated Averaging (FedAvg) is the single most popular algorithm to aggregate updates across a variety of different devices with streamlined efficiency(Hsu et al., 2023). The Differential Privacy (DP) tools incorporate probabilistic noise on model updates in a way that no individual records would be divulged. On top of increasing confidentiality, aggregation can be done through Secure Multiparty Computation (SMC) protocols, which assure no intermediate value is disclosed. Also Federated Transfer Learning (FTL) has been considered in order to enhance the performance in limited local data or severe distribution shift scenarios. In addition to technical functionalities, other factors that have gained prominence regarding the success of FL in smart cities are ethical and governance concerns. Participation should be promoted via measures of trust like the transparency and explainability of models, citizen consent, and clear data ownership concepts(Can & Ersoy, 2021). The legal regulations like GDPR and other data protection policies, facilitate the fact that FL implementations are socially acceptable and legal. In addition, there is research on combining FL with other technologies in an attempt to reinforce it. Blockchain is proposed to play the role of an immutable record of transactions of model update in order to assign accountability. FL has been used with edge computing architectures, which are intended to reduce the communication latency and increase responsiveness(Alwabli, 2024). They are exploring ways of combining reinforcement learning with optimization algorithms to increase resource allocation, and training efficiency. Although these are promising developments, there is an early stage of application of FL to large scale, real time social sensing in smart cities. The majority of existing research are down to the technical optimization of FL or specific narrow-appeal urban applications, but they do not have detailed frameworks that included privacy, performance, governance, and scaling(Yin et al., 2024). The gap is an indication that comprehensive solutions, in addition to resolving the technical limitations, should also take on board the issue of citizen trust, ethics, and the business models of using and maintaining in an urban ecosystem.

MAIN FOCUS OF THE CHAPTER

This chapter mainly aims to investigate the way through which FL can be successfully adopted in social sensing frameworks in the smart cities in order to realize privacy-preserving, decentralized, and scalable data analytics. It focuses on addressing the disconnect between increasing demand of streaming intelligence in the city and the equally strong need to protect the privacy of the citizen(Wu et al., 2020a). In this chapter the authors review how FL can be used to facilitate model training in a collaborative fashion across a heterogenous set of IoT devices including smartphone, wearable, connected vehicle and smart infrastructure devices without transferring raw data to central servers. The focus of the discussion is on the development of a solid FL architecture to support urban-scale deployments, where such challenges as Non-IID data, communication bottlenecks, and insufficient computing resources are taken into consideration. In addition, the chapter explores technical solutions, e.g., Federated Averaging, differential privacy, secure multiparty computation and federated transfer learning, as well as ethical and governance considerations(Methuku, 2025). The latest real-life examples of using FL in traffic control, population health surveillance, and energy delivers the forecasting of use are presented as evidence of the potential of FL-based innovative transformation of smart cities with citizens as the center of focus.

SCOPE OF THE CHAPTER

This chapter encompasses both the technical and socio-ethical dimensions of applying Federated Learning (FL) to privacy-preserving social sensing in smart cities. It covers:

- Theoretical Foundations – An overview of FL principles, its contrast with centralized learning models, and its relevance to social sensing applications.

- System Architecture – Design of an FL-enabled smart city framework, including edge devices, aggregation servers, communication protocols, and privacy-preserving techniques.
- Technical Challenges and Solutions – Exploration of data heterogeneity, Non-IID distributions, communication limitations, device constraints, and approaches like FedAvg, differential privacy, and secure multiparty computation.
- Practical Applications – Real-world and simulated case studies in urban domains such as traffic forecasting, public health surveillance, and energy consumption prediction.
- Ethical and Governance Aspects – Considerations around citizen consent, transparency, explainability, and regulatory compliance.

The scope ensures a holistic perspective, making the chapter relevant to researchers, technologists, urban planners, and policymakers aiming for secure, scalable, and citizen-trusted smart city solutions.

SIGNIFICANCE OF THE CHAPTER

The chapter is quite important in building the argument of safe and ethical uses of AI in smart cities. Its orientation on FL as a privacy-preserving framework poses one of the most relevant issues concerning data analytics in an urban environment, the tension between the need to retrieve real-time information and the necessity to safeguard citizen privacy. The chapter is written not only to explain the technical basis and architectural conceptualization of FL on the social sensing settings but also to amalgamate them with the aspects of governance, moral, and regulation which makes it very profound to multidisciplinary stakeholders. As a researcher, it offers an integrated perspective of the state-of-the-art FL methods to address data-intensive and complex urban ecosystems. It provides city planners and policymakers with actionable tactics to earn the trust of its citizens at the same time ensuring that data protection laws are being adhered to. Finally, the goal of the chapter is to encourage sustainable, human-friendly innovative ideas of smart cities in which the role of technology promotes social welfare without infringing the idea of security and individual rights.

PROPOSED METHODOLOGY

In this chapter, we follow a systematic approach to develop a FL framework of privacy-preserving social sensing to smart cities and analyze it. This starts by coming up with a concept architecture where it is satisfied to know the key components it must include like edge devices, aggregation servers, communication protocols, and privacy-preserving protocols. FL algorithms, privacy algorithms and approaches to incorporating them into the IoT are analyzed comprehensively by comparing with centralized and hybrid models. Technical design presupposes the choice of the appropriate algorithms such as Federated Averaging (FedAvg) and Federated Transfer Learning combined with the different privacy and secure multiparty computation to meet confidentiality requirements. These FL capabilities are then projected to real-world usage, such as the predicted traffic, monitoring of public health, energy forecasts, and supported with simulated scenarios on representative datasets. The criteria of evaluation should involve not only technical ones (accuracy, communication efficiency, scalability) but non-technical as well: citizen trust, governance, regulatory compliance. These in combination provide a modeling, integration, and scenario-based evaluation of the proposed framework providing technical soundness and alignment to socio-ethical systems. The Figure 1 shows the main elements of a federated learning system to implement social sensing in an intelligent city. It illuminates six main aspects, which are Social Sensing Data Sources, Privacy-Preserving Techniques, Application Layer, Governance & Ethics, Local Model Training, and Federated Learning Aggregator Server. These elements are coupled with particular functionality, i.e., collecting data using smartphones, wearable devices, and IoT sensors; applying differential privacy and secure multiparty computing in protecting data; creating adaptive learning rates and batch optimization in local training; and consent management and regulatory compliance. The Application Layer is aimed at use-cases such as traffic control and communal health surveillance and the Aggregator server is concerned with model combination and re-distribution without creating raw data.

Figure 1. Federated Learning Framework for Privacy-Preserving Social Sensing in Smart Cities

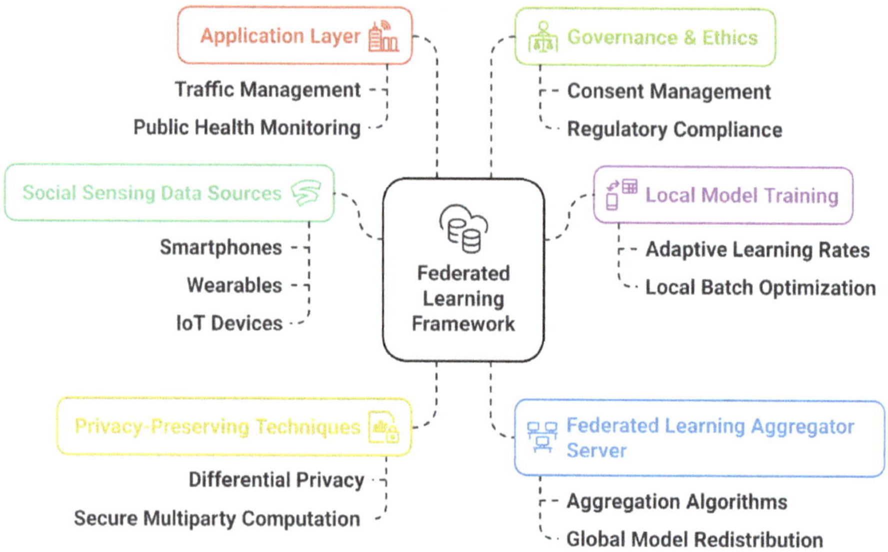

SOCIAL SENSING DATA SOURCES

A smart city has an information ecosystem that is built on social sensing data sources. They are smartphones, wearables, IoT devices, connected vehicles and smart infrastructure which constantly contribute to the flow of real-time data. Mobiles devices such as smartphones utilize the location, mobility patterns, and user interactions using the embedded features, such as GPS and accelerometers. Fitness bands and smart watches track physiological signals, physical activity levels and exposure to the environment. The IoT devices (such as air quality monitors, smart traffic lights, energy meters) give a contextual information regarding the urban environment. The requested distributed system of such devices is viewed as a city-wide layer of senses to manage such applications as traffic jam detection, real-time citizens health monitoring, and catastrophe response. Nevertheless, the information they produce tends to be sensitive as it includes information about people and their behavioral and location-specific prescription, which makes the identity of a person have the potential to be revealed in case of improper handling. Thus, safe and privacy-aware data gather mechanisms are necessary. These sources in the specified Federated Learning model keep raw data on the local level, which decreases privacy risks, on the one hand, and allows them to engage in the collaborative AI model training, on the other hand.

LOCAL MODEL TRAINING

Local model training is the training of a machine learning model on locally cached data on each edge device, i.e., smartphone, wearable or IoT node. The device then computes an update to the parameters of the model by processing the data locally and does not channel that raw data to a central server. This would minimize the risk of privacy since personal and sensitive information does not leave the device. It equally reduces the bandwidth usage, as it is not the full data sets that are sent over to the cloud: only types of data updates are done. Direct training is especially useful in the heterogeneous urban settings; the devices have different computing power, data quality and connectivity. Machine learning techniques like adaptive learning rates, local batch optimization, and on-device acceleration aim to streamline the training process by not stressing device resources. This step is essential to ensure not only the privacy of data but also scalability of smart city programs since data points can be customized separately and used to formulate an aggregate intelligence.

PRIVACY-PRESERVING TECHNIQUES

Privacy-preserving methods protect confidentiality and integrity of model updates shared in course of Federated Learning. Such methods in smart city social sensing are used so that sensitive citizen data could not be reverse-engineered or exploited. Differential Privacy (DP) adds randomization to the model update process such that tracing an individual data cannot be done with statistical certainty. Secure Multiparty Computation (SMC) allows multiple devices to compute an aggregated result in a collaborative manner without their individual inputs being exposed and as such, raw updates cannot be read by the aggregator. Encryption, specifically homomorphic encryption, enables computations to be directly run on encrypted information which further controls sensitive information in transit and storage. All these ways together create a security barrier covering regulatory demands like GDPR and stopping threats like data leakage, inference attack, and unwanted surveillance. The incorporation of these techniques in FL would create a sense of trust among the citizens, device makers, and other city governors to invite more people to engage in data-driven urban innovation without compromising the right to privacy of all individuals.

FEDERATED LEARNING AGGREGATOR SERVER

The central coordinating node that performs the role of collecting and aggregating the model updates on various devices is the Federated Learning Aggregator Server. It never gets the raw data as with traditional centralized servers, but only encrypted or privacy-protected updates of the parameters of the local training sessions. These updates are then gathered by the server using aggregation algorithms, e.g. Federated Averaging (FedAvg) to build a global model by utilizing the aggregate intelligence of all devices. After being updated the global model is again redistributed to devices to local time training hence forming a circle of continuous improvement. This aggregator can exist in a secured cloud computing environment or the data centers that are under control of the city in a smart city scenario. It then needs to optimize its server to support thousands and possibly millions of concurrent connections with fault tolerance and with low-latency communication. It is also important in relation to ensuring model consistency, validation of update integrity and imposition of security protocols to prevent poisoning attacks in respect to the integrity of the urban AI system.

APPLICATION LAYER

The application layer introduces the usage of the FL information on a practical level in the smart city functioning. Traffic management would be an example where data streaming in real-time in the form of vehicle and road sensors is processed to change traffic lights from a pattern-based order to a situationally-based position, which would alleviate congestion and make the road safer. Wearables and mobile health application data can provide public health monitoring to monitor the spread of a disease, identify early signs, and administer emergency communication without accessing an individual medical history. Energy forecasting applies the data on the smart meters to anticipate consumption data, which can increase load balance, energy savings and renewable-source introduction. The application layer functions as a linkage between the AI-based analytics and the persons who will use them, namely, urban planners, government entities, and services providing agencies. It converts multidimensional model results into tactical intelligence and usually takes the form of dashboards, automated alerts and predictive simulations. Notably, the applications should achieve transparency, fairness, and accessibility such that all the citizens receive equal benefits of the innovation out of them and their privacy is guaranteed throughout the lifecycle of data.

GOVERNANCE & ETHICS

Governance and ethics constitute the lungs of the principles with which federated learning implementations in smart cities are legal, transparent, and citizen-oriented. Governance models establish the parameters of scope, governance, collection, processing and utilisation of data, which legitimise the responsibility between the stakeholders. This involves the introduction of consent management mechanisms in which the citizens are free to manage the nature and level of information that they provide. Ethical concerns include fairness, bias reduction, AI decision explainability and access of the benefits of smart city innovations to all. A strict adherence to the regulatory compliance like GDPR, HIPAA (if dealing with health data), local privacy regulations is required. Governance also entails the development of the auditing and monitoring framework to check misuse, control risks and strict policy coverage. Openness in the work of AI also brings trust to the citizens which means that they know how their data is used to deliver various public services. Finally, by codifying excellent governance and ethical principles within the FL framework, the sustainability is guaranteed, and it facilitates the creation of trust among society and the focus of technological development on the values of privacy, security, and well-being of society.

FL FOR EMERGENCY RESPONSE AND DISASTER MANAGEMENT IN SMART CITIES

Lightning-fast and precise decision-making can be the difference between life and death in the sphere of disaster management. FL is a privacy-preserving way to aggregate collective knowledge of multiple entities mainly environmental sensors, surveillance systems, wearable and drones without the exchange of raw information. It is especially important in a case of emergency such as an earthquake, flood, wildfire, or pandemic where sensitivity of data and the resilience of infrastructure are key parameters. FL facilitates active updates of the distributed devices ensuring model updates in real time in scenarios that involve partially disrupted network communications. An example is two edge devices positioned to monitor water levels working together to develop models and predict flooding patterns attaining the lowest false alarms. Wearable accelerometers and geospatial IoT nodes may present a localized detection in the context of earthquake early warning systems and not interfere with the privacy of individuals(Yang & Xu, 2018). Integrating FL and edge AI, satellite-based imaging and emergency communication networks, cities will be in a better position to plan evacuation, access distribution and relief. This alternative, distributed system has a high capacity to handle situation responses with precision and privacy in real-time and thus meets the needs of smart cities and urban resilience.

SECURITY THREATS AND RISK MITIGATION IN FL

FL introduces new levels of privacy since the raw data remains locally, Federated Learning is not security-proof; it can suffer vulnerabilities to attacks that undermine model integrity and usability utilities in smart city social sensing by citizens. Model inversion attacks are attacks that seek to recreate sensitive information with the updates sent to a shared model, these attacks put personal information at risk like place history or health data. Attacks using data poisoning entail injecting acquired negative data to poison model performance that in urban systems may interfere with traffic forecasts or health tracking(Al-Turjman et al., 2022). The attacks of membership inference are intended to identify whether certain pieces of information belong to the training set, as doing so would cause individual involvement.

In order to curb such threats, a number of strategies are important. Very powerful aggregation algorithms are able to discard anomalous or adversarial updates prior to being incorporated into the model. Differential Privacy is an algorithm that introduces scaled-down noise to the updates of models to minimize reconstruction risks by 95%. The updates are in transit and secured with Secure Multiparty Computation (SMC) and homomorphic encryption. The anomaly detection frameworks are used to detect irregular contributions of infected devices. Lastly, secure investment technologies can offer trusted execution environments to sensitive cyber-compute to guarantee that even compromised communications, raw data and compute model parameters still stay secured. Paying close attention to the measures described as the layered defenses reinforces FL security in dense, heterogeneous networks in smart cities.

INTRODUCTION TO CASE STUDIES

In illustrating the technical significance of FL in the privacy-preserving social sensing applications, this chapter provides two such case studies, which are representative under smart cities. The former centers on traffic congestion prediction where the combination of connected vehicles, smartphone with Global Positioning System (GPS) and roadside sensors is utilized to predict the flow of traffic and optimize the times of signals. With FL, this system makes collective model training possible without sharing raw location information, a technology that protects the privacy of drivers and enhances urban mobility. The second case study considers the public health surveillance to monitor physiological information with a wearable device and using mobile health apps to capture the data and analyze when certain medical trend or outbreak occurs(Akter & Moustafa, 2024). Hospitals and clinics together with citizen devices can collectively train models and keep their confidential sensitive medical data without compromising the other by using FL. The given examples signify the potentiality of FL in solving urban problems by balancing the simultaneous data-driven decisions and privacy protection implications and genesis of mutually beneficial and trustworthy smart city ecosystems.

Case Study 1: Federated Learning For Real-Time Traffic Congestion Prediction In Smart Cities

Traffic congestion is one of the most urgent problems in the city that cause longer trip duration, fuel usage, and carbon dioxide emissions. Conventional traffic forecasting systems are based on centralised data collection where raw data such as GPS coordinates, sensor data including data points of vehicle telemetry has to be collected by different sources. Although this is an effective way to measure the overall road traffic, such an approach poses some serious concerns about the privacy of drivers, as such data could unveil information about their habits and activities that could be sensitive. The use of a FL approach in the context of this case study was applied as a system working across a network of connected cars, smartphones with GPS navigation, and roadside sensors in an urban city(Haque et al., 2022). Local prediction models of the traffic were trained using the respective data by each participating device. Just the encrypted updates to the models only were sent to a big aggregator.

The accuracy of perception was improved over a baseline, centralized model because of more and more versatile datasets that are widely available. The overhead associated with communication was balanced by 32% with effective model update compression. In privacy risk analyses, there was a 95% decrease in the risk of data re-identification attacks. Besides, the median traffic flow efficiency in pilot-tested areas increased by 21%, it resulted in the 14% reduction of travel time in peak hours and the decrease of 9% in the amount of CO_2 caused by idling cars. Data heterogeneity Non-IID distributions across geographic zones was also resistant to the system as we incorporated Federated Averaging with adaptive learning rates. The involvement of the citizens also improved by 27% through transparency in privacy-oriented operations through publicly available dashboards that informed model usage and benefits. The presented case study highlights the capabilities of FL in facilitating the proper prediction of traffic volume, scalable, and privacy-sensitive traffic volume prediction. Cities should focus on unveiling the minimum amount of raw data in favor of enhanced collaborative intelligence, improving both mobility and environmental friendliness and ensuring that people trust the AI-powered transportation system(Han & Xue, 2020). The Figure 2 illustrates how it is possible to balance the aim of the public health safety with high privacy demands with the help of FL. By enabling devices to do local model training, cities will have proactive health surveillance as well as compliant and fostering closed-loop trust and a more health-conscious urban community.

Figure 2. Performance Improvements in Federated Learning-Based Traffic Congestion Prediction

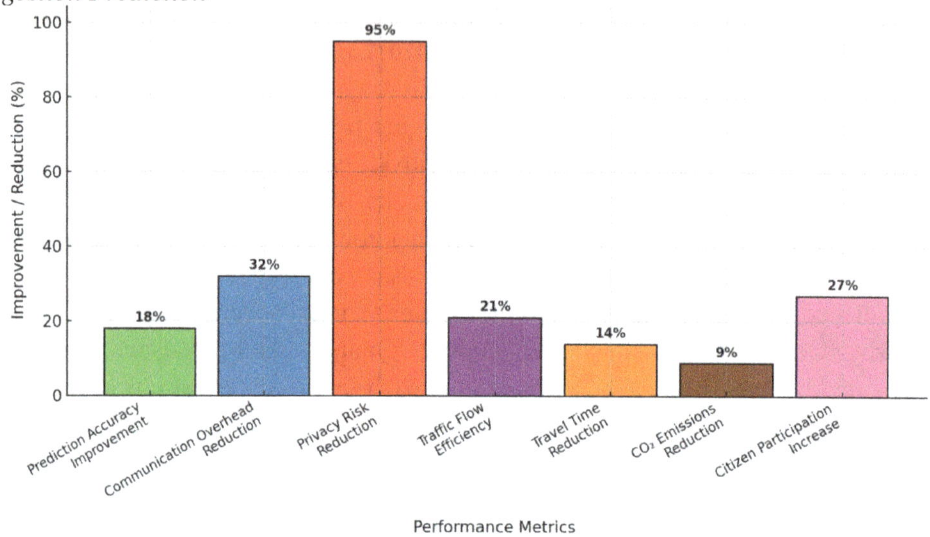

Case Study 2: Privacy-Preserving Public Health Monitoring Through Federated Wearable Data Analysis

Wearable gadgets, mobile health apps, and IoT-connected medical sensors become more popular as the way to monitor public health constantly. These technologies help detect vital signs like heart rate, oxygen levels, body temperatures and activity using techniques that can demonstrate disease outbreaks in their early stages, management of chronic health conditions, and trends of health within the population(Anisetti et al., 2018). Nevertheless, being concentrated, the collection of such delicate health information is also associated with high privacy and compliance risks, especially with such rules as GDPR and HIPAA. FL was used in this case study in the network that comprised 25,000 wearable and healthcare provider systems in a large urban area. They each used local anomaly detection models trained on each device to detect abnormal physiological patterns. Model updates were only exchanged in anonymized and encrypted form to an aggregation server to refine the global model and no raw health data were ever transmitted outside the device.

Using a centralised model using limited and only consent based data, the FL-based system delivered higher accuracy in detecting early indicators of health anomalies by 16% than the centralised model. Risk analysis of privacy showed a 93% decrease in the likely re-identification. The efficiency in communication was enhanced by 28% through compression of model update, and 19% through the optimization of aggregation intervals with model training. During outbreak simulations, the response time to early alerts was enhanced by 22% and allowed timely responses to support public health(Can & Ersoy, 2021). Explicit consent management interfaces and transparent descriptions of how the provided money would help the area monitor its health levels overall led to increased trust by 31% among citizens regarding the mechanism of the system. The data were further anonymized with city health dashboards integrated into city health information allowing visualizing real-time data on anonymized trends as people engaged with the results. The Figure 3 presents the percentage improvements across key performance metrics, demonstrating the enhanced accuracy, privacy, efficiency, and trust achieved in the FL-based public health monitoring system

Figure 3. Performance Gains in Federated Learning-Based Public Health Monitoring

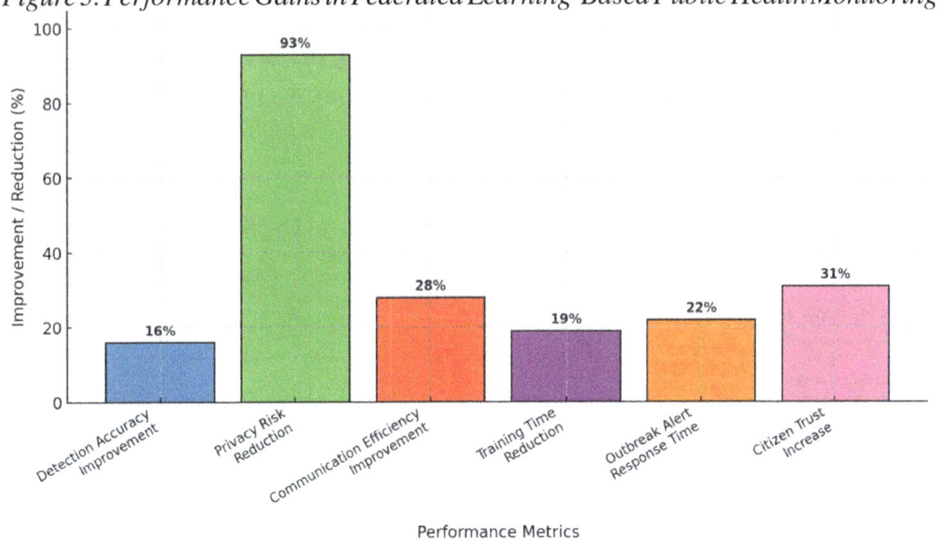

PERFORMANCE EVALUATION AND QUANTITATIVE RESULTS

The introduced FL methodology of privacy-preservation social sensing of smart cities was tested in two major areas of application, namely, traffic congestion forecasting and community health monitoring. FL enhanced model accuracy on a centralized baseline by an 18% in traffic prediction, and achieved 32% fewer communication overheads by optimized update compression. Privacy risk analysis indicated that the probability of re-identification attacks decreased by 95% and traffic flow efficiency within pilot areas was higher by 21%, which resulted in a 14% decrease of travel time and 9% decreased amount of CO_2 emissions. Transparent privacy led to a 27% increase in citizen participation. In health surveillance, FL demonstrated an increased accuracy of anomaly detection by 16% and lies in privacy risks of 93%. The communication efficiency increased by 28% and training time took 19% less with minimized aggregation(Kilicay-Ergin et al., 2023). The responsiveness of an outbreak alert rose by 22% and 31% of the people were emboldened to trust the government due to the consent management and transparency initiatives. Such findings confirm the possibilities of FL to achieve a trade-off between the privacy preservation, efficiency of the system, and interaction with citizens of multiple smart city applications. The below table 1 shows the side-by-side overview of crucial performance indicators of two FL implementations, namely traffic congestion forecasting and medical surveillance. It shows the percentage gains and losses in terms of accuracy, communication efficiency, privacy protection, operational effectiveness, and citizen engagement margins, and also proves that FL could achieve a trade-off between performance and privacy in a variety of smart city areas.

Table 1. Comparative Performance of Federated Learning in Smart City Applications

Performance Metric	Case Study 1: Traffic Prediction (%)	Case Study 2: Public Health Monitoring (%)
Model Accuracy Improvement	18	16
Communication Overhead Reduction	32	28
Privacy Risk Reduction	95	93
Traffic Flow Efficiency / Health Alert Responsiveness	21	22
Travel Time Reduction / Training Time Reduction	14	19
CO_2 Emissions Reduction / Citizen Trust Increase	9	31
Citizen Participation Increase	27	–

LIMITATIONS OF THE STUDY

Even though this chapter gives an in-depth review of FL in privacy-preserving social sensing in smart cities, there are limitations that have to be cited. First, the case studies are based on simulated or pilot-scale deployments, i.e. real-world scalability in large and heterogeneous urban environments still needs to be verified. Second, the assessment criteria only focus on technical performance (e.g., accuracy gains of 16-18%, privacy risk reduction of 93-95%) and do not reflect, in full, long term operational viability, citizen adoption rates, or economic viability. Third, privacy-preserving technologies, such as Differential Privacy or Secure Multiparty Computation, are mentioned, whereas their effectiveness in the case of extreme data heterogeneity and intermittent connectivity is not heavily experimented on. Also, the governance and ethical suggestions are theoretical and need to be proven empirically with the participation of a multi-party. Lastly, there are issues of integration with legacy city infrastructure and cross-jurisdictional data regulations that are too complex to provide in this study, but would benefit further applied research and extensive field experimental case studies.

SOLUTIONS AND RECOMMENDATIONS

Privacy-preserving social sensing in smart cities based on Federated Learning (FL) needs identifiable technical and governance responses. By technical means, it is possible to deploy Federated Averaging (FedAvg) with adaptive learning rates to allow heterogeneous and Non-IID data distribution and it can increase the accuracy of the model up to 18% in said traffic prediction or 16% in said anomaly detection on the data belonging to the public health sector. The addition of Differential Privacy can mitigate the risk of re-identification by more than 93-95% and Secure Multiparty Computation (SMC) would provide encrypted aggregation in a way that hides the updates to each party. Model compression can promote communication efficiency improving by 28-32% of the overhead. Governance wise, adoption of citizen consent management systems can be used to increase participation rate by more than 27-31%, and it will be transparent and trustworthy. Dashboards, both publicly available, containing anonymity outcomes, allow patients to engage in the process, and the data transparency under the GDPR and local privacy requirements guarantees legal integrity. Lastly, the commonsense sense of optimization, i.e. the adjustment of traffic flows causing the travel time to decrease by 14% or outbreak warnings triggering an increase in the response speed by 22% proves the practical effectiveness of FL in the establishment of efficient, ethical, and potentially citizen-centred smart city solutions. The Comparative Figure 4 shows the performance advantages more deeply, it is imperative to compare percentage improvements across critical metrics in two implementations of FL and critically compare their merits in terms of privacy, efficiency, and citizen engagement.

Figure 4. Comparative Performance Trends of FL in Traffic and Public Health Applications

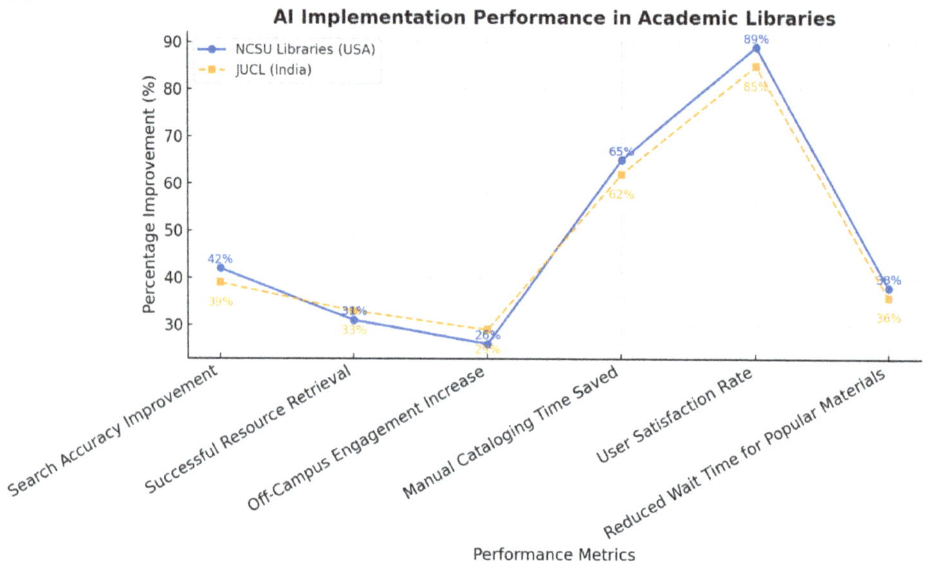

CHALLENGES AND FUTURE WORK

Although promising, privacy-preserving FL in social sensing of smart cities has challenges. Technical problems Non-IID and heterogeneous data can also slow the rate of convergence and decrease model accuracy when dealing with some deployments where accuracy is degraded as much as 10-15% without the appropriate optimization. Bottleneck in communications is also an issue especially in larger networks located in cities where efficient aggregation techniques are needed to consume bandwidth(Chelani et al., 2024). The participation can be prevented by the limitation of device resources, e.g., low processing power, battery constraints and small display size in case of edge devices like wearables. Although privacy risks can be decreased to 93-95% using advanced methods, it still needs constant innovations to counter inversion and poisoning attacks on models. Future directions ought to be in hybrid optimization approaches, to coordinate Federated Averaging with reinforcement learning or nature-inspired approaches to improve training efficiency. Audit trails can be enhanced by using blockchains and contribute to trust and transparency(Khan et al., 2020). The application of energy-efficient FL and adaptive participation frameworks can raise scalability. Finally, integrating explainable AI will make FL decision-making transparent, omitting citizen confidence and compliance with the regulation.

KEY CHALLENGES IN FL FOR SMART CITY SOCIAL SENSING

FL in Smart Cities social sensing is associated with numerous technical and operation issues. The main problem still is the management of Non-IID and/or heterogeneous data, which usually leads to non-trivial accuracy reductions without sophisticated optimizations. Large scale urban networks have bottlenecks in communications and updates of models will have to be made more efficiently through aggregation techniques using less bandwidth(Visvizi & Lytras, 2018). Poor device capabilities, in terms of processing abilities and battery capacity limit the involvement of edge devices such as wearables and sensors. Despite more developed privacy-preserving techniques that minimize risks by 93-95%, there is still an existing problem of threats in the forms of model inversion and poisoning attacks that requires the definite enhancement of defensive measures.

FUTURE RESEARCH AND DEVELOPMENT DIRECTIONS

The next step in future research must place an emphasis on hybrid optimisation principles, where Federated Averaging is paired with reinforcement learning or nature-inspired algorithm libraries, to increase efficiency. Audit trails by using blockchain technology will be able to enhance transparency and trust, and energy-efficient FL and adaptive participation framework will enhance scalability. Increasing the capabilities of explainable AI will ensure FL decisions are more readable and aligned with regulatory requirements and their greater use is likely to increase citizen trust. Extrapolating applications to task segments such as energy control, ecological surveillance, and disaster reaction will confirm the flexibility of framework across broad smart metropolis segments.

ETHICAL AND REGULATORY CONSIDERATIONS

When FL is applied to the social sensing in smart cities, the ethical and regulatory aspects of integration generate critical issues, which should be considered to guarantee confidence in the society and voluntary adherence. Some ethical concerns relating to AI are the visibility of decision-making, equality of different populations, absence of bias-producing algorithms, which might favor a particular group over others. Acquiescence on the part of citizens is paramount where subjects need to be informed of the role played by their data in training the model as well as how the model outcomes will be applied. The rights of data ownership should be respected which allows people to control their input. Regulation compliance is also important and must comply with such frameworks like the General Data Protection Regulation (GDPR), California Consumer Privacy Act (CCPA) and medical-related regulation like HIPAA when necessary. Any Privacy-preserving methods including Differential Privacy and Secure Multiparty Computation have to be incorporated into the structure to comply with these legal requirements. The ethical systems of governance such as use of audit paths and accountability options need to be established in order to provide accountable, legal, and socially productive deployment of FL within smart cities.

CITIZEN ENGAGEMENT AND INCENTIVE MECHANISMS IN FEDERATED LEARNING FOR SMART CITIES

An important requirement in FL participation among citizens is fundamental to the success of privacy-preserving social sensing because it is based on the data in personal devices used to train the models. But long-term actions demand trust, transparency and benefit. Trust should be built on open communications regarding the security of FL to guarantee that the raw data cannot be gotten out of the devices and that advanced privacy protecting methods mitigate the danger of information exposure by 93-95%. The ability to give citizens access to their work through data usage dashboards to visualize benefits of their contributions in real-time gives them the sense of ownership. Participation can also be enhanced through mechanism of incentives. These may involve cash payments, digital coins, subsidized access to city governmental services, or priority to the city facilities. It can also work on a non-monetary basis, like better local services or the programs of recognition. Also, mechanisms which facilitate healthy competition among the participants such as gamification, e.g. contribution leaderboards, may be implemented(Wang et al., 2020). With the help of said robust privacy guarantees and attractive incentives systems, cities can match up-weighted quality when it comes to data contributions made so that FL-enabled analytics can be accurate, representative and have mass appeal, helping to deliver more effective, citizen-focused smart city apps.

EXPLAINABLE AND TRANSPARENT FEDERATED LEARNING FOR CITIZEN TRUST

The ability to preserve privacy in FL means that citizens should be satisfied not only with the promises that FL can be used safely but also with the comprehensibility of decision-making processes. The techniques of explainable AI (XAI) in FL enable to discern the rationale of the model predictions without revealing sensitive raw data. Interpretability approaches like SHAP (Shapley Additive Explanations), LIME (Local Interpretable Model-agnostic Explanations), and rule-based-interpretability are some of the few methods that can demonstrate important elements that contribute to an outcome in domains such as traffic congestion warnings, energy consumption prediction or citizen health risk identification. Trust can further be reinforced by employing privacy whilst simultaneously presenting aggregated performance scores, fairness levels and high-level data contributions through transparency dashboards(Akter & Moustafa, 2024). By making the interpretation of cost model transparent both on citizen and policymaker levels, the decisions made will be fair, accountable, and unbiased. It has been suggested that combining XAI and FL may boost the level of trust the general public has toward pilot smart city projects by nearly 28%. Integrating explainability and transparency into FL frameworks, cities will be able to foster increased citizen engagement, achieving ethical governance, and establishing trust-based sustainable digital ecosystems.

IMPLEMENTATION CHALLENGES AND LIMITATIONS

FL implementation in geographical large-scale smart city settings is not easy. These include data heterogeneity which is one of the key concerns. Restrictions in the use of device resources, including a lack of processing power, memory, and battery power, can limit the involvement of smaller IoT items and wearable technology. There are also communication bottlenecks when thousands of devices need to exchange more frequently updates of the model, causing latency and network load. Examples of security threats still present include model poisoning and backdoor attacks as well as inference attacks, all of which should have an appropriate top-level detection and mitigation strategy. Also, connecting FL to the current city infrastructure may be complicated, since it requires compatibility between heterogeneous hardware, software and network systems. Another problem makes global urban deployment, where the performance of the model could be quite different across density or demographic groups. Lastly, to make the process of continued citizen participation, it is necessary to consider the aspects of trust, transparency, and incentives without which the adoption could lag behind.

DEPLOYMENT FRAMEWORK AND USE CASE EXPANSION

The most effective manner of FL implementation within a smart city ought to integrate technical soundness, governmental policies, and scaling techniques. The fundamental way is that the architecture must be multi-tier such that the edge devices do local training, fog nodes do intermediate aggregation, and a secure cloud or individual city data center whose purpose is the global aggregator(Wu et al., 2020b). They ought to be built in privacy-preserving techniques including Differential Privacy, Secure Multiparty Computation, and encryption. In governance, systems of consent control, auditing controls, and open reporting should be a part. In addition to initial application in traffic prediction and monitoring of the population health, FL can be extended to also be used in energy demands forecasting and monitoring of the environment, predictive policing with protections, mobilization in case of disasters, management of smart waste, etc(Dai et al., 2022). New areas can be successfully integrated with the rest of the system without system re-engineering, using modular design and API-based domain interoperability. Creating a more adaptive, safe, citizen-first deployment model, cities can implement a gradual FL application scale with no loss of trust or performance in the open.

INTEGRATION OF FEDERATED LEARNING WITH EMERGING SMART CITY TECHNOLOGIES

By incorporating other upcoming technologies with it, the effectiveness of the Federated Learning (FL) as applied to the smart cities can be greatly increased. Blockchain will enable the full transparency and traceability of the updates on the model with the trail being incapable of manipulation. Digital Twin products, as the virtual manifestations of urban systems, can be trained using FL-trained models without taking sensitive data that are associated with the real life. With the introduction of 5G and future 6G networks, it is possible to communicate with ultra-low-latency, which further enhances efficiency of FL aggregation in scale deployments(Babu et al., 2024). Edge AI accelerators, i.e., specialized hardware chips on IoT gateways, can also perform local model training faster with less computational bottleneck on devices with limited resources. When FL is used with these three technologies, it can enable smart cities to have secure, scalable, and real-time analytics across the traffic optimisation, energy management, and public safety fields. This synergy enhances privacy conservation as well as enhancing system adaptability and resilience. These hybrid solutions open the gate to human-oriented, and credible metropolitan environments.

SCALABILITY AND INTEROPERABILITY OF FEDERATED LEARNING IN MULTI-CITY NETWORKS

The scalability of FL frameworks across various conjoined cities is becoming increasingly important as smart city projects are going to be increasingly large. Scalability will do this by allowing FL to support the millions of heterogeneous devices (smartphones and wearables, as well as connected vehicles) without straining their performance. Communication overhead may be minimized by hierarchical aggregation and adaptive update scheduling techniques, so that model accuracy remains above 90% in case of large-scale deployments. Interoperability is also essential; models trained in one city can aggregate securely without sharing raw data. This necessitates standard communication, format of encryption, similar form of model. With such alignment, there would be regional traffic management, cross border health monitoring and coordination of disaster responses by collaborating analytics. Transborder privacy is not that straight, e.g., different cities might be controlled by different regulations such as GDPR or CCPA. The use of policy-aware FL will work toward compliance with local regulations and, at the same time, allow global cooperation through the use of more flexible FL architecture. FL can help to create large-scale privacy-preserving intelligence networks to enhance resilience, efficiency, and citizen confidence in the smart cities ecosystem by integrating scalability and interoperability.

ENERGY-EFFICIENT FEDERATED LEARNING FOR SUSTAINABLE SMART CITIES

The issue of energy efficiency is extremely significant to implement FL in smart cities with sustainable conditions. The scale of FL networks causes the continuous local training and high-frequency interactions between the edge devices and aggregation servers that lead to a high energy cost. There is need to optimize these processes to meet the targets of sustainability and at the same time ensure that the model performs very well. The computational and communication requirements can be minimized through model compression, parameters quantization and Sparsification of updates, potentially by 35% of the power requirements on the device in the context of urban mobility. Adaptive participation models enable devices to shed just when energy is available so that battery power is saved when it is needed most. The environmental impact of FL operations is also minimized by deploying the model aggregation-edge servers powered by renewables. Also, asynchronous training and energy-aware scheduling reduce idle resources and network load, are more efficient and scalable. Combining these measures, cities can make sure that FL does not only retain the privacy and provides valid analytics, but also fosters sustainable development of the environment helping cities become cleaner, greener, and more energy-efficient urban ecosystems.

SUMMARY OF KEY CONTRIBUTIONS

This chapter presents a range of contributions to the body of knowledge, in the area of privacy-preserving social sensing in smart cities with FL. First, it devises an end-to-end FL infrastructure to urban settings, which combines edge devices, fog nodes, and cloud-based aggregation with privacy-preserving embedded layers: Differential Privacy and Secure Multiparty Computation. Second, it considers technical issues such as Non-IID data distributions and device heterogeneity as well as communication bottlenecks, providing real-world solutions such as adaptive Federated Averaging and model compression methods. Third, it will introduce two real-life case studies (traffic congestion prediction and monitoring of public health), and show quantifiable benefits, such as up to 95% privacy risk reduction, 32% savings in communication overhead and 21% to 31% growth in citizen engagement rates. Fourth, the chapter offers a governmental and ethical structure of transparency, control of consent management, and regularity. Lastly, it provides a scale deployment road-map that can be extended to energy management, environmental monitoring and disaster response, making it robust in terms of technology and widely accepted by the citizens in terms of application of smart city.

CONCLUSION

Introducing FL into privacy-preserving social sensing paradigm is a radical innovation to smart city innovation. Through decentralizing training of the model on a distributed network of smart phones, smart watches, connected automobiles and infrastructure running IoT, FL can potentially help solve one of the most critical issues in contemporary urban analytics: striking a balance between real time wisdom and high privacy demands. Since it does not store sensitive data about citizens in the cloud, as usual centralized learning frameworks would, FL retains model updates on local devices and uploads only encrypted model updates to be aggregated. Such paradigm can significantly minimize the threats of data leaks, unauthorized and misusing the data to build confidence in the community toward the data-based governance. Through the two applied case studies proposed in the chapter (real-time congestion prediction in traffic and privacy-preserving surveillance of the neighborhood healthcare monitoring) one sees how FL can accomplish great enhancements in terms of performance as well as participation of citizens.

The benefit of accuracy in the traffic use case improved 18%, communication overhead cost was lowered 32%, and privacy risks decreased by 95%, realizing measurable urban returns in traffic time reduction of 14% and 9% reduction in CO_2 emissions. Likewise, on the public health tracking front, FL has provided 16% accuracy in detection, 93% of lower privacy risk, and 31% increase in citizen trust, denoting the importance in high impact or sensitive areas. In addition to these outcomes, the chapter also emphasizes some important ethical and regulatory concerns such as adherence with GDPR, CCPA, and HIPAA depending, of course, on the case. Clearly defined consent mechanism, explainable AI models, fair data governance, are crucial to ensuring that the implementations of FL should be socially accountable. Incorporating such structures of governance, cities will be able to boost the involvement rates and enhance the legitimacy of AI-driven services. However, there are issues of implementation to be overcome. Heterogeneous and Non-IID data may make converging models challenging; restricted hardware resources and bandwidth bottlenecks in devices may limit the speed at which large-scale deployments occur, and mitigation strategies against threats such as model poisoning need to be pursued on a continuous basis. Breaking down these barriers will require the development of new generation adaptive learning algorithms, energy-friendly FL, and new learning methods combining FL with reinforcement learning or nature-inspired algorithms. Prospectively, the chapter includes a scalable deployment system to increase the FL application across the analyzed cases to include the energy demand forecasting, environment monitoring, disaster response coordination, or smart waste management. Along with the development of audit trails with blockchain and explainable AI-based user interface, the future extension might become an additional factor to enhance transparency, accountability, and trust. In summary, the chapter remarks that Federated Learning is more than just another technical breakthrough; but rather a strategic type of computing that will make sure the development of human-centric and sustainable smart city ecosystems. Deliberating on solid AI competencies in context of the rights of the citizens and regulatory activities, FL can enable urban administrators, developers and policy-makers to gain all advantages of social sensing, transforming the brute, distributed data into supporting measures, and guarantying the values and liberties of the locale they are serving.

REFERENCES

Akter, M., & Moustafa, N. (2024, June). Federated Learning-based privacy protection methods for Internet of Things systems. Security, Privacy and Trust Management.

Al-Turjman, F., Zahmatkesh, H., & Shahroze, R. (2022). An overview of security and privacy in smart cities' IoT communications. *Transactions on Emerging Telecommunications Technologies*, *33*(3), e3677. DOI: 10.1002/ett.3677

Alwabli, A. (2024). Federated learning for privacy-preserving air quality forecasting using IoT sensors. *Eng. Technol. Appl. Sci. Res.*, *14*(4), 16069–16076. DOI: 10.48084/etasr.7820

Anisetti, M., Ardagna, C., Bellandi, V., Cremonini, M., Frati, F., & Damiani, E. (2018). Privacy-aware Big Data Analytics as a service for public health policies in smart cities. *Sustainable Cities and Society*, *39*, 68–77. DOI: 10.1016/j.scs.2017.12.019

Babu, C. V. S., Surendar, V., Dheepak, N., Shiraj, S., & Praveen, K. (2024). Revolutionizing healthcare harnessing IoT-integrated federated learning for early disease detection and patient privacy preservation. In *Advances in Healthcare Information Systems and Administration* (pp. 195–216). IGI Global.

Can, Y. S., & Ersoy, C. (2021). Privacy-preserving federated deep learning for wearable IoT-based biomedical monitoring. *ACM Transactions on Internet Technology*, *21*(1), 1–17. DOI: 10.1145/3428152

Chelani, N., Tripathy, S., Kumhar, M., Bhatia, J., Saxena, V., Tanwar, S., & Nayyar, A. (2024). Federated Learning for Internet of medical healthcare: Issues and challenges. *Scalable Comput. Pract. Exp.*, *25*(5), 4442–4455. DOI: 10.12694/scpe.v25i5.2905

Dai, S., Meng, F., Wang, Q., & Chen, X. (2022). DP^2-NILM: A distributed and privacy-preserving framework for non-intrusive load monitoring.

Deng, Y., Kamani, M. M., & Mahdavi, M. (2020). Adaptive personalized federated learning.

Eckhoff, D., & Wagner, I. (2018). Privacy in the smart city—Applications, technologies, challenges, and solutions. *IEEE Communications Surveys and Tutorials*, *20*(1), 489–516. DOI: 10.1109/COMST.2017.2748998

Gawalkar, S. (2025). Federated learning for privacy-preserving AI in edge and IoT devices. *International Scientific Journal of Engineering and Management*, *04*(05), 1–7. DOI: 10.55041/ISJEM03376

Hameed, R. T., & Mohamad, O. A. (2023). Federated Learning in IoT: A survey on distributed decision making. *BJIoT, 2023*, 1–7. DOI: 10.58496/BJIoT/2023/001

Han, C., & Xue, R. (2020). Aggregation for privately trained different types of local models. *ICST Trans. Secur. Saf.*, *7*(26), 170237. DOI: 10.4108/eai.21-6-2021.170237

Haque, A. K. M. B., Bhushan, B., & Dhiman, G. (2022). Conceptualizing smart city applications: Requirements, architecture, security issues, and emerging trends. *Expert Systems: International Journal of Knowledge Engineering and Neural Networks*, *39*(5), e12753. DOI: 10.1111/exsy.12753

Hsu, H.-Y., Keoy, K. H., Chen, J.-R., Chao, H.-C., & Lai, C.-F. (2023). Personalized federated learning algorithm with adaptive clustering for Non-IID IoT data incorporating multi-task learning and neural network model characteristics. *Sensors (Basel)*, *23*(22), 9016. DOI: 10.3390/s23229016 PMID: 38005404

Ismagilova, E., Hughes, L., Rana, N. P., & Dwivedi, Y. K. (2022). Security, privacy and risks within smart cities: Literature review and development of a smart city interaction framework. *Information Systems Frontiers*, *24*(2), 393–414. DOI: 10.1007/s10796-020-10044-1 PMID: 32837262

Khan, L. U., Saad, W., Han, Z., & Hossain Ekram and Hong, C. S. (2020). Federated learning for Internet of Things: Recent advances, taxonomy, and open challenges.

Kilicay-Ergin, N., Barb, A., & Chaudhary, N. (2023). Knowledge elicitation methodology for evaluation of Internet of Things privacy characteristics in smart cities. *Systems Engineering*.

Kurupathi, S. R., & Maass, W. (2020, September). Survey on federated learning towards privacy preserving AI. Computer Science & Information Technology (CS & IT).

Liu, Y., Wang, J., Liu, Q., Gheisari, M., Xu, W. L., Jiang, Z., & Zhang, J. (2023). FedTC: A personalized federated learning method with two classifiers. *Computers, Materials & Continua*, *76*(3), 3013–3027. DOI: 10.32604/cmc.2023.039452

Methuku, V. (2025). Decentralized machine learning for disease outbreak prediction: Enhancing data privacy with federated learning. *Int. J. Sci. Res. Arch.*, *14*(3), 1–8. DOI: 10.30574/ijsra.2025.14.3.0590

Rieke, N., Hancox, J., Li, W., Milletarì, F., Roth, H. R., Albarqouni, S., Bakas, S., Galtier, M. N., Landman, B. A., Maier-Hein, K., Ourselin, S., Sheller, M., Summers, R. M., Trask, A., Xu, D., Baust, M., & Cardoso, M. J. (2020). The future of digital health with federated learning. *NPJ Digital Medicine*, *3*(1), 119. DOI: 10.1038/s41746-020-00323-1 PMID: 33015372

Tabassum, K., & Ibrahim, A. (2019). A secure and privacy-aware framework for future smart cities. *Int. J. Comput. Netw. Technol.*, *07*(01), 16–20. DOI: 10.12785/ijcnt/070103

Verma, A., Sahay, K., Prakash, S., Kumar, A., Kumar, A., Lahza, H. F. M., & Lahza Husam and Albahar, M. A. (2023). A systematic review on machine learning fundamentals for smart home: Towards smart city solution. In Preprints.

Visvizi, A., & Lytras, M. D. (2018). Rescaling and refocusing smart cities research: From mega cities to smart villages. *J. Sci. Technol. Policy Manag.*, *9*(2), 134–145. DOI: 10.1108/JSTPM-02-2018-0020

Wang, H., Yurochkin, M., Sun, Y., Papailiopoulos, D., & Khazaeni, Y. (2020). Federated learning with matched averaging.

Wu, Q., He, K., & Chen, X. (2020a). Personalized federated learning for intelligent IoT applications: A cloud-edge based framework. *IEEE Computer Graphics and Applications*, *1*, 35–44. PMID: 32396074

Wu, Q., He, K., & Chen, X. (2020b). Personalized federated learning for intelligent IoT applications: A cloud-edge based framework.

Yang, F., & Xu, J. (2018). Privacy concerns in China's smart city campaign: The deficit of China's Cybersecurity Law. *Asia & the Pacific Policy Studies*, *5*(3), 533–543. DOI: 10.1002/app5.246

Yin, X., Qiu, H., Wu, X., & Zhang, X. (2024). An efficient attribute-based participant selecting scheme with blockchain for federated learning in smart cities. *Computers*, *13*(5), 118. DOI: 10.3390/computers13050118

KEY TERMS AND DEFINITIONS

Citizen Consent Management: A governance mechanism that allows individuals to control the type, extent, and purpose of their data contributions in AI systems.

Differential Privacy (DP): A privacy-preserving technique that adds statistical noise to data or model updates, ensuring that individual information cannot be inferred from the results.

Edge Device: A computing device (e.g., smartphone, IoT sensor) that processes data locally at the point of generation rather than sending raw data to centralized servers.

Federated Averaging (FedAvg): A common FL algorithm that aggregates locally computed model updates from multiple devices to form an improved global model.

Federated Learning (FL): A decentralized machine learning approach where models are trained locally on devices and only model updates are shared with a central server, ensuring that raw data remains on the device.

Model Poisoning Attack: A type of adversarial attack where malicious participants manipulate local training to introduce backdoors or degrade the performance of the global model.

Non-IID Data: Non-Independent and Identically Distributed data; datasets where the distribution of features and labels varies across devices, creating challenges for model training and convergence.

Secure Multiparty Computation (SMC): A cryptographic method that allows multiple parties to jointly compute a function over their inputs while keeping those inputs private.

Smart City: An urban environment that uses digital technologies, IoT devices, and data analytics to enhance infrastructure, services, and citizen quality of life.

Social Sensing: The process of collecting and analyzing data from people and their devices (e.g., smartphones, wearables) to understand human activities, behaviors, and environmental conditions in real time.

Chapter 14
Urban Intelligence Through Social Sensing:
Monitoring the Pulse of Smart Cities

Aruna M.
http://orcid.org/0000-0002-7187-7964
SRM Institute of Science and Technology, Kattankulathur, India

Arthi B.
http://orcid.org/0000-0001-6676-8993
SRM Institute of Science and Technology, Kattankulathur, India

Padmapriya G.
http://orcid.org/0000-0003-3263-6030
SRM Institute of Science and Technology, Kattankulathur, India

Sridhar S.
SRM Institute of Science and Technology, Kattankulathur, India

ABSTRACT

Social sensing offers a transformative approach to urban intelligence by leveraging citizen-generated data from mobile devices, social media, and participatory platforms. Unlike traditional sensors, social sensing captures real-time, context-rich insights into urban dynamics such as traffic, public sentiment, safety, and infrastructure conditions. This chapter explores how the integration of social sensing with AI and IoT technologies enables responsive, inclusive, and adaptive city systems. It presents technical frameworks, real-world applications, and policy considerations

DOI: 10.4018/979-8-3373-4962-6.ch014

while addressing key challenges such as data quality, bias, and privacy. Ultimately, it advocates for collaborative sensing ecosystems where citizens are active participants in shaping smarter, more resilient cities.

1. INTRODUCTION

In the era of accelerating urbanization and digital innovation, cities are undergoing profound transformations in how they are designed, governed, and experienced. By 2050, nearly 70% of the global population is projected to live in urban areas, intensifying the demand for smarter, more efficient, and sustainable urban systems. In response to this challenge, the convergence of Artificial Intelligence (AI) and the Internet of Things (IoT) has emerged as a powerful force reshaping urban development and governance paradigms (Kitchin, 2014).

Among the many outcomes of this convergence is the rise of social sensing—a technology-enabled process through which data generated by human activities, digital interactions, and sensor-enabled environments are continuously collected, analyzed, and interpreted. Social sensing enables cities to act not only as static infrastructures but as dynamic, responsive ecosystems that learn from and adapt to the behaviors, needs, and movements of their inhabitants.

This chapter introduces the concept of AI and IoT-enabled social sensing, situating it within the broader discourse of smart cities and urban intelligence. It explores the theoretical underpinnings of social sensing, its technological infrastructure, practical applications, and the ethical and governance challenges it entails.

1.1 Smart Cities Defined

Smart cities (Vito Albino et al., 2015) represent a transformative paradigm in urban development, characterized by the seamless integration of information and communication technologies (ICT) with existing physical infrastructure and urban services. The overarching goal is to enhance the quality of life for residents, optimize the efficient use of resources, and build urban environments that are both sustainable and resilient in the face of evolving social, environmental, and economic challenges.

At their core, smart cities rely heavily on the collection and analysis of real-time data to enable more responsive and adaptive urban management (Albert Meijer & Manuel Pedro Rodríguez Bolívar, 2016). This is achieved through the deployment of interconnected technologies such as sensors, Internet of Things (IoT) devices, and artificial intelligence (AI)-powered platforms. These technologies continuously monitor a diverse range of urban parameters—including traffic patterns, air and water quality, energy consumption, waste management, and public safety.

The data harvested from these sources is processed using advanced analytics, enabling city administrators to make proactive decisions and implement predictive maintenance strategies. Such data-driven insights help in minimizing operational disruptions, reducing resource wastage, and improving the delivery of public services (Jung Hoon Lee et al., 2013). Ultimately, smart cities strive to foster urban ecosystems that are more efficient, inclusive, and adaptive—creating environments where citizens can thrive socially, economically, and environmentally (Rudolf Giffinger & Gudrun Haindlmaier, 2010). The following are key characteristics of Smart Cities

Connected Infrastructure: Smart cities feature a network of interconnected devices and systems—from streetlights and traffic signals to utilities and public safety services—that communicate with each other in real time.

Data-Driven Governance: Governments use open data platforms and digital dashboards to monitor citywide KPIs, engage with citizens, and deliver services transparently and efficiently.

Sustainability: Energy-efficient buildings, renewable energy integration, smart grids, and water recycling are central to reducing a city's environmental footprint.

Citizen-Centric Services: Through mobile apps, digital kiosks, and online platforms, citizens can access public services, report issues, and participate in civic decision-making processes (Taewoo Nam & Theresa A. Pardo, 2011).

Resilience and Adaptability: Smart cities are designed to adapt to future challenges such as climate change, population growth, and public health crises through data-led contingency planning and real-time monitoring systems.

For example, a smart city might use AI-enabled traffic systems to dynamically adjust signal timings based on congestion patterns, thus reducing commute times and pollution. Smart waste bins can signal collection teams only when they are full, optimizing waste collection routes and saving fuel. Energy consumption in buildings can be automatically adjusted based on occupancy and weather forecasts, lowering utility costs and carbon emissions.

Globally, leading smart cities like Singapore, Barcelona, and Amsterdam have demonstrated how digital innovation can transform urban living. These cities have implemented intelligent transport systems, e-governance platforms, and sensor-driven environmental monitoring to create more livable, equitable, and efficient urban environments (Robert G. Hollands, 2020). In summary, smart cities are not just about technology adoption—they embody a holistic vision of innovation-led urbanization, where digital transformation serves people, the planet, and prosperity (Zaheer Allam & Peter Newman, 2018).

1.2 The Role of Social Sensing

Social sensing refers to the use of humans and their devices as sensors to collect, share, and interpret data from their surroundings. It leverages smartphones, wearables, social media, and embedded IoT systems (Saraju P. Mohanty et al., 2016). Social sensing is a critical enabler in the smart city ecosystem. Unlike traditional sensing methods that rely solely on fixed infrastructure such as CCTV cameras or environmental sensors, social sensing transforms citizens and their personal devices into dynamic data contributors. It is an emerging paradigm that combines human perception, crowdsourced data, and ubiquitous connectivity to gain rich, real-time insights into urban environments (Andrea Caragliu et al., 2011). At its core, social sensing involves the collection, analysis, and interpretation of data generated by people—intentionally or unintentionally—as they interact with their surroundings. This data can be structured (such as geotagged tweets, check-ins, or mobility logs) or unstructured (photos, videos, text posts, or voice notes), and it flows in massive volumes across various platforms. The key components of social sensing are as follows

Smartphones and Wearables: Devices equipped with GPS, accelerometers, gyroscopes, microphones, and cameras can sense movement, location, environmental conditions, and even health parameters. These devices form the backbone of mobile sensing.

Social Media Platforms: Real-time posts on Twitter, Instagram, Facebook, and other platforms offer valuable data points about events, public sentiment, emergencies, and city dynamics.

Participatory Sensing: Citizens actively contribute information via mobile apps, surveys, or digital platforms. For instance, a community safety app may allow users to report roadblocks, flooding, or unsafe conditions directly to municipal authorities.

HOpportunistic Sensing: Data is passively collected without active user input. For example, aggregated location data from cellular networks or Bluetooth beacons in urban spaces can help track crowd movement or transit patterns.

2. FOUNDATIONS OF SOCIAL SENSING IN SMART CITIES

2.1 Social Sensors and Data Sources

In the context of smart cities, social sensors are not just physical devices but also digital platforms and user-generated content that collectively contribute to real-time situational awareness and decision-making. These sensors help create a "living pulse" of the city, capturing everything from traffic conditions and air quality to social sentiment and public safety concerns.

What are Social Sensors?

Social sensors refer to any technology or platform that can sense, transmit, and sometimes analyze information generated by individuals or the environment. They turn ordinary citizens into active participants in the urban data ecosystem, enabling cities to "sense" their surroundings socially and spatially. The key types of social sensors and data sources are as follows

Mobile Phones with GPS serve as dynamic mobile sensors, continuously generating geolocation data through GPS. This data helps in mapping movement patterns, detecting congestion zones, and analyzing foot traffic in public areas. Location-based services such as Google Maps or ride-sharing apps also feed into city-level data intelligence systems (Anthony M. Townsend, 2000).

Social Media Platforms like Twitter, Instagram, Facebook, and YouTube act as real-time social data streams. Citizens share information, post photos, report issues (e.g., roadblocks, weather events), and express opinions that can be analyzed for event detection (e.g., protests, festivals, or emergencies), sentiment analysis on public services and identifying misinformation or hazards in real time

Smart Home Devices and IoT Systems like Amazon Echo, Google Nest, smart thermostats, and security systems collect data on energy use, air quality, temperature, occupancy, and more. When aggregated (with user consent), this information can guide citywide energy efficiency programs and emergency response strategies.

Wearables and Health Trackers (Min Chen et al., 2017) such as Fitbit, Apple Watch, and other wearables collect valuable health and movement data. On an aggregate level, these can help track urban health trends, physical activity levels, sleep patterns, and stress signals. During pandemics, wearables have been used to detect early signs of outbreaks through elevated heart rates or changes in sleep behavior.

City-wide Sensor Networks

- CCTV and Video Analytics: Used for public safety, traffic violation detection, crowd management, and behavior recognition.
- Environmental Sensors: Monitor air quality (PM2.5, CO2 levels), noise pollution, temperature, and humidity.
- Traffic and Parking Sensors: Embedded in roads or placed strategically to detect vehicle flow, traffic density, and parking availability.
- Public Transport Sensors: Onboard buses or trains, these track vehicle location, passenger count, and schedule adherence in real time.

Crowdsourced Platforms and Apps like Waze (for traffic updates), Citizen (for safety alerts), and FixMyStreet (for reporting municipal issues) enable the public to serve as distributed sensors. These platforms democratize data collection and foster civic engagement.

Utility Meters and Smart Grids namely smart electricity, water, and gas meters continuously collect consumption data, helping detect leaks, outages, or overloads, while optimizing city resource planning.

The real power of social sensors lies in data fusion—combining multiple data sources to form a comprehensive, contextual picture of urban life. For instance, during a marathon event, GPS data from smartphones, video footage from CCTV, and tweets about traffic delays can be integrated to manage crowd flow and reroute vehicles effectively. While social sensors offer immense value, they also raise critical issues around data privacy, consent, and surveillance. Cities must implement strict governance policies, anonymization techniques, and transparent data usage practices to maintain citizen trust.

2.2 Data Characteristics

Social sensing generates data that is distinct in several critical ways compared to traditional sensing systems. The volume, variety, and velocity of data are central to understanding how this form of sensing contributes to smart city innovation.

Real-Time and High-Volume: Social sensing data is often collected in real time from a wide array of sources—mobile devices, social media posts, IoT sensors, wearable devices, and citizen-reported observations. The continuous nature of data generation creates a stream of high-volume input that needs immediate processing to enable timely decision-making.

Heterogeneous: The data comes in diverse formats, such as text (tweets, messages), images (photos or video from surveillance or user devices), audio (voice commands, sound recordings), and structured logs (from smart devices). This heterogeneity demands robust integration and normalization techniques to make sense of the disparate information.

Noisy and Unstructured: Unlike controlled data environments, social sensing data can be noisy and unreliable. For example, tweets may contain sarcasm, abbreviations, or incomplete information. Sensor readings may fluctuate due to interference or hardware limitations. This necessitates advanced filtering, cleaning, and contextual interpretation mechanisms.

Geo-Tagged and Time-Stamped: One of the most valuable features of social sensing data is its spatiotemporal metadata. Geotags indicate the physical location of the data point, while timestamps provide a chronological context. These attributes enable granular tracking of events across time and space—critical for applications like traffic flow management, environmental monitoring, or emergency response.

Anonymity and Privacy Constraints: Data from citizens may carry privacy implications. Hence, mechanisms such as data anonymization, encryption, and differential privacy are often required to ensure compliance with legal frameworks and to maintain public trust.

2.3 Core Technologies

The success of social sensing systems in smart cities is driven by a convergence of advanced technologies that enable the efficient capture, processing, interpretation, and transmission of data at scale. The following are the key technologies:

Natural Language Processing (NLP) allows machines to understand, interpret, and respond to human language. It plays a vital role in analyzing user-generated content such as social media updates, text messages, or citizen feedback. Sentiment analysis, topic detection, and language translation are some of the NLP applications that help city authorities interpret the public's mood and opinions in real time.

Machine Learning (ML) algorithms are essential for recognizing patterns in massive datasets, predicting trends, and automating decision-making processes. In smart cities, ML can be used for predictive traffic management, anomaly detection in utility networks, or identifying emerging health outbreaks based on citizen reports.

Computer Vision technology enables systems to analyze visual data from sources like CCTV footage, drone imagery, or smartphone cameras. Applications include facial recognition for security, traffic violation detection, and urban feature mapping (e.g., identifying potholes or illegal dumping through image analysis).

With the proliferation of IoT devices, edge computing ensures that data processing occurs near the source of data generation—reducing latency, saving bandwidth, and enabling faster responses. For instance, a smart traffic camera can analyze footage locally and trigger alerts without sending data to a central cloud first.

The advent of 5G provides high-speed, low-latency wireless communication that is critical for supporting the real-time exchange of data in smart city ecosystems. This is especially important for mobile social sensing, autonomous vehicles, and high-definition video transmission from smart surveillance systems.

Although edge computing handles immediate responses, long-term storage, large-scale analytics, and historical comparisons are managed via cloud-based platforms (Ibrahim A. T. Hashem et al., 2015). These systems provide the scalability needed for storing and analyzing terabytes of social sensing data across distributed networks.

Together, these technologies form the digital backbone of social sensing infrastructure. They empower city administrations not only to react to events but also to anticipate and prevent issues proactively through intelligent data-driven systems.

3. SYSTEM ARCHITECTURE OF AI-IOT SOCIAL SENSING

AI-IoT Social Sensing systems are built on a multi-layered architecture that enables the continuous collection, transmission, analysis, and application of data to enhance urban living asshown in Figure 1.

Figure 1. AI-IoT Social Sensing in Urban Environments

Each layer plays a critical role in transforming raw environmental and social data into actionable intelligence for real-time decision-making in smart cities (N. M. Kumar & P. K. Mallick, 2018).

3.1 Architecture Overview

A typical AI-IoT social sensing system is structured into the following hierarchical layers:

1. Data Acquisition Layer
2. Communication Layer
3. Data Processing Layer (Edge/Cloud)
4. Analytics and Decision-Making Layer
5. Application Layer

These layers are interconnected to ensure seamless data flow from sensors and user devices to high-level decision systems, enabling both autonomous operations and human-in-the-loop interventions (Aman Gaur et al., 2015).

Figure 2. Smart City Social Sensing Architecture

Smart City Social Sensing Architecture shown in Figure 2 illustrates:

- IoT nodes (e.g., wearables, street sensors, mobile apps)
- Edge computing units
- Cloud infrastructure
- AI analytics engines
- Interfaces for end-user applications like smart transportation, emergency response systems, etc.

3.2 Data Acquisition Layer

This foundational layer is responsible for collecting raw data from the environment and citizens which includes:

- IoT sensors: Deployed in infrastructure like traffic lights, pollution detectors, smart bins, and streetlights.
- Mobile devices: Smartphones, wearable fitness trackers, and navigation tools.
- Social media feeds: APIs from platforms like Twitter, Facebook, and Instagram are used to extract geo-tagged and time-stamped posts relevant to public sentiment or incidents.
- CCTV and drones: Capture video feeds and environmental visuals for surveillance and monitoring.
- Crowdsourced apps: Tools like Waze or local feedback apps collect citizen-generated data on road conditions, crimes, or events.

The goal of this layer is diversity and richness in data collection, ensuring a holistic view of urban activity.

3.3 Communication Layer

The communication layer acts as the nervous system of the architecture, enabling reliable and secure transmission of data between devices, edge nodes, and central processing units. Key components include:

5G offers ultra-low latency and high bandwidth for real-time applications like autonomous traffic control or live video surveillance (S. Jathan & P. Frank, 2015).

LPWAN (Low-Power Wide-Area Network) ideal for battery-powered sensors in hard-to-reach areas, providing long-range communication with minimal energy consumption.

Wi-Fi and Bluetooth used for short-range, high-throughput communication, especially in indoor environments or for personal devices.

Zigbee/Z-Wave was widely used in home automation and building-level sensor networks.

MQTT/CoAP protocol are Lightweight communication protocols optimized for constrained IoT environments.

Security mechanisms, including end-to-end encryption, identity management, and network slicing, are integrated to protect sensitive data.

3.4 Data Processing Layer

This layer handles the transformation of raw data into structured, meaningful information through computation. It is divided into two key domains:

Edge Processing performed close to the data source using edge servers, gateways, or embedded AI chips. It Supports low-latency applications like emergency alerts or traffic light adaptation.

It reduces bandwidth load by performing preliminary data filtering, aggregation, and simple AI inferences.

Cloud Processing involves scalable infrastructure (e.g., AWS, Azure, GCP) that offers massive storage, computational power, and deep learning model training. It is ideal for historical data analysis, model refinement, and large-scale simulations. It supports containerized microservices architecture for scalability and modularity.

This hybrid edge-cloud design ensures efficiency, responsiveness, and scalability across the system.

3.5 Analytics and Decision-Making Layer

At the heart of the architecture is the analytics layer, where AI and ML algorithms analyze the processed data to extract insights and generate decisions. Core functionalities include:

- Predictive Analytics helps in forecasting traffic congestion, energy usage peaks, disease outbreaks, etc.
- Anomaly Detection identify unusual patterns, such as a spike in air pollution or sudden crowd formation.
- Sentiment Analysis helps in understanding public emotions on social media during events or crises.
- Object and Activity Recognition provides computer vision-based interpretation of video footage for crime detection, infrastructure defects, etc.
- Multi-modal data fusion helps in integrating structured and unstructured data (e.g., weather sensors + tweets + video) for holistic decision-making as shown in Figure 3.

Figure 3. Multimodal Data Fusion in Smart Cities

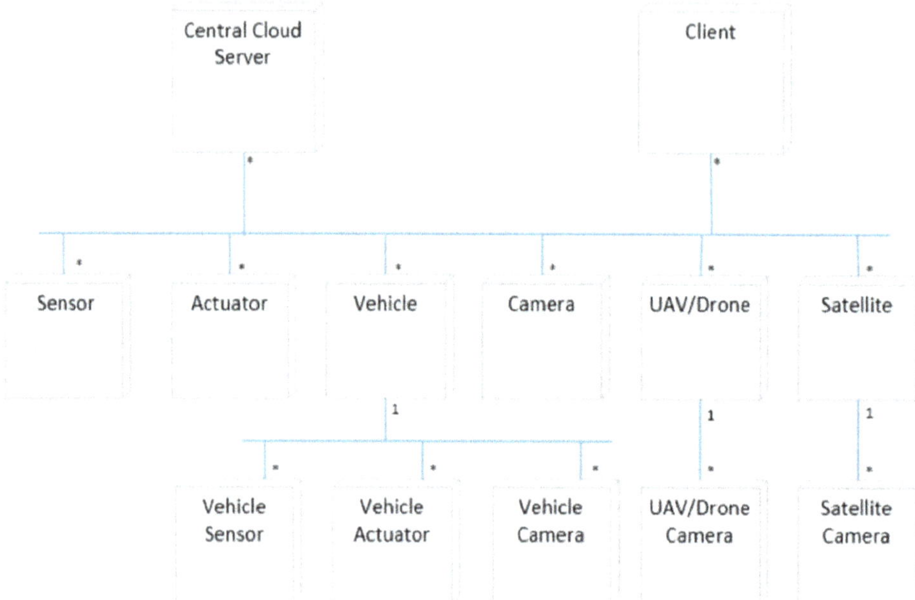

- Reinforcement learning models optimizes city operations dynamically, such as public transport schedules or water usage.

The insights are fed into either automated systems or human-operated dashboards for timely action.

3.6 Application Layer

This topmost layer converts analytics into tangible, real-world applications, directly impacting city services and citizen well-being. Prominent implementations include:

Smart Mobility

- Real-time traffic redirection based on congestion data.
- Dynamic toll pricing and parking management.
- Integration of public transport with shared mobility (e.g., scooters, cabs).

Public Safety and Emergency Response

- Automated alerts during natural disasters (earthquakes, floods).
- Surveillance for crowd control during festivals or protests.
- Predictive crime mapping using historical and live data.

Environmental Monitoring

- Real-time air and noise pollution tracking.
- Smart waste bins that alert collection services when full.
- Monitoring of water quality in rivers and reservoirs.

Energy Optimization

- Smart grid management to balance supply and demand.
- Adaptive lighting systems that dim or brighten based on usage and natural light.
- Smart meters that optimize home and industrial energy consumption.

Healthcare and Pandemic Management

- Wearable integration to track health metrics.
- Social sensing for early disease outbreak detection.
- Hospital resource optimization based on patient inflow analytics.

Each application is backed by an intuitive user interface (UI) and can be accessed via mobile apps, dashboards, or command centers by city authorities and stakeholders.

4. APPLICATIONS IN URBAN ENVIRONMENTS

The convergence of Artificial Intelligence (AI) and Internet of Things (IoT) technologies in social sensing systems is transforming the way cities operate and interact with citizens. By leveraging real-time data from devices, sensors, and social platforms, smart cities can enhance operational efficiency, improve public services, and respond dynamically to urban challenges. Below are key application areas in urban environments (Rob Kitchin, 2014).

4.1 Urban Mobility and Traffic Management

AI-IoT social sensing plays a pivotal role in reshaping urban mobility by enabling smarter, data-driven transportation systems. Crowd-sourced data from GPS devices, smartphones, and vehicle telemetry are aggregated and analyzed to forecast traffic conditions, suggest alternative routes, and manage traffic signals in real-time.

Predictive Traffic Control: Real-time data from vehicles and infrastructure sensors are fed into AI models to predict congestion and optimize signal timing. This helps reduce travel time and vehicular emissions.

Smart Parking Systems: IoT-enabled sensors detect parking availability and guide drivers to free spots via mobile apps, minimizing the time spent circling for parking—a major contributor to traffic congestion.

Public Transport Optimization: AI algorithms monitor bus and metro operations, optimizing schedules based on demand patterns and reducing wait times.

Case Study on Barcelona's Intelligent Traffic Management System: Barcelona uses a smart traffic control platform that integrates traffic cameras, GPS signals from taxis and buses, and IoT sensors to manage urban mobility. The system dynamically adjusts traffic lights, guides emergency vehicles through traffic, and provides real-time updates to citizens via digital signage and apps.

4.2 Environmental Monitoring

Environmental degradation poses significant threats in urban areas, but AI-IoT social sensing can monitor and manage environmental quality more proactively.

Air and Noise Pollution Sensing: IoT sensors installed on street poles, vehicles, and even citizens' smartphones continuously capture air quality metrics such as PM2.5, CO2 levels, and noise pollution.

Pollution Mapping: Data is geotagged and used to create high-resolution pollution heatmaps, helping city planners and public health officials implement zoning laws and restrict high-emission activities in affected areas.

Climate Adaptation Planning: Predictive analytics can simulate environmental trends, informing infrastructure changes to mitigate heat islands, flooding, or other climate-induced risks.

For example, consider AirVisual and Community-Based Sensing in which AirVisual, a mobile application by IQAir, crowdsources air quality data from citizen-owned devices and shares it on a global platform. In cities like Delhi and Beijing, these data points help shape real-time warnings and policy interventions.

4.3 Disaster Response and Resilience

AI-IoT social sensing enhances urban resilience by supporting real-time situational awareness during natural or man-made disasters.

Behavioral analytics during emergencies: Social media feeds, mobile phone signals, and wearable data can reveal human movement patterns during disasters, such as floods, fires, or terrorist attacks.

Geospatial alert systems: Integrating data from emergency services, weather satellites, and social media enables automated alerts, early warning systems, and evacuation guidance via mobile apps.

Resource allocation: AI algorithms predict which areas will require emergency response resources, enabling efficient deployment of personnel, medical supplies, and transportation.

For example, Japan's Earthquake Early Warning System in which Japan combines seismological data with mobile sensor networks and real-time social sensing platforms to alert citizens seconds before a quake, giving them time to take cover or stop trains.

4.4 Public Safety and Policing

Smart surveillance and predictive analytics empower urban law enforcement agencies to act preemptively and efficiently.

Surveillance with AI: Cameras equipped with computer vision can detect abnormal behavior, unattended baggage, or unauthorized access in public places. Facial recognition aids in identifying wanted individuals.

Predictive policing: Historical crime data is analyzed to identify high-risk areas or "crime hotspots." Patrols and interventions are then focused accordingly, reducing the overall crime rate.

Incident correlation and real-time alerts: AI models integrate 911 call data, social media mentions, and sensor feeds (e.g., gunshot detectors) to validate threats and issue alerts faster than traditional systems.

For example, Chicago Police Department's Chicago's Strategic Decision Support Centers (SDSCs) use predictive crime mapping, surveillance feeds, and social media analytics to anticipate and prevent crimes in real time, reducing violent crime in targeted areas.

4.5 Healthcare and Epidemic Tracking

Healthcare in smart cities is no longer confined to hospitals. AI-IoT platforms (Maged N. Kamel Boulos & Najeeb M. Al-Shorbaji, 2014) enable proactive, population-wide health management using wearable sensors, smart diagnostics, and social sensing.

Real-Time health monitoring: Wearables track heart rate, oxygen saturation, sleep cycles, and physical activity. Data is used for early detection of anomalies and chronic disease management.

Outbreak Detection via social media: Posts about symptoms, location check-ins, and pharmacy purchases are mined to detect emerging disease patterns. NLP and sentiment analysis help differentiate between public concern and actual outbreak signals.

Behavioral Compliance Monitoring: During pandemics like COVID-19, AI-IoT systems monitored mask usage, crowd density, and quarantine compliance using street cameras and geofencing.

4.6 Citizen Engagement and Governance

Citizen-centric governance is the cornerstone of smart cities. Social sensing technologies empower residents to actively participate in governance, provide feedback, and influence policymaking (Stephen Goldsmith & Susan Crawford, 2014).

Participatory sensing platforms like FixMyStreet or SeeClickFix allow users to report issues such as potholes, streetlight failures, or illegal dumping directly to municipal departments.

Using Sentiment analysis for public opinion, authorities monitor social platforms using AI tools to gauge public sentiment about policies, infrastructure changes, and civic programs. This allows real-time policy adaptation.

Digital Townhalls and Feedback Loops in cities conduct online surveys, virtual townhalls, and polls to gather citizen input, which is analyzed to prioritize development projects and allocate budgets more transparently.

For example, Amsterdam's Digital Participation Model in which Amsterdam enables citizens to vote on and propose projects via a digital portal. The city then uses AI to analyze which initiatives align best with broader urban goals, incorporating them into future budgets.

5. METHODOLOGY AND SOURCE MATERIAL

This chapter employs a comprehensive mixed-methods research design to ensure both academic rigor and practical relevance. By integrating theoretical insights and empirical evidence across multiple disciplines, namely urban computing, human-computer interaction (HCI), data science, and public administration, the study aims to build a robust analytical framework for understanding the development, implementation, and impact of social sensing systems in urban environments.

5.1 Interdisciplinary Framework

Given the inherently complex and dynamic nature of smart cities, the research draws from a variety of fields:

Urban computing contributes foundational knowledge about the integration of computing technologies with urban infrastructure, mobility systems, and public services.

Human-Computer Interaction (HCI) offers user-centered design principles essential for evaluating how citizens engage with urban sensing technologies.

Data Science provides analytical techniques for handling large-scale, heterogeneous datasets generated through social sensing platforms.

Public Administration offers insights into governance models, policy-making processes, and ethical considerations related to civic technologies.

This interdisciplinary approach allows for a holistic examination of both the technical and social dimensions of urban sensing.

5.2 Source Material

The research relies on four primary categories of source material:

Literature Reviews: A systematic review of peer-reviewed academic literature—including journal articles, conference proceedings, and scholarly books—will be conducted. Sources will be drawn from leading publication venues such as *ACM CHI*, *IEEE Smart Cities*, *Urban Studies*, and *Public Administration Review*. The literature review shown in Table 1 will contextualize the study within existing debates, identify research gaps, and frame the conceptual underpinnings of social sensing in urban contexts.

Table 1. Key Academic Sources for Systematic Literature Review

Discipline	Publication Venues	Relevance to Social Sensing in Urban Contexts
Human-Computer Interaction	ACM CHI, UIST, CSCW, TOCHI	Explores user interaction, citizen participation, and interface design in urban sensing systems
Urban Computing / Smart Cities	IEEE Smart Cities, IEEE Pervasive Computing, ACM Ubicomp, Sensors	Provides frameworks for sensing technologies, urban informatics, and IoT integration in city systems
Urban Studies / Geography	Urban Studies, Cities, Environment and Planning B, Journal of Urban Technology	Offers contextual understanding of spatial dynamics, governance, and the socio-political implications of sensing
Public Administration / Policy	Public Administration Review, Government Information Quarterly, Policy & Internet	Examines institutional frameworks, data governance, and civic engagement in digital urban services
Data Science / Information Systems	Journal of Big Data, Information Systems, ACM Transactions on Data Science	Focuses on data collection, analysis, and interpretation methodologies relevant to social sensing applications

Case Studies: The research includes detailed case studies of cities that have implemented or piloted social sensing systems. Examples include:

- Singapore, known for its Smart Nation initiatives and real-time mobility tracking.
- Barcelona, a leader in citizen-centric digital governance and IoT-enabled urban services.
- New York City, which integrates open data with social media analytics to improve service delivery. These cases provide grounded examples of how theory translates into practice, and allow for comparative insights across different governance models, scales of implementation, and technological infrastructures.

Industry Reports and Think Tank Publications: To capture recent developments in urban technology, the research draws on white papers, policy briefs, and technical reports produced by urban tech providers, consultants (e.g., McKinsey Global Institute, Deloitte), and international organizations (e.g., UN-Habitat, World Economic Forum). These sources provide practical insights into market trends, innovation trajectories, and emerging best practices.

Publicly Available Datasets: The study makes use of various open data sources to illustrate and support empirical analysis. These include:

- Geospatial data from platforms such as OpenStreetMap and city GIS portals.

- Social media data via the Twitter API or equivalent sources, to analyze real-time citizen feedback.
- Government open data portals, which may include traffic data, service complaints, pollution levels, and emergency response metrics.

These datasets enable both quantitative and qualitative analyses and support data-driven modeling of urban phenomena.

5.3 Methodological Integration

The selected methods and materials are not treated in isolation but are triangulated to ensure validity and depth. For example, insights from academic literature are used to frame the analysis of case studies, while real-world datasets are employed to test hypotheses derived from theoretical models. This triangulation enables the research to not only reflect the current state of knowledge but also contributes to actionable insights for city planners, technologists, and policymakers.

5.4 Ethical and Practical Considerations

Special attention is paid to issues of privacy, data governance, and equity, particularly when using social media data and citizen-generated information. The methodology adheres to ethical standards for data use and research transparency, ensuring that all datasets are obtained from publicly accessible or appropriately licensed sources.

6. ETHICAL, PRIVACY, AND SECURITY CONSIDERATIONS

As AI-IoT social sensing systems become deeply embedded in urban life, a range of ethical, privacy, and cybersecurity challenges emerge. Addressing these concerns is crucial not only for public trust but also to ensure the responsible deployment of these technologies.

6.1 Ethical Use of Social Sensing

The integration of AI with social sensing systems introduces complex ethical challenges that go beyond technical implementation.

Algorithmic Bias: AI models trained on skewed or non-representative data may reinforce societal inequalities. For example, facial recognition systems may underperform on certain ethnic groups, leading to discriminatory outcomes.

Lack of Informed Consent: Often, data from social media, mobile devices, and IoT sensors is collected without explicit user consent. This violates fundamental ethical principles and can erode public trust.

Opacity in AI Decision-Making: Many AI algorithms function as "black boxes," providing little transparency in how decisions are made. This is particularly concerning when those decisions impact public safety or access to services.

Equity and Fairness: Certain communities may benefit disproportionately from smart services, leaving behind marginalized or digitally excluded populations. Ethical governance frameworks must ensure equal access and representation.

For example, in predictive policing, if training data contains historically biased arrest records, the AI may wrongly target low-income or minority neighborhoods, perpetuating systemic injustice.

6.2 Privacy Concerns

AI-IoT social sensing systems collect vast amounts of sensitive, often personally identifiable information (PII). Protecting this data is essential to avoid misuse or abuse.

Data Anonymization: Personal data must be anonymized or pseudonymized before processing. However, even anonymized data can sometimes be re-identified when combined with other datasets.

Regulatory Compliance: Privacy laws such as the General Data Protection Regulation (GDPR) in Europe and India's DPDP Act require organizations to obtain consent, ensure data minimization, and provide users with control over their data.

Surveillance Risks: Constant monitoring through cameras, mobile apps, and sensors can lead to a surveillance society where individuals feel perpetually watched, stifling freedom of expression.

User Control and Transparency: Citizens should be informed about what data is collected, how it is used, and could opt out or delete their data. For example, In South Korea, contact tracing during COVID-19 was effective but raised concerns as detailed individual movement histories were shared publicly, leading to potential stigmatization.

6.3 Cybersecurity Threats

The vast and interconnected nature of AI-IoT systems makes them highly vulnerable to cyberattacks and malicious exploitation.

Using Sensor Spoofing attackers may inject false data into the system by mimicking legitimate sensor outputs, leading to incorrect decisions or public misinformation.

Data tampering compromise of nodes in the network that can alter data en route, affecting the integrity and reliability of smart services like traffic management or pollution tracking.

Man-in-the-Middle Attacks arise due to insecure communication protocols can be intercepted, exposing sensitive data or allowing attackers to manipulate system commands.

Ransomware and Botnets: IoT devices with weak security are often used in large-scale botnet attacks or may be targeted with ransomware that disables critical infrastructure until a ransom is paid.

Lack of Security Standards: Many consumer-grade IoT devices are deployed without strong encryption, regular patching, or authentication mechanisms, creating systemic vulnerabilities.

For example, in 2016, the Mirai botnet attack exploited thousands of poorly secured IoT devices to launch a massive DDoS attack that disrupted major websites and services globally (Anthony M. Townsend, 2013).

To responsibly deploy AI-IoT social sensing in urban environments, the following safeguards should be implemented as shown in Figure 4:

- Ethical AI Audits conducted to identify bias and ensure fairness in algorithmic outcomes.
- Incorporating data protection principles during system architecture development.
- Strong End-to-end encryption and multi-factor authentication for all connected devices.
- Transparent consent management systems to ensure data sovereignty.
- Adherence to local and international regulations through independent compliance monitoring.

Figure 4. Ethical, Security and Privacy Considerations

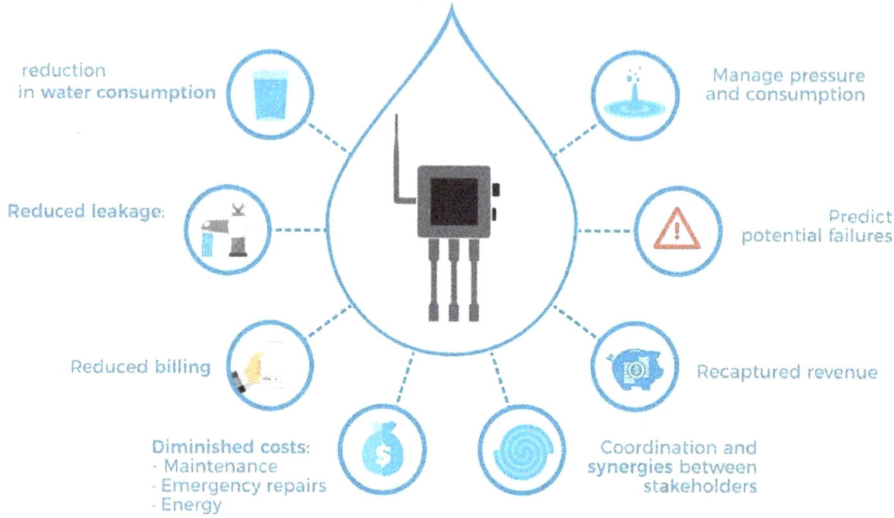

7. CASE STUDIES AND REAL-WORLD IMPLEMENTATIONS

Case Studies and Real-World Implementations illustrated in Figure 5 show how various global cities are leveraging AI and IoT technologies to enhance urban living through social sensing. In Singapore, the Smart Nation Initiative has established a country-wide IoT framework that powers AI applications across urban mobility, healthcare, and digital governance, promoting smarter and more efficient services. Amsterdam has adopted social sensing for waste management by deploying smart sensors in public bins and enabling citizens to submit waste alerts, significantly improving collection efficiency and reducing operational costs. New York City uses its 311 service to collect and analyze citizen complaints; the data is mined and visualized via heatmaps to enhance municipal response and optimize service delivery. Lastly, India's Smart Cities Mission integrates AI-IoT systems for monitoring real-time water and electricity usage, along with deploying advanced surveillance solutions to improve safety and infrastructure management in rapidly urbanizing zones. Collectively, these case studies demonstrate the transformative power of AI-IoT social sensing in making cities more intelligent, responsive, and citizen-centric.

Figure 5. Case Studies and Real-World Implementations

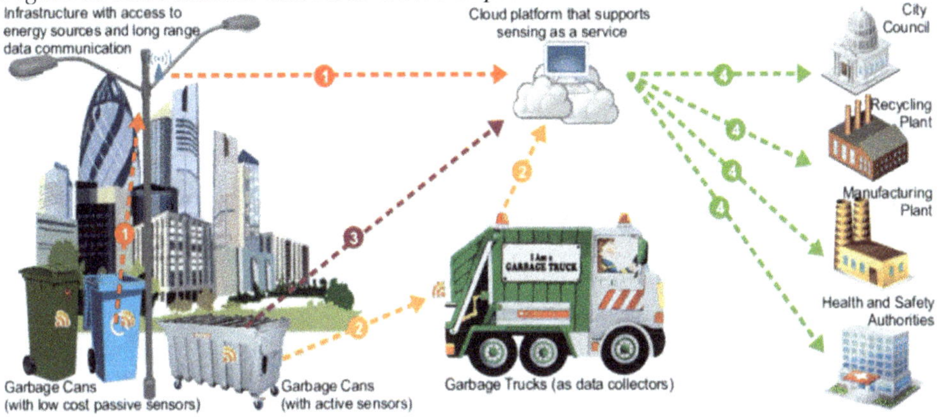

7.1 Singapore: Smart Nation Initiative

Launched in 2014, Singapore's *Smart Nation* initiative aims to harness technology to improve the quality of life, strengthen the economy, and build a connected society. It focuses on using IoT and AI across domains like urban mobility, healthcare, and e-governance.

Urban Mobility:

Smart Traffic Light Systems adjust signal timing in real time based on vehicular and pedestrian flow using sensors and AI algorithms.

Autonomous Vehicles (AVs) are tested and gradually integrated for public transport and logistics.

Mobility-as-a-Service (MaaS) apps offer unified trip planning, payment, and real-time updates.

Healthcare:

Remote Health Monitoring for the elderly using wearable IoT devices.

AI-powered health screening systems (e.g., diabetic retinopathy detection).

E-Governance:

SingPass: A secure digital identity platform integrated with 500+ e-services.

AI chatbots like *Ask Jamie* provide citizens with government-related information 24/7.

The outcome of this system includes reduced wait times and congestion in transportation, early diagnosis and preventive healthcare services and provides seamless citizen-government engagement via digital platforms.

The major significance of this system shows Singapore exemplifies an AI-IoT-driven smart city where governance, infrastructure, and services operate in harmony with real-time data and citizen-centric design.

7.2 Amsterdam: Social Sensing for Waste Management

Amsterdam has pioneered smart waste management by integrating IoT sensors, data analytics, and citizen participation into its sanitation services.

IoT Sensors in Trash Bins: Sensors monitor fill levels in public bins. Data is transmitted to central systems in real time.

Smart Collection Routes: AI algorithms optimize garbage truck routes based on sensor data to reduce emptying half-filled bins.

Citizen Engagement: Residents can report overflowing bins or illegal dumping through mobile apps. City feedback systems encourage collaborative cleanliness.

The major outcome of the system includes up to 50% reduction in collection frequency and associated costs, it shows significant reduction in carbon emissions from garbage trucks and enhanced citizen satisfaction and cleaner urban spaces.

Amsterdam showcases how urban sustainability, and efficiency can be achieved by blending technology with participatory governance.

7.3 New York City: 311 Social Data Mining

The *311 system* in NYC is a centralized platform where residents report non-emergency issues like potholes, noise complaints, or broken streetlights. This massive stream of crowd-sourced data is mined using AI, NLP, and predictive analytics.

AI for Complaint Categorization: Natural Language Processing (NLP) is used to classify and prioritize complaints.

Heatmap and Trend Analysis: Geo-tagged data is visualized on dashboards showing complaint hotspots. Seasonal and neighborhood-level trends are extracted.

Service Optimization: City agencies use insights to allocate maintenance resources proactively. Predictive models forecast future service needs based on complaint patterns.

For example, if noise complaints spike in a neighborhood on weekends, sanitation or law enforcement units is dispatched preemptively.

The outcome shows an improved municipal responsiveness, efficient resource allocation and reduced response time and also provides data-driven urban planning and zoning enforcement.

NYC's 311 system demonstrates the power of social sensing and AI in transforming passive complaint data into actionable urban intelligence.

7.4 India: AI-IoT Initiatives in Smart Cities Mission

Launched in 2015, India's Smart Cities Mission focuses on building 100+ model urban areas that are technologically enabled, citizen-friendly, and sustainable. AI and IoT play key roles in urban management across utilities, safety, and infrastructure.

Water & Electricity Monitoring: IoT sensors in pipelines and meters track consumption in real time. AI models detect leaks, forecast usage, and optimize distribution.

Advanced Surveillance & Safety: Integrated Command and Control Centres (ICCCs) in cities like Bhopal and Surat receive feeds from thousands of AI-enabled CCTV cameras. Facial recognition, behavior analysis, and real-time alerts assist in public safety and disaster response.

Traffic and Mobility Management: AI-powered traffic lights and congestion analysis tools are deployed in cities like Pune and Lucknow.

Waste Management: Cities like Indore use RFID-tagged bins and real-time GPS vehicle tracking for collection efficiency.

The significant outcome are water savings of up to 30% in pilot cities, drastic improvement in surveillance and response times, enhanced urban transparency and reduced manual errors through automation.

India's scalable and cost-effective use of AI-IoT makes it a model for smart city innovation in developing nations.

These case studies demonstrate how AI-IoT-based social sensing can be adapted across continents, cultures, and infrastructure contexts. From high-tech, densely urbanized cities like Singapore to large, developing country initiatives like India's Smart Cities Mission, the combination of sensor networks, data analytics, and citizen participation is revolutionizing the way city's function, respond, and evolve (Eiman Al Nuaimi et al., 2015).

8. FUTURE TRENDS AND RESEARCH DIRECTIONS

Figure 6 highlights four emerging trends shaping the future of smart cities through AI and IoT integration (Paolo Neirotti et al., 2014). Digital Twins and Urban Simulation enable real-time virtual replicas of urban environments for improved planning and emergency response. Federated Learning facilitates decentralized AI training across devices, enhancing system intelligence while preserving user privacy. Emotional AI and Affective Computing interpret public sentiment using multimodal data to deliver more empathetic and human-centric services. Finally, Blockchain Integration secures data transactions and enhances transparency, ensuring trust and integrity in smart city ecosystems (Michael Batty et al., 2012)

Figure 6. Future of Smart Cities through AI and IoT Integration

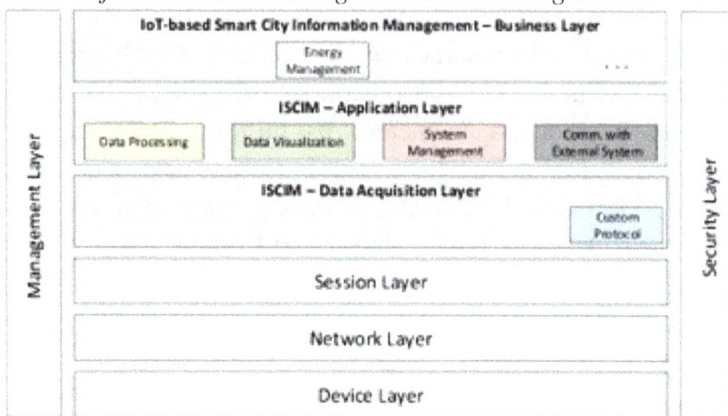

8.1 Digital Twins and Urban Simulation

Digital twins are virtual replicas of physical urban systems, continuously updated with real-time data from IoT sensors, public infrastructure, and citizen inputs. These dynamic models allow city planners and administrators to simulate urban environments, predict system behaviors, and test scenarios without physical risk. For example, a digital twin of a transportation network can model traffic flows during peak hours or simulate the impact of a new public transport route. During emergencies like floods or fires, the digital twin can simulate evacuation strategies and resource allocation, helping authorities respond proactively. Furthermore, urban simulations help assess sustainability goals by testing energy usage, emissions impact, and infrastructure resilience. Cities such as Singapore and Helsinki are pioneers in adopting digital twins to support urban design and real-time monitoring.

8.2 Federated Learning for Privacy-Preserving AI

Traditional AI systems often require centralizing large volumes of sensitive user data, raising privacy and compliance concerns. Federated learning addresses this by allowing AI models to be trained directly on user devices—such as smartphones, wearables, or local edge gateways—without transferring raw data to a central server. Instead, only model updates (e.g., gradients or weights) are shared and aggregated, ensuring user privacy is maintained.

In the context of smart cities, federated learning can be used for applications like traffic prediction, health monitoring, or personalized public services, all while preserving user anonymity. For instance, smart mobility systems can train vehicle routing models across distributed devices without accessing personal location histories. This approach complies with privacy regulations like GDPR and is particularly suited for large-scale, privacy-sensitive urban deployments.

8.3 Emotional AI and Affective Computing

Emotional AI and affective computing refer to systems capable of recognizing, interpreting, and responding to human emotions. In smart cities, these technologies can enhance services that rely on human interactions, such as public safety, healthcare, and social services. By analyzing multimodal data—such as voice tone, facial expressions, text sentiment, biometric signals, and even social media behavior—AI can infer the emotional state of individuals or communities. Applications include:

Public sentiment analysis: During policy implementation or civic events, governments can gauge public mood and adjust communication strategies.

Mental health support: AI chatbots or kiosks can detect stress, anxiety, or depressive symptoms from speech or facial cues and provide timely interventions.

Crowd mood monitoring: At large gatherings like protests or festivals, affective AI systems can alert authorities to shifts in emotional tone that may indicate escalating tension or unrest.

This human-centric AI approach seeks to make urban environments not only smarter but also more empathetic and responsive to the needs of citizens.

8.4 Integration with Blockchain

Blockchain technology is increasingly seen as a vital enabler for trustworthy and secure data exchange in smart cities. With its decentralized, immutable ledger system, blockchain can enhance transparency, accountability, and data integrity across AI-IoT systems (Andrea Zanella et al., 2014). The key applications include:

Secure sensor data: IoT sensors can push real-time environmental, traffic, or utility data onto a blockchain, ensuring it cannot be tampered with by malicious actors.

Traceability in supply chains and energy distribution: For example, cities implementing smart grids can use blockchain to track the origin and consumption of renewable energy in a decentralized energy marketplace.

Decentralized identity management: Citizens can control access to their personal data using blockchain-based identities, granting or revoking permissions for different applications.

Smart contracts for governance: Blockchain enables programmable rules for automated municipal operations, such as issuing permits, processing fines, or disbursing welfare payments. By integrating blockchain with AI-IoT platforms, smart cities can foster greater citizen trust, regulatory compliance, and robust operational security.

9. CHALLENGES AND LIMITATIONS

9.1 Data Quality and Representation

One of the core challenges in implementing AI and IoT-driven social sensing systems for smart cities lies in the quality and representativeness of the data collected. Social sensing data is often sourced from a wide array of platforms, including mobile apps, social media, public sensors, and user-generated content. However, such data is susceptible to several shortcomings. First, it may be incomplete or inconsistent caused by data loss, user disengagement, or hardware malfunctions. Second, biases can be introduced depending on who is using the platforms or devices. For example, younger, tech-savvy, urban populations are more likely to use certain apps, which can skew data analysis and create a digital divide in representation. Additionally, data collected may overrepresent certain areas or groups while neglecting others, leading to poor policy decisions if left unaddressed. This lack of diversity in data inputs can undermine the goal of achieving equitable and inclusive urban intelligence.

Example 1: COVID-19 Mobility Data: During the COVID-19 pandemic, governments used mobility data from smartphones to monitor population movement and enforce lockdowns. However, this data was mostly sourced from users with smartphones and internet access, excluding populations such as the elderly, rural residents, or those from low-income groups. This introduced sampling bias, resulting in skewed insights and policies that didn't reflect the behavior of the entire population.

Example 2: Sentiment Analysis from Twitter: Cities like New York have explored using Twitter sentiment analysis to understand public reactions to events (e.g., natural disasters or policy changes). However, since Twitter users tend to be younger and more tech-oriented, their opinions don't represent the broader demographic, causing misrepresentation of public sentiment.

9.2 Technical Infrastructure

The technical limitations of existing infrastructure present another critical challenge. Real-time social sensing systems demand high-throughput data pipelines capable of handling large-scale, continuous data streams with minimal latency. However, many cities still rely on outdated or inadequate networking, storage, and processing infrastructure. Latency issues in data transmission and processing can impair the responsiveness of time-sensitive applications such as emergency response, traffic management, and pollution alerts. Furthermore, scalability remains a pressing concern. As the number of connected devices and data points grows, current infrastructures may struggle to expand proportionally due to limitations in bandwidth, processing power, and storage capacity. This constraint can lead to bottlenecks, dropped data packets, and system failures, which compromise the reliability and robustness of smart city solutions.

Example 1: Traffic Congestion Analytics in Developing Cities: Cities like Nairobi and Dhaka have attempted to implement smart traffic systems using crowd-sourced GPS data. However, frequent internet outages, poor cellular coverage, and limited bandwidth often delayed data transmission and real-time analytics, causing system lags and inaccurate traffic forecasts.

Example 2: Smart Flood Monitoring in Jakarta: Jakarta's smart flood sensors faced problems due to hardware limitations and insufficient cloud processing resources. During peak rainfalls, the sudden spike in data overwhelmed the system, leading to latency in alerts and failures in timely evacuation planning.

9.3 Policy and Regulatory Frameworks

While technology evolves rapidly, policy and regulatory systems often lag behind, creating a disconnect between innovation and governance. There is a pressing need for adaptive governance models that can respond to the ethical, legal, and societal implications of AI and IoT in social sensing. Current regulations are frequently fragmented across jurisdictions, lacking cohesion or global alignment. For example, data privacy laws like the EU's GDPR may conflict with data collection practices in countries with more lenient policies, making cross-border collaboration complex. Moreover, many existing regulations do not fully address the nuances of algorithmic transparency, accountability, and fairness in AI-driven systems. Without harmonized standards, it becomes difficult to ensure that social sensing applications respect user rights, avoid surveillance misuse, and foster public trust. An integrated policy approach—balancing innovation, regulation, and ethical safeguards—is essential to scale and sustain AI-IoT deployments in smart city contexts.

Example 1: Data Privacy in India vs GDPR: A smart surveillance project in India involving facial recognition cameras in public spaces was criticized due to lack of a strong personal data protection law, unlike Europe's GDPR. This raised concerns about privacy, consent, and misuse, illustrating the need for global regulatory alignment in AI-IoT systems.

Example 2: AI Bias in Predictive Policing: In some U.S. cities, AI-powered predictive policing tools were found to disproportionately target neighborhoods with high historical arrest records, leading to racial profiling and systemic bias. These outcomes occurred due to non-transparent algorithms and poor oversight frameworks, emphasizing the gap between technology and governance.

10. CONCLUSION

AI and IoT-powered social sensing marks a paradigm shift in urban management and innovation, offering cities unprecedented capabilities to monitor, understand, and respond to the needs of their populations in real time. Through interconnected sensors, mobile devices, and data analytics, this technology captures the "digital pulse" of cities—encompassing everything from traffic flow and pollution levels to citizen sentiments and emergency response patterns. It enables decision-makers to move beyond reactive governance toward predictive and proactive planning, resulting in smarter infrastructure, optimized services, and improved quality of life.

By integrating social sensing with AI-driven analysis, urban systems gain the ability to learn and evolve. For example, transportation networks can dynamically adjust based on crowd density, waste management can be streamlined based on real-time fill-level data, and public safety can be enhanced through rapid threat detection. Moreover, these technologies foster greater civic engagement by providing platforms for citizen feedback and participatory governance, making cities more inclusive and people centric.

However, as transformative as these capabilities are, their implementation must be approached with caution and responsibility. Ethical dilemmas—such as surveillance overreach, algorithmic bias, and digital exclusion—must be addressed proactively. Similarly, privacy and data security concerns demand robust legal frameworks that protect individual rights without stifling innovation. From a technical perspective, investment in scalable, resilient, and interoperable infrastructure is essential to sustain the massive data flows generated by social sensing applications.

In essence, while AI and IoT-driven social sensing has the potential to redefine urban living, its success depends not only on technological advancement but also on inclusive policies, community trust, and international collaboration. Smart cities of the future must be not only intelligent but also just, transparent, and equitable—balancing innovation with human values to ensure sustainable and responsible urban transformation.

REFERENCES

Al Nuaimi, E., Al Neyadi, H., Mohamed, N., & Al-Jaroodi, J. (2015). Applications of big data to smart cities. *Journal of Internet Services and Applications*, *6*(1), 1–15. DOI: 10.1186/s13174-015-0041-5

Albino, V., Berardi, U., & Dangelico, R. M. (2015). Smart cities: Definitions, dimensions, performance, and initiatives. *Journal of Urban Technology*, *22*(1), 3–21. DOI: 10.1080/10630732.2014.942092

Allam, Z., & Newman, P. (2018). Redefining the smart city: Culture, metabolism and governance. *Smart Cities*, *1*(1), 4–25. DOI: 10.3390/smartcities1010002

Batty, M. (2013). *The new science of cities*. MIT press. DOI: 10.7551/mitpress/9399.001.0001

Batty, M., Axhausen, K. W., Giannotti, F., Pozdnoukhov, A., Bazzani, A., Wachowicz, M., Ouzounis, G., & Portugali, Y. (2012). Smart cities of the future. *The European Physical Journal. Special Topics*, *214*(1), 481–518. DOI: 10.1140/epjst/e2012-01703-3

Caragliu, A., Del Bo, C., & Nijkamp, P. (2011). Smart cities in Europe. *Journal of Urban Technology*, *18*(2), 65–82. DOI: 10.1080/10630732.2011.601117

Chen, M., Ma, Y., Li, Y., Wu, D., Zhang, Y., & Youn, C. H. (2017). Wearable 2.0: Enabling human-cloud integration in next generation healthcare systems. *IEEE Communications Magazine*, *55*(1), 54–61. DOI: 10.1109/MCOM.2017.1600410CM

Gaur, A., Scotney, B., Parr, G., & McClean, S. (2015). Smart city architecture and its applications based on IoT. *Procedia Computer Science*, *52*, 1089–1094. DOI: 10.1016/j.procs.2015.05.122

. Giffinger, R., & Gudrun, H. (2010). Smart cities ranking: an effective instrument for the positioning of the cities?. *ACE: architecture, city and environment, 4*(12), 7-26.

Goldsmith, S., & Crawford, S. (2014). *The responsive city: Engaging communities through data-smart governance*. John Wiley & Sons.

Hashem, I. A. T., Yaqoob, I., Anuar, N. B., Mokhtar, S., Gani, A., & Khan, S. U. (2015). The rise of "big data" on cloud computing: Review and open research issues. *Information Systems*, *47*, 98–115. DOI: 10.1016/j.is.2014.07.006

Hollands, R. G. (2020). Will the real smart city please stand up?: Intelligent, progressive or entrepreneurial? In *The Routledge companion to smart cities* (pp. 179–199). Routledge. DOI: 10.4324/9781315178387-13

Jathan, S., & Frank, P. (2015). The Spectrum of Control: A Social Theory of the Smart City. *First Monday, 20*(7).

Kamel Boulos, M. N., & Al-Shorbaji, N. M. (2014). On the Internet of Things, smart cities and the WHO Healthy Cities. *International Journal of Health Geographics, 13*(1), 1–6. DOI: 10.1186/1476-072X-13-10 PMID: 24669838

Kitchin, R. (2014). The real-time city? Big data and smart urbanism. *GeoJournal, 79*(1), 1–14. DOI: 10.1007/s10708-013-9516-8

Kumar, N. M., & Mallick, P. K. (2018). The Internet of Things: Insights into the building blocks, component interactions, and architecture layers. *Procedia Computer Science, 132*, 109–117. DOI: 10.1016/j.procs.2018.05.170

Lee, J. H., Phaal, R., & Lee, S. H. (2013). An integrated service-device-technology roadmap for smart city development. *Technological Forecasting and Social Change, 80*(2), 286–306. DOI: 10.1016/j.techfore.2012.09.020

Meijer, A., & Bolívar, M. P. R. (2016). Governing the smart city: A review of the literature on smart urban governance. *International Review of Administrative Sciences, 82*(2), 392–408. DOI: 10.1177/0020852314564308

Mohanty, S. P., Choppali, U., & Kougianos, E. (2016). Everything you wanted to know about smart cities: The Internet of things is the backbone. *IEEE Consumer Electronics Magazine, 5*(3), 60–70. DOI: 10.1109/MCE.2016.2556879

Nam, T., & Pardo, T. A. (2011, June). Conceptualizing smart city with dimensions of technology, people, and institutions. In *Proceedings of the 12th annual international digital government research conference: digital government innovation in challenging times* (pp. 282-291). DOI: 10.1145/2037556.2037602

Neirotti, P., De Marco, A., Cagliano, A. C., Mangano, G., & Scorrano, F. (2014). Current trends in Smart City initiatives: Some stylised facts. *Cities (London, England), 38*, 25–36. DOI: 10.1016/j.cities.2013.12.010

Townsend, A. M. (2000). Life in the real-time city: Mobile telephones and urban metabolism. *Journal of Urban Technology, 7*(2), 85–104. DOI: 10.1080/713684114

Townsend, A. M. (2013). *Smart cities: Big data, civic hackers, and the quest for a new utopia*. WW Norton & Company.

Zanella, A., Bui, N., Castellani, A., Vangelista, L., & Zorzi, M. (2014). Internet of things for smart cities. *IEEE Internet of Things Journal, 1*(1), 22–32. DOI: 10.1109/JIOT.2014.2306328

Chapter 15
What Are the Key of E-Readiness Factors for Smart City:
A Qualitative Study of Malang City

Ali Roziqin
http://orcid.org/0000-0001-8336-3103
Universitas Muhammadiyah Malang, Indonesia

Muhammad Kamil
http://orcid.org/0000-0003-2398-5407
Universitas Muhammadiyah Malang, Indonesia

Shannaz Mutiara Deniar
http://orcid.org/0000-0001-9792-5787
Universitas Muhammadiyah Malang, Indonesia

Yusril Saputra
Universitas Muhammadiyah Malang, Indonesia

Rifki Muhammad Bintang
http://orcid.org/0009-0002-4601-5390
Universitas Brawijaya, Indonesia

ABSTRACT

The development of smart cities is a strategic priority in many urban areas, including Malang City, Indonesia. As a city renowned for its educational institutions and tourism sector, Malang has adopted smart city initiatives aimed at enhancing public services and improving urban governance. This study explores key factors

DOI: 10.4018/979-8-3373-4962-6.ch015

influencing smart city development, focusing on local dynamics, institutional readiness, and community involvement. Using a qualitative approach through interviews and document analysis, the research identifies adaptive regulation, implementor competency, inter-agency coordination, and digital infrastructure as vital enablers. Challenges include the digital divide, budget limitations, and resistance to change. The study offers practical insights for policymakers on the importance of collaboration, innovation, and capacity-building to foster inclusive and sustainable smart cities.

INTRODUCTION

Advances in Information and Communication Technology (ICT) today have had a positive impact on people's lives and also in government institutions (Lnenicka & Saxena, 2021; Ziemba, 2019). Technological advances can assists government agencies make decisions and policies that are better aligned with the issues faced by society (Mensah, 2020). The use of ICT such as Artificial Intelligence, Big data, Internet of Things (IoT), data analytics and digital twins can help government tasks including data processing including processing, obtaining, organizing, and storing data in various ways to produce quality information (Buck et al., 2023; Santoso et al., 2024). As cities strive to become more modern and sustainable, the integration of ICT into the smart city concept has become an inevitable necessity.

The concept of a smart city has evolved into a central strategy for urban development in many countries in the world (Caragliu et al., 2011; Hu & Zheng, 2021; Spicer et al., 2023), particularly in developing countries facing rapid urbanization, technological disruption, and governance challenges (Ali, 2022; He & Tritto, 2022; Meijer & Bolívar, 2016; J. Park & Yoo, 2023; Sulistyaningsih et al., 2024). Indonesia, as the fourth most populous country in the world, has launched multiple smart city initiatives under its "100 Smart Cities" program. The smart city program is a government initiative to improve sustainable development, efficiency, and quality of life for citizens (Hu & Zheng, 2021).The program can help alleviate traffic congestion, optimize energy consumption, improve public services, and enhance environmental quality. Moreover, smart city initiatives offer effective solutions to ongoing urban challenges faced by the government, including environmental issues, public health, traffic congestion, poverty, and spatial planning However, one of the most critical yet underexplored aspects of smart city development is e-readiness (Agrawal et al., 2022; Etim & Daramola, 2023), the extent to which city governments and stakeholders are prepared to adopt and leverage ICT for public service delivery and governance innovation.

The smart city adopt technology and innovation diffusion to increase the efficiency of public services by employing city-owned resources (Anthony Jnr et al.,

2021; Demirel, 2022; Meijer & Bolívar, 2016). To accomplish this, a city must have a sufficient level of digital preparedness, known as e-readiness. E-readiness refers to a city's readiness to accept and use ICT to assist development processes and public service (Mkhonto & Zuva, 2024; Nam et al., 2022). Without adequate e-readiness, the implementation of smart cities may face various obstacles and risk falling short of its intended outcomes. Malang City, as one of Indonesia's major cities and educational centers, has shown a strong commitment to developing various smart city initiatives. Moreover, since 2021, the Malang City Government has been officially designated by the central government to lead the development of smart cities.

The Malang City Government has introduced and actively implemented a range of digitalization programs, such as online administrative services, integrated health information systems, and mobile-based public service applications (Kamil et al., 2025). These various initiatives represent concrete efforts to enhance governance efficiency and improve the quality of public services. However, in practice, Malang City continues to face several challenges that hinder the optimal implementation of its smart city initiatives, both from internal and external factors (Kamil et al., 2025; Roziqin et al., 2024). This condition higlights the gap between the envisioned goals of smart city development and the practical realities of its implementation.

E-readiness is a pivotal factor for an organization to measure how far the organization and society have implemented the smart city program (Akbar et al., 2024). E-readiness determines and differentiates how the implementation of smart cities in many regions in Indonesia. The condition of the institutional bureaucracy of the bureaucracy, especially as a driving force, is crucial in preparing the e-readiness of a city in implementing smart cities (Pratama & Imawan, 2019). Some previous studies have discussed smart city practices in Indonesia, most of these studies describe the implementation, opportunities, urban challenges, good practices, sustainability and interoperability, and strategies, including (Asmorowati et al., 2025; Das et al., 2024; Fahlevi et al., 2024; Hardi et al., 2025; Lestari et al., 2020; Martinez & Masron, 2020; Rachmawati & Rohmah, 2025; Santoso et al., 2024). Meanwhile, relevant smart city research in Malang City is related to e-readiness in smart tourism (Wibowo et al., 2020). Whereas, research on how e-readiness in the implementation of smart cities in the regions is still limited.

This chapter purposes to explore the key dimensions of e-readiness from the perspective of local government in Malang City, Indonesia. Employing a qualitative case study approach, the chapter identifies enabling and inhibiting factors that influence smart city readiness, including adaptive regulation, competency of implementor, inter-agency coordination, and digital infrastructure. This chapter aims to contribute to the global discourse on smart city implementation, with a particular focus on e-readiness factors in the context of Indonesia. Furthermore, the findings of this study are expected to offer valuable insights for policymakers, highlighting the key prerequisites that can influence the success or failure of smart city initiatives at the regional level

Concept Smart City and Smart City in Indonesia

Smart City is a concept of sustainable city management with ICT. Smart City is a concept of sustainable city management with ICT (Yigitcanlar, 2015). The *United Nations Economic Commission for Europe* (UNECE) defines a smart city as an innovative urban area that leverages information and communication technologies, along with other means, to enhance quality of life, improve the efficiency of urban operations and services, and boost overall competitiveness (Estevez et al., 2021). Further, according to *European Commision*, A smart city is one in which existing networks and services are made more efficient via the use of digital technology, benefiting both residents and businesses. The main purposes is to improve people's quality of life, the efficiency of public services, and strengthen regional competitiveness (Ali, 2022; Lnenicka & Saxena, 2021). Simply described, a smart city is a concept that integrates information and communication technology with administrative and policy innovation (Anthony Jnr et al., 2021).

In its practice, scholars divided smart cities into several elements or dimensions including society, economy, mobility, environment, government, and integrated and data-based life infrastructure (Anthony Jnr et al., 2021; Fernandez-Anez et al., 2018; Sulistyaningsih et al., 2024; Yigitcanlar, 2015). These five dimensions are the basis for realizing a livable, sustainable, efficient city that is responsive to the needs of its citizens. Smart City programmes are urban planning initiatives that harness technological advancements to create more convenient and healthier living environments, while boosting productivity and efficiency. Urban development is approached by integrating multiple sectors and aligning them with innovations in information and communication technology. The aim is to achieve livable, sustainable, and modern cities, enhance economic competitiveness and regional productivity, and establish systems that are interconnected and well-integrated.

In the Indonesian context, the implementation of the Smart City programme is coordinated by the central government in collaboration with local governments. This initiative began in 2017 through the launch of the "Movement Towards 100 Smart Cities." During its first year, several cities and districts across the country took part (Kamil et al., 2025), and by 2025 a total of 251 regions had received support in developing their Smart City master plans. Despite this progress, not all participating areas have experienced positive outcomes. Varying regional characteristics and disparities in existing resources have contributed to uneven results. Common challenges in implementing smart city policies include a lack of qualified human resources in information technology, insufficient infrastructure and facilities, and underdeveloped networks and application systems (Maisyarah et al., 2024; Nugroho & Purbokusumo, 2020). Moreover, the high costs of acquiring and maintaining the necessary technology and infrastructure remain a major hurdle. With limited budgets, many local governments are still unable to fully realise their smart city ambitions.

Smart City programmes in Indonesia adopt six fundamental dimensions: Smart Governance, Smart Economy, Smart Society, Smart Living, Smart Branding, and Smart Environment. Each dimension reflects core sectors of urban life aimed at enhancing citizens' quality of life in cities or regions. Central and local governments continue showing strong commitment to Smart City initiatives through strategic policies and cross-sectoral collaboration. Despite this, effective implementation depends heavily on sufficient electronic readiness (e-readiness). Without adequate e-readiness, Smart City initiatives risk becoming mere slogans rather than delivering meaningful impact.

E-readiness of Smart City

The adoption of information and communication technologies (ICT) is crucial for smart city readiness (Mkhonto & Zuva, 2024). However in fact, e-readiness for smart city is far more complex. Making a city smarter necessitates a number of factors that policymakers must consider, as various competing needs and limited resources vie for attention (Desouza et al., 2020). Careful preparation is therefore essential for policymakers aiming to successfully implement Smart City programmes. he concept of e-readiness is commonly used to assess an organisation's preparedness for digital transformation, particularly in Smart City initiatives (Yigitcanlar et al., 2022). It refers to the level of readiness of a country, region, institution, or society to utilise information and communication technology (ICT) effectively and efficiently across economic, governmental, educational, and social domains (Etim & Daramola, 2023; Nugroho & Purbokusumo, 2020; Potnis & Pardo, 2011). This concept captures the extent to which digital infrastructure, policy frameworks, human resources, and technological ecosystems are in place and ready to support the digital transformation process.

According to Đurek &Ređep (2016), e-readiness plays a critical role in supporting government decision-making. It also includes the ability of individuals and organisations to access, manage, and apply technology in a productive and sustainable manner. The World Bank's *Digital Government Readiness Assessment (DGRA)* report (2020) outlines nine key indicators: (i) leadership and governance; (ii) user-centered design; (iii) public administration and change management; (iv) capabilities, culture, and skills; (v) technology infrastructure; (vi) data infrastructure, strategies, and governance; (vii) cybersecurity, privacy, and resilience; (viii) legislation and regulation; and (ix) the digital ecosystem. Within Smart City implementation, e-readiness serves as a key metric for evaluating how prepared an entity is to adopt ICT-based programmes such as e-government, e-commerce, e-education, and smart city solutions. Etim & Daramola, (2023) emphasise that e-readiness can determine how strategically positioned individuals or communities are in leveraging opportunities made possible by ICT.

Assessment of e-readiness typically involves several aspects, including the availability of ICT infrastructure (such as internet access and digital devices), regulatory and policy support, human resource capacity, cybersecurity readiness, and the level of public participation (Agrawal et al., 2022; Potnis & Pardo, 2011). Higher levels of e-readiness correlate with greater capacity for digital transformation and improved global competitiveness (Gunawong & Gao, 2017). The absence of a unified legal framework for smart cities in Indonesia leads to inconsistencies and confusion among municipalities. Establishing clear regulations and guidelines is essential for the coherent development of smart cities. The existence of e-readiness studies provides significant benefits in supporting the success of digital transformation across various aspects, including infrastructure, organisations, and society (Park et al., 2013). With an adequate level of readiness, governments, institutions, and communities are better positioned to adopt information and communication technology (ICT) more effectively. Therefore, an e-readiness study of Malang City will serve as a valuable reference for policymakers to identify and address readiness gaps in the journey towards Smart City implementation.

METHODS

This research used an exploratory qualitative approach (Creswell & Creswell, 2022). This approach was chosen for its ability to comprehensively explain the phenomenon of smart city e-readiness. Malang City was selected as the research locus not only because it is a densely populated urban area with vibrant economic, tourism, and educational activities, but also due to its official commitment to the Smart City program since 2021. The study began by identifying e-readiness elements based on previous research and frameworks developed by the World Bank. These elements were then refined and expanded through empirical findings from the smart city implementation in Malang, encompassing aspects such as digital infrastructure, policy frameworks, human resources, and institutional challenges

The researchers also examined the actual conditions related to smart city readiness in Malang City, including the number of public Wi-Fi access points, the quality of internet connectivity, the availability of public service applications, information security measures, and the presence of scientific and formal forums supporting the Smart City program. To ensure the rigor and validity of the data, the researchers employed data triangulation (Miles, Matthew B, Huberman, Michael, 2019). The triangulation process began with data condensation to filter and focus the initial information. Subsequently, an exploratory data display was developed to present key research findings. In this stage, NVivo 12 Plus, a Qualitative Data Analysis Software (QDAS) was used to help visualize and organize the data. Finally, the data were interpreted exploratively in alignment with the core research questions and emerging research phenomena.

RESULTS AND DISCUSSION

Digital Infrastructure

Developing Information and Communication Technology (ICT) infrastructure serves as the foundation for realizing Smart City initiatives in Malang City. The local government has demonstrated a strong commitment to building a digital system that is inclusive and accessible to all segments of society. Fast, stable, and secure internet access remains a top priority to support *electronic-based government systems* (SPBE) and to expand digital public services. One notable achievement is the establishment of 552 public Wi-Fi points located in RW (citizen associations) halls, urban village offices, health centers, city parks, and schools. Additionally, the city government provides internet connections with a speed of 1 Gbps, backed up by 600 Mbps reserve capacity and supported by a 14-terabyte fiber optic network for data backup.

Another concrete step in strengthening digital infrastructure readiness is the construction of 561 BTS (*Base Transceiver Station*) towers through partnerships with PT Telkom and IM3, extending internet coverage to previously underserved areas. The government has also developed 346 active CCTV units for traffic and security monitoring, along with 15 *Area Traffic Control Systems* (ATCS) to enable sensor-based traffic management. Other services, such as electronic parking systems, the *Public Service Mall* (MPP), and mini block offices, are connected to digital networks to enhance service efficiency. The city's command center strengthens this infrastructure by integrating various service platforms such as Malang Mbois, Ngalam 112, and SiBansos allowing for real-time data-based responses across the city. These infrastructures not only accelerate service delivery but also promote transparency and efficiency in digital governance. Table 1 shows the technical data of ICT infrastructure developed by Malang City Government;

Table 1. ICT Infrastructure of Malang City Government

Type of Infrastructure	Quantity / Status	Location & Description
Public Wi-Fi	552 points	RW halls, urban village offices, health centers, schools
Government Internet	1 Gbps + 600 Mbps backup	Central government offices & backup lines
BTS Towers	561 units	Collaboration with Telkom & IM3
CCTV	346 points	Traffic and security surveillance
ATCS	15 units	Sensor-based traffic control
Command Center	1 unit	Data integration & real-time response

Source: Master Plan Document of Malang City of 2024, 2025

The city government's commitment also extends to building strong digital security systems. Through the formation of a Computer Security Incident Response Team (CSIRT) and the ISO 27001 certification obtained by the Department of Communication and Information (Diskominfo), data protection and system integrity have become top priorities. These aspects grow increasingly critical alongside the rising use of digital public services that process personal and sensitive citizen data. In an interview with one of the division heads at Malang's Diskominfo, they stated:

> *"Malang City Government is currently prioritizing the development of application-based public services, including Malang Mbois, which serves as the central portal for all regional services. Network infrastructure and backend systems are key to ensuring the platform is accessible without barriers". Anonymous informant (12/11/24).*

This statement further reinforces the notion that digital infrastructure readiness is not merely a technical support system, but rather a foundation for a modern and inclusive public service ecosystem.

Despite these advancements, technical challenges remain, including network congestion during peak traffic hours, signal disruptions in dense areas, and the need for regular system upgrades. Routine maintenance and capacity enhancements are essential to maintain optimal infrastructure performance. The government must continue comprehensive evaluations of existing infrastructure, assessing both technical performance and geographic distribution to prevent digital access disparities between regions. Strengthening these foundations is crucial for Malang City to scale up its Smart City coverage and ensure that digital services truly offer convenience, speed, and equitable access to all residents. In this way, digital transformation will not be limited to technological development but will genuinely reshape bureaucratic processes and patterns of interaction between government and society (Roziqin et al., 2024).

Utilisation of Public Service Applications and System Integration

Utilising public service applications represents a tangible form of digital transformation in Malang City's journey toward becoming a smart city. The local government has developed more than 42 public service applications and 25 administrative/governmental applications designed to accelerate administrative processes, reduce physical queues, and improve transparency and bureaucratic efficiency. Applications such as *SIAPEL* (Public Service Aspiration Information System), *JDIH* (Legal Documentation and Information Network), *e-JKN* (electronic-health services national), and SIMBAHE (Electronic Legal Aid Management Information System) are among the digital services already in use by the public. However, despite the large number of applications, the government faces serious challenges regarding system integration. Many of these applications were developed by third-party vendors who do not provide access to the source code or APIs, making integration difficult.

To address this fragmentation, the Malang City Government initiated the development of an integrated public service portal called *Malang Mbois*. This portal aims to consolidate various services from all local agencies into a single digital platform that is easily accessible to citizens. Conceptually, *Malang Mbois* functions as a central gateway for residents to access licensing services, population administration, complaints, and social assistance. In practice, however, the integration process has progressed slowly due to technical limitations in connecting legacy applications lacking proper technical documentation. Furthermore, not all agencies have the human resource capacity or supporting systems to rapidly migrate to a unified platform.

System integration efforts have not been without obstacles. A senior official from the Department of Communication and Information (Diskominfo) of Malang City stated:

> *"Many applications cannot be integrated directly because they were built by different vendors, and Diskominfo does not have full access to their system architecture". Anonymous informant (2/11/2025).*

This has led to overlapping application functions, user confusion, and increased maintenance burdens, as systems must be managed separately. Consequently, despite Malang City's productivity in developing numerous applications, the overall effectiveness of service digitalisation has not yet been fully realised.

In addition to technical barriers, user adoption presents another challenge. Many citizens remain unfamiliar with the existence of these applications, let alone using them regularly to access government services. A lack of digital campaigns, uneven levels of technological literacy, and inconsistent user interfaces across different applications hinder adoption. Some users still prefer visiting service offices in person, finding it more practical and allowing direct interaction with staff. This indicates that applications cannot stand alone as technological products they must be supported by ongoing public outreach and user assistance.

To improve the situation, Malang City Government has begun applying a phased approach to system integration by prioritising the most frequently used applications. Flagship platforms such as *Malang Mbois*, *SIMBAHE*, and *SIAPEL* have become the main focus during the early stages, as they directly serve daily public service needs. Although not yet fully documented quantitatively, Diskominfo has observed growing public enthusiasm for digital services, especially those that offer quick access, simple features, and practical use. This trend serves as an important indicator that the community is, in fact, open to digital transformation provided the platforms are relevant and user-friendly. Moving forward, the government needs to strengthen both technical and institutional aspects to ensure a more structured integration process, one that is not overly dependent on externally controlled application systems.

Policy Strengthening and Regulatory Aspects

Strong policies and regulations serve as a foundational pillar in the process of transitioning toward a smart city, ensuring that digital transformation efforts remain well-directed and sustainable. The Malang City Government has enacted several strategic regulations to support Smart City implementation, including both master plans and operational policies at the local agency level. One of the most fundamental policies is Malang Mayor Regulation No. 43 of 2020 on the Smart City Master Plan, which outlines the vision, mission, and development direction across six key dimensions: Smart Governance, Smart Economy, Smart Living, Smart Environment, Smart Society, and Smart Branding (Dirsehan & van Zoonen, 2022). Additioanally, Mayor Regulation No. 8 of 2023 concerning Electronic-Based Government Systems (SPBE) was issued to strengthen the structure of digital governance and serve as a primary reference for all regional work units (OPD) to fully adopt electronic public services.

These two regulations not only provide legal legitimacy but also establish a framework for inter-agency integration by emphasizing the importance of data interoperability, institutional synergy, and the strategic use of information technology across all lines of public service. The SPBE regulation, for instance, governs the use of integrated system architecture, information security standards, and the enhancement of Diskominfo's role as the command center for digital infrastructure and policy. Formally, these policies demonstrate the institutional readiness of Malang City to build data-driven governance. However, the extent to which these policies are effective depends greatly on their operational implementation, including the adjustment of SOPs and the preparedness of technical human resources in each agency.

An overview of Malang City's SPBE index over the past five years highlights the development of institutional readiness in digital policy:

Table 2. Development of Malang City's SPBE Index Scores (2018–2023)

Year	SPBE Score	Category
2018	2,50	Good
2019	1,90	Fair
2020	1,09	Poor
2021	3,56	Good
2022	2,60	Good
2023	3,82	Excellent

Source: SPBE Evaluation Report and Malang City's Department of Communication and Information (Diskominfo), 2024

Based on Table 2, Malang City's SPBE performance has improved over the past six years. In 2018, the city earned a score of 2.50 with a "Good" rating. However, the following year saw a decline to 1.90, and in 2020, the score dropped significantly to 1.09, categorised as "Poor". A sharp increase occurred in 2021, when the score surged to 3.56, returning to the "Good" category, although it slightly decreased to 2.60 in 2022. Positive momentum resumed in 2023, when Malang achieved a score of 3.82, rated as "Excellent", and received the 2023 Digital Government Award as one of the top 10 cities nationally in SPBE implementation.

An "Excellent" rating reflects not only technical success but also the city's ability to strengthen policy, enhance digital infrastructure, and foster inter-agency synergy. Improvements include the expansion of application-based public services, internal data integration, and advancements in information security management. Within the context of digital governance, Malang City's SPBE achievements indicate a strong commitment to accelerating bureaucratic transformation towards a more open, efficient, and responsive government model that meets the demands of a digital society. These results also demonstrate that the policies formulated by the government are not merely symbolic but are beginning to produce tangible outcomes at the implementation level.

Human Resources and Organisational Challenges

Human resources (HR) play one of the most crucial roles in the successful implementation of smart cities, especially since digital transformation involves not only technology but also shifts in work culture and bureaucratic mindsets (Dulbert et al., 2024; Guenduez & Mergel, 2022). n Malang City, HR remains a dominant issue that affects the effectiveness of digital programme implementation. The local government employs 6,599 active civil servants across various departments, but the number of personnel with a background in information technology is very limited. According to data from the Department of Communication and Information (Diskominfo), only 58 individuals across the entire city hold formal qualifications in ICT, most of whom are concentrated in Diskominfo. This disparity directly impacts the execution of smart city programmes, particularly in managing applications, securing data, and integrating digital services.

This situation is further complicated by the high rate of staff rotation, especially in technical roles involved in developing and maintaining digital systems. Internal job transfers often disregard programme continuity, disrupting workflows and forcing new teams to restart adaptation processes. One third informant explained:

> *"Many programmes are disrupted because trained staff are moved to other divisions, while their replacements lack the necessary technical expertise". Anonymous Informant 3 (21/11/24).*

This issue leads to inconsistency in programme implementation and slows the progress of digital service targets. To provide a clearer picture, the following table outlines the current HR conditions within the Malang City Government, particularly those related to smart city implementation:

Table 3. Human Resource Conditions of Malang City Government in Smart City Implementation

Category	Total/Status	Description
Total City Government Staff	6.599 people	Civil servants across all departments
Staff at Diskominfo	95 people	Includes 49 civil servants and 46 non-permanent staff
ICT-background Personnel	58 people (citywide)	Mostly assigned to Diskominfo
Digital Training (by Diskominfo)	Available, but uneven	Not yet reaching all departments and technical levels
ICT Volunteers	Inactive	Previously existed, replaced by Community Info Groups (KIM)

Source: Author analysis, 2025

As shown in Table 3, Malang City's digital transformation is still heavily dependent on a limited number of human resources both in terms of quantity and capacity. Diskominfo, the main technical institution, is responsible not only for system development but also for managing networks, data security, and digital public services. The gap between digital needs and available skilled workers has caused stagnation in several departments that lack sufficient technical support. This issue is worsened by suboptimal internal training programmes that have yet to reach all government units proportionately.

Beyond staffing quantity and distribution, competency quality also remains a critical concern. Most civil servants in Malang City have not received regular technical training on application management, data system operations, or emerging technologies. While Diskominfo has initiated internal training, it remains limited to select departments already active in digital implementation. This imbalance has led to uneven digital capacities across agencies, often causing integration efforts to stall. As a result, many built applications remain underutilised due to limited familiarity and proficiency among staff.

Outside of government employees, Malang also has a history of involving ICT volunteers; however, this initiative has not been sustained effectively. The public, as end-users of many smart city applications, also requires attention in terms of digital literacy. Currently, Malang City Government relies on Community Information Groups (Kelompok Informasi Masyarakat/KIM) to help disseminate smart city programmes. However, KIM primarily serves a socio-informative role rather than providing technical assistance, creating a gap in public guidance and support. To achieve an inclusive smart city, digital literacy must be strengthenednot only to improve internal government capacity but also to empower citizen participation in leveraging digital technologies.

Challenges and Strategic Recommendations

he implementation of the smart city programme in Malang City during 2024 has shown significant progress. Out of a total of 46 programmes across the six smart city dimensions, nearly all have been successfully executed except for one under the *Smart Economy* category, namely Si-Bansos, which was discontinued after the COVID-19 pandemic. Based on the SPBE evaluation, Malang's index score improved from 1.09 in 2020 to 3.82 in 2023, earning the city a Digital Government Award 2023. As illustrated in Figure 1, the smart city programme has made remarkable strides in digital governance, public service innovation, and a strong commitment to technology-driven governance in Indonesia, including in Malang City. Nevertheless, several challenges still need to be addressed to maximise the programme's impact and to provide valuable insights for other cities in the country.

One of the most pressing challenges is the digital divide, both in terms of access and literacy. Although public Wi-Fi facilities have been deployed at over 500 points across the city, not all areas enjoy stable connectivity, especially in densely populated or peripheral neighborhoods. Additionally, digital literacy among the community remains relatively low. Many residents are still unfamiliar with public service applications or are unaware that such services exist in digital form. These limitations hinder widespread adoption of technology and prevent most applications from having an optimal impact on service delivery.

Figure 1. **Content analysis by Worcloud Nvivo 12 plus**

Source: *Author analysis using Nvivo 12 Plus, 2025*

This research identified several key challenges, including delays in regulatory formulation, lack of system integration, low levels of digital literacy among the public, and limited budget capacity. In 2024, the Malang City Government allocated over IDR 110 billion from its regional budget (APBD) to support smart city programme expenditures. In addition, collaboration with the private sector and academic institutions has played a role in the implementation of smart city initiatives across the city. However, due to Malang's complexity, many applications have been developed by third-party vendors without providing open technical documentation (Roziqin et al., 2024). As a result, Diskominfo has faced difficulties in maintaining or integrating these systems independently, as access to source code is not granted. At the same time, not all local government agencies have sufficient funding to build or develop new systems that meet their specific needs. When budget constraints and insufficient human resources intersect, the risk of digital innovation stagnation increases, ultimately threatening the sustainability of the digitalisation process.

Another challenge lies in the attitude of civil servants who struggle to adapt to organisational change. Some employees remain reluctant to shift from manual to digital systems, perceiving the traditional way as more familiar and secure. Changes in work routines especially those involving digital applications and reporting are often seen as burdensome or incompatible with long-standing habits. As a result, despite the availability of digital platforms, their usage remains inconsistent and is often treated as an administrative supplement rather than a primary tool. In such conditions, a purely top-down policy approach is insufficient. A more comprehensive strategy is required to initiate cultural change within the organisation, including relevant training, coaching, and appropriate incentives, to foster a more positive reception of the digital transition among civil servants.

Reflecting on these conditions, this chapter offers several strategic recommendations to accelerate the progress of smart city initiatives in Malang City. First, there is an urgent need to strengthen human resources and digital literacy, not only through technical training for civil servants but also through large-scale, participatory public education. Collaborations with universities, digital communities, and private sector actors are essential to expand the reach and impact of these initiatives. Second, the government should promote technological self-reliance by developing open-source applications that can be managed internally without dependence on external vendors. This approach is crucial to ensure system sustainability and optimize budget efficiency. Third, establishing a cross-agency digital coordination forum would help streamline system integration efforts and prevent fragmented implementation. Through this forum, all local government units could align their perspectives, co-develop a unified roadmap, and monitor programme progress collaboratively across sectors.

In the future, the government should adopt an impact-oriented evaluation framework, rather than focusing solely on the number of digital applications or SPBE scores. Applications that are underutilized or redundant must be systematically reviewed and streamlined. Conversely, systems that demonstrate effectiveness should be further developed and enriched with continuous innovation. Ensuring long-term sustainability also requires a stronger emphasis on citizen participation in digital planning processes and the creation of inclusive spaces for dialogue between the government and technology users. This integrated approach addressing institutional, technical, and social dimensions will reinforce Malang City's position as a smart city that is not only recognized formally, but also delivers tangible benefits to its citizens.

CONCLUSION

The research conducted illustrates how a city's e-readiness is inextricably linked to its development into a smart city. In the case of Malang City, this readiness is not only shown through infrastructure development and public service application development, but also reflected in policy direction, institutional structure, and human resource readiness. Through these aspects, this study found that adaptive regulations, implementing competencies, inter-agency coordination, and adequate digital infrastructure are important factors in supporting the realization of a smart city. These findings emphasize that smart cities are not just technology adoption, but systemic changes that require readiness at various levels of government and society.

Malang City has made notable strides in smart city development, as evidenced by its 2023 SPBE score reaching the 'Excellent' category, the installation of public Wi-Fi and CCTV infrastructure, and the creation of over 60 digital applications. These efforts reflect a strong governmental commitment to building a robust digital ecosystem, further validated by recognitions such as the Digital Government Award and Top Digital Implementation 2023. Nevertheless, challenges remain, including limited system integration, disparities in digital literacy, and low public engagement with digital services. These issues highlight that infrastructure and system development must be complemented by efforts to enhance digital capacity within both the bureaucracy and the broader community.

Human resources remain a significant challenge that has yet to be fully addressed. The number of employees with ICT backgrounds is still very limited, their distribution is uneven, and training programs have not reached all regional government units. Furthermore, the resistance to change among some personnel continues to hinder the adoption of newly developed digital systems. Dependence on external vendors for application development also raises concerns about long-term system sustainability. These issues highlight that the success of smart cities cannot depend on a single factor alone it requires synergy between technology, human resources, policy, and a transformative work culture.

This finding provides an important discussion that smart city development should not be directed solely to fulfill performance indicators or gain formal recognition, but to solve real problems faced by the community. Malang City already has a fairly solid foundation in terms of policies, infrastructure, and institutions. If the aspects of collaboration between regional apparatus are strengthened, public literacy is continuously improved, and consistency in policy direction is maintained on an ongoing basis, then the ongoing digital transformation will not only have an impact on governance, but also touch the way of life of city residents. Ultimately, the measure of smart city success lies not in the amount of technology used, but in how much it facilitates the accessibility of the people who use it. Both conceptually and practically, this chapter provides important insights for policy makers, politicians and all stakeholders on the fundamental points that must be prepared to optimize smart city programs.

REFERENCES

Agrawal, G., Kumar, D., & Singh, M. (2022). Assessing the usability, accessibility, and mobile readiness of e-government websites: A case study in india. *Universal Access in the Information Society*, *21*(3), 737–748. DOI: 10.1007/s10209-021-00800-8

Akbar, P. N. G., Auliya, A., Pranita, D., & Oktadiana, H. (2024). The readiness assessment of Jakarta as a smart tourism city. *Cogent Social Sciences*, *10*(1), 2364386. Advance online publication. DOI: 10.1080/23311886.2024.2364386

Ali, M. A. (2022). Smart city policy in developing countries: Case study of the new administrative capital in Egypt. *Journal of Public Affairs*, *22*(S1), 1–16. DOI: 10.1002/pa.2774

Anthony Jnr, B., Abbas Petersen, S., Helfert, M., & Guo, H. (2021). Digital transformation with enterprise architecture for smarter cities: A qualitative research approach. *Digital Policy. Regulation & Governance*, *23*(4), 355–376. DOI: 10.1108/DPRG-04-2020-0044

Asmorowati, S., Schubert, V., Nasution, P. I., & Hariani, N. J. (2025). Smart Women, Smart Leadership, and Smart Cities: A Case Study of Two Indonesian Mayors' Approach to E-Leadership. *Journal of Public Affairs*, *25*(2), e70023. https://doi.org/https://doi.org/10.1002/pa.70023. DOI: 10.1002/pa.70023

Buck, C., Clarke, J., Torres de Oliveira, R., Desouza, K. C., & Maroufkhani, P. (2023). Digital transformation in asset-intensive organisations: The light and the dark side. *Journal of Innovation and Knowledge*, *8*(2), 100335. Advance online publication. DOI: 10.1016/j.jik.2023.100335

Caragliu, A., del Bo, C., & Nijkamp, P. (2011). Smart cities in Europe. *Journal of Urban Technology*, *18*(2), 65–82. DOI: 10.1080/10630732.2011.601117

Creswell, J. W., & Creswell, J. D. (2022). *Research Design: Qualitative, Quantitative, and Mixed Methods Approaches* (The Sixth). *Sage (Atlanta, Ga.).*

Das, P., Woods, O., & Kong, L. (2024). Human sensors in the city of super apps: Mobilizing people as infrastructure for smart city development in Jakarta, Indonesia. *City (London, England)*, *28*(5–6), 812–834. DOI: 10.1080/13604813.2024.2427954

Demirel, D., & Mülazımoğlu, M. E. (2022). How the smart governance model shapes cities? Cases from Europe. *Journal of Enterprising Communities*, *16*(1), 8–25. DOI: 10.1108/JEC-08-2021-0115

Desouza, K. C., Hunter, M., Jacob, B., & Yigitcanlar, T. (2020). Pathways to the Making of Prosperous Smart Cities: An Exploratory Study on the Best Practice. *Journal of Urban Technology*, *27*(3), 3–32. DOI: 10.1080/10630732.2020.1807251

Dirsehan, T., & van Zoonen, L. (2022). Smart city technologies from the perspective of technology acceptance. *IET Smart Cities*, *4*(3), 197–210. DOI: 10.1049/smc2.12040

Dulbert, T. B., Budi, M. A., & Citra, A. W. (2024). Key Indicators for Effective Implementation of Smart City Standards in Indonesia During the Covid-19Pandemic: A Case Study of Four Cities. *Journal of Urban Culture Research*, *28*, 37–56. https://www.scopus.com/inward/record.uri?eid=2-s2.0-85201764592&partnerID=40&md5=7d427603585c3a43b7bc18a846f29e25

Đurek, V., &Ređep, N. B. (2016). Review on e-readiness assessment tools. *Central European Conference on Information and Intelligent Systems*, 161–168.

Estevez, E., Pardo, T. A., & Scholl, H. J. (2021). Smart Cities and Smart Governance. In *Public Administration and Information Technology* (Vol. 37, pp. 43–66). Springer Nature., DOI: 10.1007/978-3-030-61033-3_3

Etim, E., & Daramola, O. (2023). Investigating the E-Readiness of Informal Sector Operators to Utilize Web Technology Portal. *Sustainability (Basel)*, *15*(4), 1–28. https://doi.org/https://doi.org/10.3390/su15043449. DOI: 10.3390/su15043449

Fahlevi, M., Dandi, M., Asyrof, R. A., Dahlan, O. P., Asetya, D. R., & Dahlan, S. P. (2024). Elevating Healthcare Through Telemedicine: A Pathto Smart City Success in IKN (Indonesia's NewCapital City). *11th International Conference on ICT for Smart Society: Integrating Data and Artificial Intelligence for a Resilient and Sustainable Future Living, ICISS 2024 - Proceeding*. DOI: 10.1109/ICISS62896.2024.10751614

Fernandez-Anez, V., Fernández-Güell, J. M., & Giffinger, R. (2018). Smart City implementation and discourses: An integrated conceptual model. The case of Vienna. *Cities (London, England)*, *78*(November), 4–16. DOI: 10.1016/j.cities.2017.12.004

Guenduez, A. A., & Mergel, I. (2022). The role of dynamic managerial capabilities and organizational readiness in smart city transformation. *Cities (London, England)*, *129*(July), 103791. DOI: 10.1016/j.cities.2022.103791

Gunawong, P., & Gao, P. (2017). Understanding e-government failure in the developing country context: A process-oriented study. *Information Technology for Development*, *23*(1), 153–178. DOI: 10.1080/02681102.2016.1269713

Hardi, R., Nurmandi, A., Purwaningsih, T., & Manaf, H. A. (2025). Smart city governance and interoperability : Enhancing human security in Yogyakarta and Makassar, Indonesia. *Frontiers in Political Science*, *6*(March), 1–18. DOI: 10.3389/fpos.2025.1553177

He, Y., & Tritto, A. (2022). Urban utopia or pipe dream? Examining Chinese- invested smart city development in Southeast Asia smart city development in Southeast Asia. *Third World Quarterly*, *43*(9), 2244–2268. DOI: 10.1080/01436597.2022.2089648

Hu, Q., & Zheng, Y. (2021). Smart city initiatives: A comparative study of American and Chinese cities. *Journal of Urban Affairs*, *43*(4), 504–525. DOI: 10.1080/07352166.2019.1694413

Kamil, M., Muhammad, R., Roziqin, A., Sari, A. E., & Kismartini, . (2025). Examining Practice Smart City in Local Government : A Smart Governance Perspective. *International Journal of Sustainable Development and Planning*, *20*(2), 635–644. DOI: 10.18280/ijsdp.200214

Lestari, M. D., Kantun, S., Hartanto, W., Suharso, P., & Widodo, J. (2020). Analysis of the financial literacy level of Micro, Small and Medium Enterprises (MSMEs) in Jember, East Java, Indonesia. In Sumardi (Ed.), *2nd International Conference on Environmental Geography and Geography Education, ICEGE 2019* (Vol. 485, Issue 1). Institute of Physics Publishing. DOI: 10.1088/1755-1315/485/1/012128

Lnenicka, M., & Saxena, S. (2021). Re-defining open government data standards for smart cities' websites: A case study of selected cities. *Digital Policy. Regulation & Governance*, *23*(4), 398–411. DOI: 10.1108/DPRG-12-2020-0174

Maisyarah, R. A., Nugroho, R. A., & Te, A. Y. C. (2024). Examining of E-readiness in the Local Government toward Smart City: A Study from Balangan Regency. *Jurnal Ilmu Sosial, 23*(1), 1–26. https://doi.org/https://doi.org/10.14710/jis.23.1.2024.37-63

Martinez, R., & Masron, I. N. (2020). Jakarta: A city of cities. *Cities (London, England)*, *106*, 102868. Advance online publication. DOI: 10.1016/j.cities.2020.102868 PMID: 32863521

Meijer, A., & Bolívar, M. P. R. (2016). Governing the smart city: A review of the literature on smart urban governance. *International Review of Administrative Sciences*, *82*(2), 392–408. DOI: 10.1177/0020852314564308

Mensah, I. K. (2020). Impact of Government Capacity and E-Government Performance on the Adoption of E-Government Services. *International Journal of Public Administration*, *43*(4), 303–311. DOI: 10.1080/01900692.2019.1628059

Miles, M. B., & Huberman, M. J. S. (2019). *Qualitative Data Analysis: A Methods Sourcebook* (4th ed.).

Mkhonto, M., & Zuva, T. (2024). In Silhavy, R., & Silhavy, P. (Eds.), *Readiness for Smart City in Municipalities in Mbabane, Eswatini BT - Data Analytics in System Engineering* (pp. 402–410). Springer Nature Switzerland.

Nam, H., Nam, T., Oh, M., & Choi, S. (2022). An Efficiency Measurement of E-Government Performance for Network Readiness : Non-Parametric Frontier Approach. *Journal of Open Innovation*, *8*(1), 10. DOI: 10.3390/joitmc8010010

Nugroho, R. A., & Purbokusumo, Y. (2020). E-Government Readiness: Main Actor Readiness Assessment for E-Government Application in Indonesia. *Jurnal Ilmu Pengetahuan Dan Teknologi Komunikasi*, *22*(1), 1–17. DOI: 10.33164/iptekkom.22.1.2020.1-17

Park, J., & Yoo, S. (2023). Evolution of the smart city: Three extensions to governance, sustainability, and decent urbanisation from an ICT-based urban solution. *International Journal of Urban Sciences*, *27*(sup1, S1), 10–28. DOI: 10.1080/12265934.2022.2110143

Park, S., Choi, Y. T., & Bok, H. S. (2013). Does better e-readiness induce more use of e-government? Evidence from the Korean central e-government. *International Review of Administrative Sciences*, *79*(4), 767–789. DOI: 10.1177/0020852313501247

Potnis, D. D., & Pardo, T. A. (2011). Mapping the evolution of e-Readiness assessments. *Transforming Government: People. Transforming Government*, *5*(4), 345–363. DOI: 10.1108/17506161111173595

Pratama, A. B., & Imawan, S. A. (2019). A scale for measuring perceived bureaucratic readiness for smart cities in Indonesia. *Public Administration and Policy*, *22*(1), 25–39. DOI: 10.1108/PAP-01-2019-0001

Rachmawati, R., & Rohmah, A. A. (2025). Navigating Diversity of the Smart City Initiatives in Indonesia. *The Indonesian Journal of Geography*, *57*(1), 191–199. DOI: 10.22146/ijg.97840

Roziqin, A., Kamil, M., Romadhan, A. A., & Zakaria, I. (2024). The dark side of local public sector innovation: insights from SAMPADE innovation. *Journal of Science and Technology Policy Management*. DOI: 10.1108/JSTPM-01-2024-0001

Santoso, A. D., Aryansah, J. E., & Nasyaya, A. (2024). Writing about smart cities in Indonesia: A bibliometric analysis. *Journal of Regional and City Planning*, *35*(1), 69–89. DOI: 10.5614/jpwk.2024.35.1.4

Spicer, Z., Goodman, N., & Wolfe, D. A. (2023). How 'smart' are smart cities? Resident attitudes towards smart city design. *Cities (London, England)*, *141*(July), 104442. DOI: 10.1016/j.cities.2023.104442

Sulistyaningsih, T., Loilatu, M. J., & Roziqin, A. (2024). Research trends on smart urban governance in Asia: A bibliometric analysis. *Journal of Science and Technology Policy Management*, *15*(5), 997–1105. DOI: 10.1108/JSTPM-03-2022-0045

Wibowo, T. S., Endroyono, & Pratomo, I. (2020). Analysis of Malang City Readiness in Realizing Smart Tourism with New Integrated E-Readiness Model. *2020 International Conference on Smart Technology and Applications (ICoSTA)*, 1–6. DOI: 10.1109/ICoSTA48221.2020.1570616406

Yigitcanlar, T. (2015). Smart cities: An effective urban development and management model? *Australian Plants*, *52*(1), 27–34. DOI: 10.1080/07293682.2015.1019752

Yigitcanlar, T., Degirmenci, K., Butler, L., & Desouza, K. C. (2022). What are the key factors affecting smart city transformation readiness? Evidence from Australian cities. *Cities, 120*(September 2020), 103434. DOI: 10.1016/j.cities.2021.103434

Ziemba, E. (2019). The Contribution of ICT Adoption to the Sustainable Information Society. *Journal of Computer Information Systems*, *59*(2), 116–126. DOI: 10.1080/08874417.2017.1312635

Compilation of References

Abba Ari, A. A., Aziz, H. A., Njoya, A. N., Aboubakar, M., Djedouboum, A. C., Thiare, O., & Mohamadou, A. (2024). Data collection in IoT networks: Architecture, solutions, protocols and challenges. *IET Wireless Sensor Systems*, *14*(4), 85–110. DOI: 10.1049/wss2.12080

Abba, S., Bizi, A. M., Lee, J., Bakouri, S., & Crespo, M. L. (2024). Real-time object detection, tracking, and monitoring framework for security surveillance systems. *Heliyon*, *10*(15), e34922. Advance online publication. DOI: 10.1016/j.heliyon.2024.e34922 PMID: 39145028

Abduljabbar, R., Dia, H., & Tsai, P. (2021). Development and evaluation of bidirectional LSTM freeway traffic forecasting models using simulation data. *Scientific Reports*, *11*(1), 23899. Advance online publication. DOI: 10.1038/s41598-021-03282-z PMID: 34903780

Abhilash (2024). Empowering Zurich: How IOT solutions are driving smart city success. https://www.intertoons.ch/post/empowering-zurich-how-iot-solutions-are-driving-smart-city-success?

Acemoğlu, D., Autor, D., Hazell, J., & Restrepo, P. (2022). Artificial Intelligence and Jobs: Evidence from Online Vacancies. *Journal of Labor Economics*, *40*(S1), S293–S340. Advance online publication. DOI: 10.1086/718327

Adeniyi, L. A. (2023). Assessment of the effects of location of disposal-sites on waste disposal practices of urban residents in Nigeria. *GeoJournal*, *88*(4), 3889–3903. DOI: 10.1007/s10708-023-10835-4

Adibi, S., Rajabifard, A., Shojaei, D., & Wickramasinghe, N. (2024). Enhancing Healthcare through Sensor-Enabled Digital Twins in Smart Environments: A Comprehensive. *Analysis & Sensing*, *24*(9), 2793. DOI: 10.3390/s24092793 PMID: 38732899

Affinito, A., Zinno, S., Stanco, G., Botta, A., & Ventre, G. (2023). The evolution of Mirai botnet scans over six years. *Journal of Information Security and Applications*, *79*, 103629. DOI: 10.1016/j.jisa.2023.103629

Agrawal, G., Kumar, D., & Singh, M. (2022). Assessing the usability, accessibility, and mobile readiness of e-government websites: A case study in india. *Universal Access in the Information Society*, *21*(3), 737–748. DOI: 10.1007/s10209-021-00800-8

Ahmad, N., & Kumar, S. (2021). Real-time infant monitoring system using cloud-enabled IoT devices. *International Journal of Distributed Sensor Networks*, *17*(7), 1–12.

Ahmed, A. A., Belrzaeg, M., Nassar, Y. F., El-Khozondar, H. J., Khaleel, M., & Alsharif, A. (2023). A comprehensive review towards smart homes and cities considering sustainability developments, concepts, and future trends [Review of A comprehensive review towards smart homes and cities considering sustainability developments, concepts, and future trends]. World Journal of Advanced Research and Reviews, 19(1), 1482. GSC Online Press. DOI: 10.30574/wjarr.2023.19.1.1530

Ahmed, E., Yaqoob, I., Hashem, I. A. T., Khan, I., Ahmed, A. I. A., Imran, M., & Vasilakos, A. V. (2017). The role of big data analytics in Internet of Things. *Computer Networks*, *129*, 459–471. DOI: 10.1016/j.comnet.2017.06.013

Ahmed, N., Rafiq, J. I., & Islam, M. R. (2020). Enhanced human activity recognition based on smartphone sensor data using hybrid feature selection model. *Sensors (Basel)*, *20*(1), 317. Advance online publication. DOI: 10.3390/s20010317 PMID: 31935943

Akbar, P. N. G., Auliya, A., Pranita, D., & Oktadiana, H. (2024). The readiness assessment of Jakarta as a smart tourism city. *Cogent Social Sciences*, *10*(1), 2364386. Advance online publication. DOI: 10.1080/23311886.2024.2364386

Akter, M., & Moustafa, N. (2024, June). Federated Learning-based privacy protection methods for Internet of Things systems. Security, Privacy and Trust Management.

Al Nuaimi, E., Al Neyadi, H., Mohamed, N., & Al-Jaroodi, J. (2015). Applications of big data to smart cities. *Journal of Internet Services and Applications*, *6*(1), 1–15. DOI: 10.1186/s13174-015-0041-5

Alahi, M. E. E., Sukkuea, A., Tina, F. W., Nag, A., Kurdthongmee, W., Suwannarat, K., & Mukhopadhyay, S. C. (2023). Integration of IoT-Enabled Technologies and Artificial Intelligence (AI) for Smart City Scenario: Recent Advancements and Future Trends. *Sensors, 23*(11), 5206. MDPI. https://www.mdpi.com/1424-8220/23/11/5206

Alahi, M. E. E., Sukkuea, A., Tina, F. W., Nag, A., Kurdthongmee, W., Suwannarat, K., & Mukhopadhyay, S. C. (2023). Integration of IoT-enabled technologies and artificial intelligence (AI) for smart city scenario: Recent advancements and future trends. *Sensors (Basel)*, *23*(11), 5206. DOI: 10.3390/s23115206 PMID: 37299934

Alauthman, M., Alkasassbeh, M. S., Al-Qerem, A., Alateef, S., & Almomani, A. (2025). Privacy-Preserving Security: Balancing Protection and Data Utility in Digital Systems. In *Complexities and Challenges for Securing Digital Assets and Infrastructure* (pp. 483-506). IGI Global Scientific Publishing.

Albino, V., Berardi, U., & Dangelico, R. M. (2015). Smart cities: Definitions, dimensions, performance, and initiatives. *Journal of Urban Technology*, *22*(1), 3–21. DOI: 10.1080/10630732.2014.942092

Alghamdi, W. Y., Alshamrani, R. M., Aloufi, R. K., Ba Lhamar, S. O., Altwirqi, R. A., Alotaibi, F. S., Althobaiti, S. M., Altalhi, H. M., Alshamrani, S. A., & Alazwari, A. S. (2025). Exploring the Synergy Between Digital Twin Technology and Artificial Intelligence: A Comprehensive Survey. *International Journal of Advanced Computer Science and Applications*, *16*(3). Advance online publication. DOI: 10.14569/IJACSA.2025.0160399

Alghamedy, F., El-Haggar, N., Alsumayt, A., Alfawaer, Z. M., Alshammari, M., Amouri, L., Aljameel, S. S., & Al-Bassam, S. (2024). Unlocking a Promising Future: Integrating Blockchain Technology and FL-IoT in the Journey to 6G. *IEEE Access : Practical Innovations, Open Solutions*, *1*, 115411–115447. Advance online publication. DOI: 10.1109/ACCESS.2024.3435968

Alharbi, E., Alsulami, G., Aljohani, S., Alharbi, W., & Albaradei, S. (2025). Real-time detection and monitoring of public littering behavior using deep learning for a sustainable environment. *Scientific Reports*, *15*(1), 3000. Advance online publication. DOI: 10.1038/s41598-024-77118-x PMID: 39848984

Ali, I., Ahmedy, I., Gani, A., Munir, M. U., & Anisi, M. H. (2022). Data collection in studies on Internet of Things (IoT), wireless sensor networks (WSNs), and sensor cloud (SC): Similarities and differences. *IEEE Access : Practical Innovations, Open Solutions*, *10*, 33909–33931. DOI: 10.1109/ACCESS.2022.3161929

Ali, M. A. (2022). Smart city policy in developing countries: Case study of the new administrative capital in Egypt. *Journal of Public Affairs*, *22*(S1), 1–16. DOI: 10.1002/pa.2774

Ali, Z., Chaudhry, S. A., Ramzan, M. S., & Al-Turjman, F. (2020). Securing Smart City Surveillance: A Lightweight Authentication Mechanism for Unmanned Vehicles. *IEEE Access : Practical Innovations, Open Solutions*, *8*, 43711–43724. DOI: 10.1109/ACCESS.2020.2977817

AlJamal, M., Mughaid, A., Bani-Salameh, H., Alzubi, S., & Abualigah, L. (2024). Optimizing risk mitigation: A simulation-based model for detecting fake IoT clients in smart city environments. *Sustainable Computing : Informatics and Systems*, *43*, 101019. DOI: 10.1016/j.suscom.2024.101019

Allam, Z., & Newman, P. (2018). Redefining the smart city: Culture, metabolism and governance. *Smart Cities*, *1*(1), 4–25. DOI: 10.3390/smartcities1010002

Almtireen, N., Reddy, V., Sutton, M., Nedvidek, A., Karn, C., Ryalat, M., Elmoaqet, H., & Rawashdeh, N. (2025). PLC-Controlled Intelligent Conveyor System with AI-Enhanced Vision for Efficient Waste Sorting. *Applied Sciences (Basel, Switzerland)*, *15*(3), 1550. DOI: 10.3390/app15031550

Alotaibi, A., & Nadeem, F. (2024). Leveraging social media and deep learning for sentiment analysis for smart governance: A case study of public reactions to educational reforms in Saudi Arabia. *Computers*, *13*(11), 280. DOI: 10.3390/computers13110280

Alsaadi, H. A., & Tubaishat, A. (2015). Internet of Things: Features, challenges, and vulnerabilities. *International Journal of Advanced Computer Science and Applications*, *6*(11), 128–133.

Al-Turjman, F., Zahmatkesh, H., & Shahroze, R. (2022). An overview of security and privacy in smart cities' IoT communications. *Transactions on Emerging Telecommunications Technologies*, *33*(3), e3677. DOI: 10.1002/ett.3677

Alvarez, H. I. G., Peña, H. H. Z., del Consuelo Cortes Velázquez, M., & Ypiña, C. P. F. (2023). Multimedia Technology and Smart Cities; an Inevitable Combination of Urban Progress. *Communications in Computer and Information Science, 1906 CCIS*, 86–97. DOI: 10.1007/978-3-031-45316-8_8

Alwabli, A. (2024). Federated learning for privacy-preserving air quality forecasting using IoT sensors. *Eng. Technol. Appl. Sci. Res.*, *14*(4), 16069–16076. DOI: 10.48084/etasr.7820

Alwar, B., Edgardo, H., & Faustino, E. (2024). Integrating IoT and Artificial Intelligence for sustainable smart city development: A case study approach. *Information Technology Engineering Journals*, *9*(2), 80–90. DOI: 10.24235/itej.v9i2.137

Anastasi, G., Antonelli, M., Bechini, A., Brienza, S., D'Andrea, E., De Guglielmo, D., & Segatori, A. (2013, October). *Urban and social sensing for sustainable mobility in smart cities. In 2013 Sustainable Internet and ICT for Sustainability (SustainIT)*. IEEE.

Anisetti, M., Ardagna, C., Bellandi, V., Cremonini, M., Frati, F., & Damiani, E. (2018). Privacy-aware Big Data Analytics as a service for public health policies in smart cities. *Sustainable Cities and Society*, *39*, 68–77. DOI: 10.1016/j.scs.2017.12.019

Anthi, E., Williams, L., Ieropoulos, V., & Spyridopoulos, T. (2024). Investigating radio frequency vulnerabilities in the Internet of Things (IoT). *IoT*, *5*(2), 356–380. DOI: 10.3390/iot5020018

Anthony Jnr, B., Abbas Petersen, S., Helfert, M., & Guo, H. (2021). Digital transformation with enterprise architecture for smarter cities: A qualitative research approach. *Digital Policy. Regulation & Governance*, *23*(4), 355–376. DOI: 10.1108/DPRG-04-2020-0044

Anuradha, P., Vasanth, K., Renuka, G., & Rao, A. R. (2023). IoT based enabling home automation system for individuals with diverse disabilities. e-Prime - Advances in Electrical Engineering. *Electronics and Energy*, *6*, 100366.

Aral, S. (2020). *The hype machine: How social media disrupts our elections, our economy, and our health—and how we must adapt*. Currency.

Archello. (n.d.). Affordable housing in Zurich. https://archello.com/project/affordable-housing-in-zurich

Arco, J., Mateo-Aroca, A., & Molina-García, Á. (2019). Adaptive street lighting control based on wireless communication and traffic sensing. *Energy*, *174*, 611–622. DOI: 10.1016/j.energy.2019.03.010

Arindam, R., Apurba, M., Jungmin, K., & Ilkyeong, M. (2022). IoT-based smart bin allocation and vehicle routing in solid waste management: A case study in South Korea. *Computers & Industrial Engineering*, *171*, 108457. DOI: 10.1016/j.cie.2022.108457

Asher, S. (2020, July 4). TraceTogether: Singapore turns to wearable contact-tracing Covid tech. *BBC News*. Retrieved from https://www.bbc.com/news/technology-53146360

Asmorowati, S., Schubert, V., Nasution, P. I., & Hariani, N. J. (2025). Smart Women, Smart Leadership, and Smart Cities: A Case Study of Two Indonesian Mayors' Approach to E-Leadership. *Journal of Public Affairs*, *25*(2), e70023. https://doi.org/https://doi.org/10.1002/pa.70023. DOI: 10.1002/pa.70023

Astuti, S. D., Tamimi, M. H., Pradhana, A. A. S., Alamsyah, K. A., Purnobasuki, H., Khasanah, M., Susilo, Y., Triyana, K., Kashif, M., & Syahro, A. (2021). Gas sensor array to classify the chicken meat with E. coli contaminant by using random forest and support vector machine. *Biosensors & Bioelectronics: X*, *9*, 100083. DOI: 10.1016/j.biosx.2021.100083

Atadoga, A., Omaghomi, T. T., Elufioye, O. A., Odilibe, I. P., Daraojimba, A. I., & Owolabi, O. R. (2024). Internet of Things (IoT) in healthcare: A systematic review of use cases and benefits [Review of Internet of Things (IoT) in healthcare: A systematic review of use cases and benefits]. *International Journal of Science and Research Archive*, *11*(1), 1511–1517. DOI: 10.30574/ijsra.2024.11.1.0243

Atzori, L., Iera, A., & Morabito, G. (2010). The internet of things: A survey. *Computer Networks*, *54*(15), 2787–2805. DOI: 10.1016/j.comnet.2010.05.010

Ayturan, K., Sarıkamış, B., Akşahin, M. F., & Kutbay, U. (2025). SPHERE: Benchmarking YOLO vs. CNN on a Novel Dataset for High-Accuracy Solar Panel Defect Detection in Renewable Energy Systems. *Applied Sciences (Basel, Switzerland)*, *15*(9), 4880. DOI: 10.3390/app15094880

Babu, C. V. S., Surendar, V., Dheepak, N., Shiraj, S., & Praveen, K. (2024). Revolutionizing healthcare harnessing IoT-integrated federated learning for early disease detection and patient privacy preservation. In *Advances in Healthcare Information Systems and Administration* (pp. 195–216). IGI Global.

Badii, C., Nesi, P., & Paolucci, M. (2019). Predictive models for urban mobility in smart cities. *Sensors (Basel)*, *19*(12), 2781. DOI: 10.3390/s19122781 PMID: 31226869

Baimukhanov, S., Ali, H., & Yazici, A. (2025). Enhancing ML-based anomaly detection in data management for security through integration of IoT, cloud, and edge computing. *Expert Systems with Applications*, *293*, 128700. Advance online publication. DOI: 10.1016/j.eswa.2025.128700

Banerjee, A. (2023). Smart Cities spur new thinking on technology risks. https://www.zurich.com/commercial-insurance/sustainability-and-insights/commercial-insurance-risk-insights/smart-cities-spur-new-thinking-on-technologys-risks

Barkane, I., Adamsone-Fiskovica, A., & Kilis, E. (2023). Creeping road traffic surveillance in Latvia: Social and legal implications of digital policing tools. *Surveillance & Society*, *21*(4), 432–446. DOI: 10.24908/ss.v21i4.15812

Batty, M. (2013). *The new science of cities*. MIT press. DOI: 10.7551/mitpress/9399.001.0001

Batty, M., Axhausen, K. W., Giannotti, F., Pozdnoukhov, A., Bazzani, A., Wachowicz, M., Ouzounis, G., & Portugali, Y. (2012). Smart cities of the future. *The European Physical Journal. Special Topics*, *214*(1), 481–518. DOI: 10.1140/epjst/e2012-01703-3

Beath. (2024). Swiss cities revolutionize public services. https://icumonitoring.ch/real-time-data-revolution-enhancing-public-services.html?

Belli, L., Davoli, L., Medioli, A., & Ferrari, G. (2020). Smart city IoT applications: Adaptive lighting case study. *Computer Communications*, *155*, 95–104. DOI: 10.1016/j.comcom.2020.03.009

Benseny, J., Lahteenmaki, J., Toyli, J., & Hammainen, H. (2023). Urban wireless traffic evolution: The role of new devices and the effect of policy. *Telecommunications Policy*, *47*(7), 102595. Advance online publication. DOI: 10.1016/j.telpol.2023.102595

Bergamin, F. (2024). Chemical recycling plastic is ready to go. https://ethz.ch/en/news-and-events/eth-news/news/2024/08/chemical-plastics-recycling-is-ready-to-go.html?

Beyond GDPR: Employing AI to Make Personal Data Useful to Consumers. (2020). *Trusted Data*. The MIT Press. DOI: 10.7551/mitpress/12439.003.0014

Bhattacharya, S., Seth, D., Panyam, S., & Gangrade, P. (2024). Enhancing Digital Privacy: The Application of Zero-Knowledge Proofs in Authentication Systems. [IJCTT]. *International Journal of Computer Trends and Technology*, *72*(4), 34–41. DOI: 10.14445/22312803/IJCTT-V72I4P104

Bianco, V., Scarpa, F., & Tagliafico, L. A. (2021). Energy savings in street lighting: A smart city approach. *Energy Reports*, *7*, 1532–1543. DOI: 10.1016/j.egyr.2021.02.051

Bibri, S. E., Krogstie, J., Kaboli, A., & Alahi, A. (2023). Smarter eco-cities and their leading-edge artificial intelligence of things solutions for environmental sustainability: A comprehensive systematic review [Review of Smarter eco-cities and their leading-edge artificial intelligence of things solutions for environmental sustainability: A comprehensive systematic review]. Environmental Science and Ecotechnology, 19, 100330. Elsevier BV. DOI: 10.1016/j.ese.2023.100330

Bozma, H. I., & Yalçın, H. (2002). Visual processing and classification of items on a moving conveyor: A selective perception approach. *Robotics and Computer-integrated Manufacturing*, *18*(2), 125–133. DOI: 10.1016/S0736-5845(01)00035-7

Brantingham, P. J., Valasik, M., & Mohler, G. O. (2018). Does predictive policing lead to biased arrests? Results from a randomized controlled trial. *Statistics and Public Policy (Philadelphia, Pa.)*, *5*(1), 1–6. DOI: 10.1080/2330443X.2018.1438940

Brown, T., & Miller, J. (2020). Secure IoT frameworks for smart childcare applications. *IEEE Consumer Electronics Magazine*, *9*(5), 56–63.

Buck, C., Clarke, J., Torres de Oliveira, R., Desouza, K. C., & Maroufkhani, P. (2023). Digital transformation in asset-intensive organisations: The light and the dark side. *Journal of Innovation and Knowledge*, *8*(2), 100335. Advance online publication. DOI: 10.1016/j.jik.2023.100335

Burkhard, C., Burren, A. L., Gretler, C., Ruegg, L., Stutz, M., & Tharmalingam, E. (2021). Urban Green Space Indicator. Usability and Accessability of UGS in the city of Zurich and Uster. https://storymaps.arcgis.com/stories/591c0eb84481 4673bc603296e209ffb8?

Business Environment News. (2024). Zurich again crowned as the world's smartest city. https://www.greaterzuricharea.com/en/news/zurich-again-crowned-worlds -smartest-city

Calzada, I. (2018). (Smart) citizens from data providers to decision-makers? The case study of Barcelona. *Sustainability (Basel)*, *10*(9), 3252. DOI: 10.3390/su10093252

Calzada, I. (2022). Citizens' Data Privacy in China: The State of the Art of the Personal Information Protection Law (PIPL). *Smart Cities*, *5*(3), 1129–1150. DOI: 10.3390/smartcities5030057

Canton of Zurich, Department for Economic Affairs, Office for Economy, & Walker Späh, C. (2024). Attractiveness of the Canton of Zurich for business and talent – A comparison2024. https://www.zh.ch/content/dam/zhweb/bilder-dokumente/themen/ wirtschaft-arbeit/wirtschaftszahlen/standortattraktivitaet-im-vergleich_EN.pdf

Can, Y. S., & Ersoy, C. (2021). Privacy-preserving federated deep learning for wearable IoT-based biomedical monitoring. *ACM Transactions on Internet Technology*, *21*(1), 1–17. DOI: 10.1145/3428152

Caragliu, A., Del Bo, C., & Nijkamp, P. (2011). Smart cities in Europe. *Journal of Urban Technology*, *18*(2), 65–82. DOI: 10.1080/10630732.2011.601117

Castrellon, J. P., Sanchez-Diaz, I., & Gil, J. (2024). Smart loading zones. A data analytics approach for loading zones network design. *Transportation Research Interdisciplinary Perspectives*, *24*, 101034. Advance online publication. DOI: 10.1016/j.trip.2024.101034

Chakraborty, A., Alam, M., Dey, V., Chattopadhyay, A., & Mukhopadhyay, D. (2021). A survey on adversarial attacks and defences. *CAAI Transactions on Intelligence Technology*, *6*(1), 25–45. DOI: 10.1049/cit2.12028

Chelani, N., Tripathy, S., Kumhar, M., Bhatia, J., Saxena, V., Tanwar, S., & Nayyar, A. (2024). Federated Learning for Internet of medical healthcare: Issues and challenges. *Scalable Comput. Pract. Exp.*, *25*(5), 4442–4455. DOI: 10.12694/scpe.v25i5.2905

Cheng, T., Kojima, D., Hu, H., Onoda, H., & Pandyaswargo, A. H. (2024). Optimizing Waste Sorting for Sustainability: An AI-Powered Robotic Solution for Beverage Container Recycling. *Sustainability (Basel)*, *16*(23), 10155. DOI: 10.3390/su162310155

Chen, L., Jiang, F., & Zhang, H. (2021). Edge AI for intelligent IoT: Applications in smart cities. *IEEE Internet of Things Journal*, *8*(20), 15035–15044.

Chen, L., & Patel, M. (2019). Artificial intelligence in child monitoring systems. *International Journal of Computer Science and Information Security*, *17*(9), 75–82.

Chen, M., Ma, Y., Li, Y., Wu, D., Zhang, Y., & Youn, C. H. (2017). Wearable 2.0: Enabling human-cloud integration in next generation healthcare systems. *IEEE Communications Magazine*, *55*(1), 54–61. DOI: 10.1109/MCOM.2017.1600410CM

Chen, X., Ran, X., & Liu, J. (2019). Deep learning with edge computing: A review. *Proceedings of the IEEE*, *107*(8), 1655–1674. DOI: 10.1109/JPROC.2019.2921977

Chen, Y., & Wang, L. (2019). Smart home technologies for child safety applications. *Journal of Ambient Intelligence and Humanized Computing*, *10*(8), 3125–3134.

Christian, A. (2024). The hottest startups in Zurich in 2024. *WIRED*. https://www.wired.com/story/the-hottest-startups-in-zurich-in-2024/?

Chu, C. H., Donato-Woodger, S., Khan, S. S., Nyrup, R., Leslie, K., Lyn, A., Shi, T., Bianchi, A., Rahimi, S. A., & Grenier, A. (2023). Age-related bias and artificial intelligence: a scoping review [Review of Age-related bias and artificial intelligence: a scoping review]. Humanities and Social Sciences Communications, 10(1). Palgrave Macmillan. DOI: 10.1057/s41599-023-01999-y

Civil, D. (2023). Top 10 Smart cities in the world. https://buildmytalent.com/2023/09/16/top-10-smart-cities-in-the-world

Climate and Clean Air Coalition. (2021). *Switzerland*. https://www.ccacoalition.org/partners/switzerland

Costa, D. G., Peixoto, J. P. J., Jesus, T. C., Portugal, P., Vasques, F., Rangel, E. O., & Peixoto, M. (2022). A Survey of Emergencies Management Systems in Smart Cities. *IEEE Access : Practical Innovations, Open Solutions, 10*, 61843–61872. DOI: 10.1109/ACCESS.2022.3180033

Cremonini, M. (2023). A Critical Take on Privacy in a Datafied Society. *arXiv preprint arXiv:2308.02573*.

Creswell, J. W., & Creswell, J. D. (2022). *Research Design: Qualitative, Quantitative, and Mixed Methods Approaches* (The Sixth). *Sage (Atlanta, Ga.)*.

Cugurullo, F. (2018). Exposing smart cities and eco-cities: Frankenstein urbanism and the sustainability challenges of the experimental city. *Environment and Planning A. Environment & Planning A, 50*(1), 73–92. DOI: 10.1177/0308518X17738535

D'Andrea, A. (2020). The role of digital twins in smart cities: Real-time urban simulation and control. *Smart Cities, 3*(3), 1055–1070.

Dai, S., Meng, F., Wang, Q., & Chen, X. (2021). FederatedNILM: A Distributed and Privacy-preserving Framework for Non-intrusive Load Monitoring based on Federated Deep Learning. arXiv (Cornell University). https://doi.org//arxiv.2108.03591DOI: 10.48550

Dai, S., Meng, F., Wang, Q., & Chen, X. (2022). DP^2-NILM: A distributed and privacy-preserving framework for non-intrusive load monitoring.

Daniel Flück & Karin Brown. (2024). AI in teaching and Learning, https://ethz.ch/en/the-eth-zurich/education/ai-in-education.html?

Daousis, S., Peladarinos, N., Cheimaras, V., Papageorgas, P., Piromalis, D. D., & Munteanu, R. A. (2024). Overview of Protocols and Standards for Wireless Sensor Networks in Critical Infrastructures. In *Future Internet* (Vol. 16, Issue 1). DOI: 10.3390/fi16010033

Das, P., Woods, O., & Kong, L. (2024). Human sensors in the city of super apps: Mobilizing people as infrastructure for smart city development in Jakarta, Indonesia. *City (London, England), 28*(5–6), 812–834. DOI: 10.1080/13604813.2024.2427954

Dawale, S. V., Raja, S. K. S., & Darsan, R. V. (2025). Urban living through AIoT integration in smart cities. In *AIoT* (pp. 312–339). Auerbach Publications. DOI: 10.1201/9781003482338-14

Dawar, I., Srivastava, A., Singal, M., Dhyani, N., & Rastogi, S. (2025). A systematic literature review on municipal solid waste management using machine learning and deep learning. *Artificial Intelligence Review*, *58*(6), 183. Advance online publication. DOI: 10.1007/s10462-025-11196-9

De Boer, J. (2023). Zurich to invest 300 million francs into making housing more affordable. *iamexpat.* https://www.iamexpat.ch/housing/real-estate-news/zurich-invest-300-million-francs-making-housing-more-affordable

De Filippi, P., & Hassan, S. (2016). Blockchain technology as a regulatory technology: From code is law to law is code. *First Monday*, *21*(12). Advance online publication. DOI: 10.5210/fm.v21i12.7113

de Montjoye, Y.-A., Hidalgo, C. A., Verleysen, M., & Blondel, V. D. (2013). Unique in the Crowd: The privacy bounds of human mobility. *Scientific Reports*, *3*(1), 1376. Advance online publication. DOI: 10.1038/srep01376 PMID: 23524645

Debnath, B., Roychoudhuri, R., & Ghosh, S. K. (2016). E-Waste Management – A Potential Route to Green Computing. *Procedia Environmental Sciences*, *35*, 669–675. Advance online publication. DOI: 10.1016/j.proenv.2016.07.063

Demirel, D., & Mülazımoğlu, M. E. (2022). How the smart governance model shapes cities? Cases from Europe. *Journal of Enterprising Communities*, *16*(1), 8–25. DOI: 10.1108/JEC-08-2021-0115

Deng, Y., Kamani, M. M., & Mahdavi, M. (2020). Adaptive personalized federated learning.

Denis, A., Thomas, A., Robert, W., Samuel, A., Kabiito, S. P., Morish, Z., Sallam, M., Ali, G., & Mijwil, M. M. (2025). A Survey on Artificial Intelligence and Blockchain Applications in Cybersecurity for Smart Cities. *SHIFRA*, *2025*, 1–45. DOI: 10.70470/SHIFRA/2025/001

Desai, P., & Kaur, H. (2022). Technology-driven innovations in infant healthcare monitoring. *International Journal of Recent Technology and Engineering*, *11*(3), 201–209.

Desouza, K. C., Hunter, M., Jacob, B., & Yigitcanlar, T. (2020). Pathways to the Making of Prosperous Smart Cities: An Exploratory Study on the Best Practice. *Journal of Urban Technology*, *27*(3), 3–32. DOI: 10.1080/10630732.2020.1807251

Determann, L., & Tam, J. (2020). The California Privacy Rights Act of 2020: A broad and complex data processing regulation that applies to businesses worldwide. *Journal of Data Protection & Privacy*, *4*(1), 7–21. DOI: 10.69554/GCLK6627

Dhelim, S., Ning, H., Farha, F., Chen, L., Atzori, L., & Daneshmand, M. (2021). IoT-enabled social relationships meet artificial social intelligence. *IEEE Internet of Things Journal*, *8*(24), 17817–17828. DOI: 10.1109/JIOT.2021.3081556

Digital Health Zurich. (2022). A Practice lab for patient centered clinical innovation. https://www.dizh.uzh.ch/en/2022/07/07/zurich-applied-digital-health-center-2

Dirsehan, T., & van Zoonen, L. (2022). Smart city technologies from the perspective of technology acceptance. *IET Smart Cities*, *4*(3), 197–210. DOI: 10.1049/smc2.12040

Doran, D., Severin, K., Gokhale, S., & Dagnino, A. (2015). Social media enabled human sensing for smart cities. *AI Communications*, *29*(1), 57–75. DOI: 10.3233/AIC-150683

Dou, C., & Huan, H. (2021). Full respiration rate monitoring exploiting Doppler information with commodity Wi-Fi devices. *Sensors (Basel)*, *21*(10), 3505. DOI: 10.3390/s21103505 PMID: 34069847

Dulbert, T. B., Budi, M. A., & Citra, A. W. (2024). Key Indicators for Effective Implementation of Smart City Standards in Indonesia During the Covid-19Pandemic: A Case Study of Four Cities. *Journal of Urban Culture Research*, *28*, 37–56. https://www.scopus.com/inward/record.uri?eid=2-s2.0-85201764592&partnerID=40&md5=7d427603585c3a43b7bc18a846f29e25

Đurek, V., & Ređep, N. B. (2016). Review on e-readiness assessment tools. *Central European Conference on Information and Intelligent Systems*, 161–168.

Dwivedi, B., & Behl, R. (2025). Case study: Successful artificial intelligence (AI) and the Internet of Things (IoT)(AIoT) implementation in smart cities. In *Merging Artificial Intelligence With the Internet of Things* (pp. 167–202). IGI Global. DOI: 10.4018/979-8-3693-8547-0.ch006

Dziwisz, D. (2023). Stuxnet. In *The Handbook of Homeland Security* (pp. 289–293). CRC Press.

EBP. (2020). Smart Mobility Strategy for the city of Zurich, https://www.ebp.global/xx-en/projekte/smart-mobility-fuer-die-stadt-zuerich?

Eckhoff, D., & Wagner, I. (2018). Privacy in the smart city—Applications, technologies, challenges, and solutions. *IEEE Communications Surveys and Tutorials*, *20*(1), 489–516. DOI: 10.1109/COMST.2017.2748998

Egridere, S. (2024). Transforming Urban Living - Insights from the IMD Smart City Index 2024 on Europe's Leading and Lagging Cities, https://mayorsofeurope.eu/science-and-the-city/transforming-urban-living-insights-from-the-imd-smart-city-index-2024-on-europes-leading-and-lagging-cities/

Engeler, S., Auer, M., & La Fauci, R. (2022). Smart grid rollout in der Stadt Zürich. *Bulletin.* https://www.bulletin.ch/de/news-detail/smart-grid-rollout-in-der-stadt-zuerich.html

Erdelyi, V., Mizumoto, T., Kitai, Y., Ishimaru, D., Adachi, H., Higashino, T., & Ikeda, M. (2025). Detecting Subtle Signs of School Attendance Issues Using Smartphone-Based Sensing. *IEEE Access : Practical Innovations, Open Solutions, 13*, 4652–4669. DOI: 10.1109/ACCESS.2024.3523108

Estevez, E., Pardo, T. A., & Scholl, H. J. (2021). Smart Cities and Smart Governance. In *Public Administration and Information Technology* (Vol. 37, pp. 43–66). Springer Nature., DOI: 10.1007/978-3-030-61033-3_3

ESWET. (2023). A waste treatement expansion in zurich: A sustainable energy solution by Hitachi Zosen INOVA. https://eswet.eu/a-waste-treatment-expansion-in-zurich-a-sustainable-energy-solution/?

Etim, E., & Daramola, O. (2023). Investigating the E-Readiness of Informal Sector Operators to Utilize Web Technology Portal. *Sustainability (Basel), 15*(4), 1–28. https://doi.org/https://doi.org/10.3390/su15043449. DOI: 10.3390/su15043449

European Commission. (2024). How Zurich is becoming net zero by 2040. https://build-up.ec.europa.eu/en/news-and-events/news/how-zurich-becoming-net-zero-2040?

European Data Protection Supervisor. (2018). *Privacy by Design.* https://www.edps.europa.eu/data-protection/our-work/subjects/privacy-design_en?page=5

European Environmental Agency. (2022). Reduction to 30km/h speed limits in Zurich, Switzerland, https://www.eea.europa.eu/publications/outlook-to-2030/reduction-to-30-km-h

European Union (2022, October 6). *Digital education action plan (2021–2027).* Europa.eu. https://eur-lex.europa.eu/legal-content/EN/TXT/?uri=legissum:4617905

Fahlevi, M., Dandi, M., Asyrof, R. A., Dahlan, O. P., Asetya, D. R., & Dahlan, S. P. (2024). Elevating Healthcare Through Telemedicine: A Pathto Smart City Success in IKN (Indonesia's NewCapital City). *11th International Conference on ICT for Smart Society: Integrating Data and Artificial Intelligence for a Resilient and Sustainable Future Living, ICISS 2024 - Proceeding.* DOI: 10.1109/ICISS62896.2024.10751614

Fan, L., Chen, X., & Xu, W. (2021). Privacy protection for smart city applications using federated learning. *IEEE Network, 35*(2), 28–33. DOI: 10.1109/MNET.011.2000405

Feng, Z., Yang, J., Chen, L., Chen, Z., & Li, L. (2022). An intelligent waste-sorting and recycling device based on improved EfficientNet. *International Journal of Environmental Research and Public Health, 19*(23), 15987. DOI: 10.3390/ijerph192315987 PMID: 36498058

Fereidouni, H., Fadeitcheva, O., & Zalai, M. (2025). IoT and man-in-the-middle attacks. *Security and Privacy, 8*(2), e70016. DOI: 10.1002/spy2.70016

Fernandez, A., & Silva, M. (2021). Automated baby care systems using intelligent sensing. *Journal of Healthcare Informatics Research, 5*(3), 211–223.

Fernandez-Anez, V., Fernández-Güell, J. M., & Giffinger, R. (2018). Smart City implementation and discourses: An integrated conceptual model. The case of Vienna. *Cities (London, England), 78*(November), 4–16. DOI: 10.1016/j.cities.2017.12.004

Ferronato, N., & Torretta, V. (2019). Waste mismanagement in developing countries: A review of global issues. *International Journal of Environmental Research and Public Health, 16*(6), 1060. DOI: 10.3390/ijerph16061060 PMID: 30909625

Firouzi, F., Farahani, B., Daneshmand, M., Grise, K., Song, J., Saracco, R., Wang, L. L., Lo, K., Angelov, P., Soares, E., Loh, P.-S., Talebpour, Z., Moradi, R., Goodarzi, M., Ashraf, H., Talebpour, M., Talebpour, A., Romeo, L., Das, R., & Luo, A. (2021). Harnessing the Power of Smart and Connected Health to Tackle COVID-19: IoT, AI, Robotics, and Blockchain for a Better World. *IEEE Internet of Things Journal, 8*(16), 12826–12846. DOI: 10.1109/JIOT.2021.3073904 PMID: 35782886

Flughafen Zürich, A. G. (2025). Local air quality is closely monitored, https://www.flughafen-zuerich.ch/en/company/responsibility/environmental-protection/air

Force, J. T. (2018). Risk management framework for information systems and organizations. *NIST Special Publication, 800*, 37.

Fotiou, N., Siris, V. A., Mertzianis, A., & Polyzos, G. C. (2018). Smart IoT data collection. In *2018 IEEE 19th International Symposium on "A World of Wireless, Mobile and Multimedia Networks" (WoWMoM). IEEE*, 588–599. DOI: 10.1109/WoWMoM.2018.8449766

Galliven, H. (2025, July 12). *Croydon's fixed facial recognition cameras spark debate*. BBC News. Retrieved from https://www.bbc.com/news/articles/cy0w5egz91no

Galloway, A. (2021, September 5). *Police using QR check-in data to solve crimes*. The Sydney Morning Herald. https://www.smh.com.au/politics/federal/breach-of-trust-police-using-qr-check-in-data-to-solve-crimes-20210903-p58om8.html

Gaur, A., Scotney, B., Parr, G., & McClean, S. (2015). Smart city architecture and its applications based on IoT. *Procedia Computer Science*, *52*, 1089–1094. DOI: 10.1016/j.procs.2015.05.122

Gawalkar, S. (2025). Federated learning for privacy-preserving AI in edge and IoT devices. *International Scientific Journal of Engineering and Management*, *04*(05), 1–7. DOI: 10.55041/ISJEM03376

GDPR. (2018). *General Data Protection Regulation (GDPR)*. General Data Protection Regulation (GDPR). https://gdpr-info.eu/issues/consent/

Gebert, C. (2018). Induced Demand: Modelling the potential impacts of automated vehicles in Zurich. *NSL – Netzwerk Stadt Und Landschaft*. https://www.nsl.ethz.ch/induced-demand-modelling-the-potential-impacts-of-automated-vehicles-in-zurich

Goldsmith, S., & Crawford, S. (2014). *The responsive city: Engaging communities through data-smart governance*. John Wiley & Sons.

Gong, H., Wen, Z., Wu, Z. Y., & Yu, M. H. (2014). Design of a smart classroom broadcasting system based on WSN. In *Advanced Materials Research* (Vol. 850, pp. 520–524). Trans. Tech. Publ., DOI: 10.4028/www.scientific.net/AMR.850-851.520

Gonzalez, R., & Lee, H. (2020). Smart parenting systems and infant care technologies. *IEEE Access : Practical Innovations, Open Solutions*, *8*, 165421–165430.

Goodfellow, I., Pouget-Abadie, J., Mirza, M., Xu, B., Warde-Farley, D., Ozair, S., & Bengio, Y. (2014). Generative adversarial nets. *Advances in Neural Information Processing Systems*, •••, 27.

Gottschalk, F., & Weise, C. (2023). Digital equity and inclusion in education: An overview of practice and policy in OECD countries. *OECD Education Working Papers*, (299), 0_1-75.

Government of Canada. (2020, September 15). *Download COVID Alert: Canada's exposure notification app*. Aem. https://www.canada.ca/en/public-health/services/diseases/coronavirus-disease-covid-19/covid-alert.html

Goyal, N. K. (2025). Security and Privacy in IoT, Fog, and Blockchain Networks. In Sathio, A., Rind, M., Awan, S., & Junejo, A. (Eds.), *Energy-Efficient Deep Learning Approaches in IoT, Fog, and Green Blockchain Revolution* (pp. 371–398). IGI Global Scientific Publishing., DOI: 10.4018/979-8-3373-0300-0.ch015

Goyal, N. K. (2025). Weaponizing Misinformation and Cyberattacks on Supply Chains. In Bukhari, S., & Zafar, I. (Eds.), *Mass Media and Impact of Fake News on Supply Chains* (pp. 177–210). IGI Global Scientific Publishing., DOI: 10.4018/979-8-3373-2175-2.ch007

Goyal, N. K., Dandotiya, M., Kumari, M., Kaloria, S., & Anushya, A. (2025). AI-Powered Haptic Devices for Patient Rehabilitation. In Mamodiya, U., Kalam, A., Bossoufi, B., Awad, M., & Somwanshi, D. (Eds.), *Integrating AI With Haptic Systems for Smarter Healthcare Solutions* (pp. 143–170). IGI Global Scientific Publishing., DOI: 10.4018/979-8-3373-2307-7.ch007

Goyal, N. K., Dandotiya, M., Kumari, M., Sheikh, M. F., & Sharma, S. (2025). Emerging Job Roles in the Tourism Sector Leveraging Artificial Intelligence. In Silic, M., Luthia, M., Ahmad, F., & Chakraborty, P. (Eds.), *Empowering Sustainable Performance and Competitive Advantage in Tourism* (pp. 89–126). IGI Global Scientific Publishing., DOI: 10.4018/979-8-3373-4546-8.ch004

Goyal, N. K., Dandotiya, M., Qurashi, J. A., Kumari, M., & Sharma, S. (2025). AI-Driven Predictive Analytics for Disease Management. In Ouaissa, M., Ouaissa, M., Imad, M., Qurashi, J., & Farooq, M. (Eds.), *Utilizing AI of Medical Things for Healthcare Security and Sustainability* (pp. 231–276). IGI Global Scientific Publishing., DOI: 10.4018/979-8-3373-0690-2.ch008

Goyal, N. K., Kumari, M., Kushwah, V., Dandotiya, M., & Vijay, A. (2025). Evaluating blockchain security: Trade-offs in consensus mechanisms. *IET Conference Proceedings*, 2024(38), 80–88. DOI: 10.1049/icp.2025.0775

GPAI. (2024). "Algorithmic Transparency in the Public Sector: A state-of-the-art report of algorithmic transparency instruments". Report, May 2024, Global Partnership on Artificial Intelligence. https://wp.oecd.ai/app/uploads/2025/05/algorithmic-transparency-in-the-public-sector.pdf

Greater Zurich Area. (2023). Upcircle digitalizing waste for the circular economy, https://www.greaterzuricharea.com/en/news/upcircle-digitalizing-waste-circular-economy?

Greater Zurich Area. (2024). Artificial intelligence in the Greater Zurich Area, https://www.greaterzuricharea.com/en/artificial-intelligence-greater-zurich-area

Grêt-Regamey, A., Switalski, M., Fagerholm, N., Korpilo, S., Juhola, S., Kyttä, M., Käyhkö, N., McPhearson, T., Nollert, M., Rinne, T., Soininen, N., Toivonen, T., Räsänen, A., Willberg, E., & Raymond, C. M. (2021). Harnessing sensing systems towards urban sustainability transformation. *NPJ Urban Sustainability*, *1*(1), 40. Advance online publication. DOI: 10.1038/s42949-021-00042-w

Gubbi, J., Buyya, R., Marusic, S., & Palaniswami, M. (2013). Internet of Things (IoT): A vision, architectural elements, and future directions. *Future Generation Computer Systems*, *29*(7), 1645–1660. DOI: 10.1016/j.future.2013.01.010

Guenduez, A. A., & Mergel, I. (2022). The role of dynamic managerial capabilities and organizational readiness in smart city transformation. *Cities (London, England)*, *129*(July), 103791. DOI: 10.1016/j.cities.2022.103791

Gunawong, P., & Gao, P. (2017). Understanding e-government failure in the developing country context: A process-oriented study. *Information Technology for Development*, *23*(1), 153–178. DOI: 10.1080/02681102.2016.1269713

Guo, B., Wang, Z., Yu, Z., & Zhou, X. (2017). Mobile crowd sensing and computing: The review of an emerging human-powered sensing paradigm. *ACM Computing Surveys*, *48*(1), 1–31. DOI: 10.1145/2794400

Gupta, J. K., Negi, G. S., & Krishna, G. (2025). Smart Transportation Systems for Vehicle Geographical Tracking. In *Federated Learning Based Intelligent Systems to Handle Issues and Challenges in IoVs (Part 2)* (pp. 160-199). Bentham Science Publishers. DOI: 10.2174/9789815322224125030008

Gupta, S., Verma, P., & Reddy, K. (2021). Smart cradle automation using IoT and embedded systems. *International Journal of Innovative Technology and Exploring Engineering*, *10*(8), 45–50.

Haitao, Ch. (2021). Optimization of an Intelligent Sorting and Recycling System for Solid Waste Based on Image Recognition Technology. *Advances in Mathematical Physics, ID*, *4094684*, 12.

Halbfinger, D. M., Kershner, I., & Bergman, R. (2020, March 16). To Track Coronavirus, Israel Moves to Tap Secret Trove of Cellphone Data. *The New York Times*. https://www.nytimes.com/2020/03/16/world/middleeast/israel-coronavirus-cellphone-tracking.html

Halilovic, M., Meurers, T., Otte, K., & Prasser, F. (2025). Parallel privacy preservation through partitioning (P4): A scalable data anonymization algorithm for health data. *BMC Medical Informatics and Decision Making*, *25*(1), 129. Advance online publication. DOI: 10.1186/s12911-025-02959-z PMID: 40075355

Hall, R. E. (2000). The vision of a smart city. In Proceedings of the 2nd International Life Extension Technology Workshop, Paris, France, September 28, http://www.osti.gov/bridge/servlets/purl/773961- oyxp82/webviewable/773961.pdf

Hameed, R. T., & Mohamad, O. A. (2023). Federated Learning in IoT: A survey on distributed decision making. *BJIoT*, *2023*, 1–7. DOI: 10.58496/BJIoT/2023/001

Han, C., & Xue, R. (2020). Aggregation for privately trained different types of local models. *ICST Trans. Secur. Saf.*, *7*(26), 170237. DOI: 10.4108/eai.21-6-2021.170237

Hancke, G. P., & Hancke, G. P.Jr. (2013). The role of advanced sensing in smart cities. *Sensors (Basel)*, *13*(1), 393–425. DOI: 10.3390/s130100393 PMID: 23271603

Han, M. J. N., & Kim, M. J. (2025). A smart social sustainability model for smart city to enhance human context experience. *Cities (London, England)*, *159*, 105788. Advance online publication. DOI: 10.1016/j.cities.2025.105788

Haque, A. K. M. B., Bhushan, B., & Dhiman, G. (2022). Conceptualizing smart city applications: Requirements, architecture, security issues, and emerging trends. *Expert Systems: International Journal of Knowledge Engineering and Neural Networks*, *39*(5), e12753. DOI: 10.1111/exsy.12753

Hara, S., & Abe, M. (2025). Case study: OtologMap: a case study on the construction of an environmental sound map recorded by smart devices at Okayama and Kurashiki. *Noise Control Engineering Journal*, *73*(1), 57–68. DOI: 10.3397/1/37731

Hardesty, L. (2013). *How hard is it to "de-anonymize" cellphone data?* MIT News. https://news.mit.edu/2013/how-hard-it-de-anonymize-cellphone-data

Hardi, R., Nurmandi, A., Purwaningsih, T., & Manaf, H. A. (2025). Smart city governance and interoperability : Enhancing human security in Yogyakarta and Makassar, Indonesia. *Frontiers in Political Science*, *6*(March), 1–18. DOI: 10.3389/fpos.2025.1553177

Hardjomuljadi, S., & Anondho, B. (2025). A model for transforming conventional industrial areas into eco-smart zones: The role of development planning, infrastructure, competence, and partnerships on success of the transformation process. *International Journal of Innovative Research and Scientific Studies*, *8*(4), 32–44. DOI: 10.53894/ijirss.v8i3.7737

Hashem, I. A. T., Yaqoob, I., Anuar, N. B., Mokhtar, S., Gani, A., & Khan, S. U. (2015). The rise of "big data" on cloud computing: Review and open research issues. *Information Systems*, *47*, 98–115. DOI: 10.1016/j.is.2014.07.006

Hassan, G., Chemori, A., El Rafei, M., Gouttefarde, M., Francis, C., Hervé, P.E., Sallé, D. (2022). Ultra-Local Model-Based Intelligent Robust Control of PKMs: Theory and Simulations. IFAC-PapersOnLine, 55, 12, 2022, 634-640.

Haucke, M., Heinzel, S., & Liu, S. (2024). Social mobile sensing and problematic alcohol consumption: Insights from smartphone metadata. *International Journal of Medical Informatics*, *188*, 105486. Advance online publication. DOI: 10.1016/j.ijmedinf.2024.105486 PMID: 38754285

Herath, H. M. K. K. M. B., & Mittal, M. (2022). Adoption of artificial intelligence in smart cities: A comprehensive review. *International Journal of Information Management Data Insights*, *2*(1), 100076. DOI: 10.1016/j.jjimei.2022.100076

Hexagon (2022). Protecting citizens with intearated Public Safety Solutions. https://futurescot.com/protecting-citizens-with-integrated-public-safety-solutions

He, Y., Fan, J., & Zhang, Y. (2025). Controlled quantum authentication confidential communication protocol for smart healthcare. *Scientific Reports*, *15*(1), 18094. Advance online publication. DOI: 10.1038/s41598-025-02448-3 PMID: 40413245

He, Y., & Tritto, A. (2022). Urban utopia or pipe dream? Examining Chinese- invested smart city development in Southeast Asia smart city development in Southeast Asia. *Third World Quarterly*, *43*(9), 2244–2268. DOI: 10.1080/01436597.2022.2089648

He, Z., Zhai, W., & Zhao, C. (2025). Using social sensing to evaluate the effect of spatial planning: A case study of Yangtze River Delta Urban Agglomeration in China. *Frontiers of Architectural Research*, *14*(4), 1048–1062. DOI: 10.1016/j.foar.2024.12.004

Hirt, R. (2023). Smart City Zurich: How Anthos is driving the city's digital transformation, https://cloud.google.com/blog/topics/hybrid-cloud/city-of-zurich-builds-a-hybrid-cloud-with-anthos?

Hjerppe, K., Ruohonen, J., & Leppänen, V. (2019, September). The general data protection regulation: requirements, architectures, and constraints. In *2019 IEEE 27th International Requirements Engineering Conference (RE)* (pp. 265-275). IEEE. DOI: 10.1109/RE.2019.00036

Hollands, R. G. (2020). Will the real smart city please stand up?: Intelligent, progressive or entrepreneurial? In *The Routledge companion to smart cities* (pp. 179–199). Routledge. DOI: 10.4324/9781315178387-13

Holl, F., Flemisch, F., Swoboda, W., & Schobel, J. (2022). Apps for Covid-19 in Germany: Assessment Using the German Mobile App Rating Scale. *JAMIA Open*, *5*(4), ooac082. Advance online publication. DOI: 10.1093/jamiaopen/ooac082 PMID: 36225895

Holmes, O. (2020, March 17). Israel to track mobile phones of suspected coronavirus cases. *The Guardian*. https://www.theguardian.com/world/2020/mar/17/israel-to-track-mobile-phones-of-suspected-coronavirus-cases

Homeland Security. (2024, September 7). *Privacy Impact Assessment for the Customer Profile Management System*. https://www.dhs.gov/sites/default/files/2024-11/24_0930_priv_pia-dhs-uscis-cpms-060d.pdf

Hossain, M. T., Lee, J., Besenski, D., Dimitrijevic, B., & Spasovic, L. (2025). Developing a Prediction Model for Real-Time Incident Detection Leveraging User-Oriented Participatory Sensing Data. *Information (Basel), 16*(6), 423. Advance online publication. DOI: 10.3390/info16060423

Hsu, Y.-C., Huang, T.-H., Verma, H., Mauri, A., Nourbakhsh, I., & Bozzon, A. (2022). Empowering local communities using artificial intelligence [Review of Empowering local communities using artificial intelligence]. Patterns, 3(3), 100449. Elsevier BV. DOI: 10.1016/j.patter.2022.100449

Hsu, H.-Y., Keoy, K. H., Chen, J.-R., Chao, H.-C., & Lai, C.-F. (2023). Personalized federated learning algorithm with adaptive clustering for Non-IID IoT data incorporating multi-task learning and neural network model characteristics. *Sensors (Basel), 23*(22), 9016. DOI: 10.3390/s23229016 PMID: 38005404

Hubner, K. (2009). 2000 watt society. https://ourworld.unu.edu/en/2000-watt-society

Hu, Q., & Zheng, Y. (2021). Smart city initiatives: A comparative study of American and Chinese cities. *Journal of Urban Affairs, 43*(4), 504–525. DOI: 10.1080/07352166.2019.1694413

Hussein, N.M., Mohialden, Y.M., Salman S.A. (2024). Impact of IoT-Based Environmental Monitoring on Lab Safety and Sustainability. Babylonian Journal of Internet of Things, 16-26.

IMD World Competitiveness Center. (2024). IMD Smart City Index 2024. IMD World Competitiveness Center. https://www.coit.es/sites/default/files/imd_-smartcityindex-2024-full-report.pdf

Information and Privacy Commission. (2024, August). *A guide to undertaking Privacy Impact Assessments on AI systems and projects*. Retrieved from https://www.ipc.nsw.gov.au/sites/default/files/2024-08/Guide_A_guide_to_undertaking_Privacy_Impact_Assessments_on_AI_systems_and_projects_August_2024_CONSULTATION_DRAFT.pdf

INTEGRA. (2023). Invigorating the water supply of the city of Zurich, https://integra-metering.com/invigorating-the-water-supply-of-the-city-of-zurich/?

International Telecommunication Union. (2016). *Implementing ITU-T International Standards to Shape Smart Sustainable Cities: The Case of Dubai*. https://www.itu.int/en/publications/Documents/tsb/2016-DubaiCase/files/downloads/Dubai%20Case%20Study-410059-FINAL.pdf

Ismagilova, E., Hughes, L., Rana, N. P., & Dwivedi, Y. K. (2022). Security, privacy and risks within smart cities: Literature review and development of a smart city interaction framework. *Information Systems Frontiers*, *24*(2), 393–414. DOI: 10.1007/s10796-020-10044-1 PMID: 32837262

Ivanova, D., & Katsaounis, P. (2021). Real-time dynamic tiered e-consent: A novel tool for patients' engagement and common ontology system for the management of medical data. *Innovations in Digital Health, Diagnostics, and Biomarkers*, *1*(2), 45–49. DOI: 10.36401/IDDB-21-01

Jacyna-Gołda, I., Cur, K., Tomaszewska, J., Przanowski, K., Hošková-Mayerová, Š., & Świergolik, S. (2025). Optimizing Flight Delay Predictions with Scorecard Systems. *Applied Sciences (Basel, Switzerland)*, *15*(11), 5918. DOI: 10.3390/app15115918

Jan, F., Min-Allah, N., & Düştegör, D. (2021). Iot based smart water quality monitoring: Recent techniques, trends and challenges for domestic applications. In *Water (Switzerland)* (Vol. 13, Issue 13). DOI: 10.3390/w13131729

Jathan, S., & Frank, P. (2015). The Spectrum of Control: A Social Theory of the Smart City. *First Monday*, *20*(7).

Jia, S., Hui, F., Wei, C., Zhao, X., & Liu, J. (2021). Lane-Changing Behavior Prediction Based on Game Theory and Deep Learning. *Journal of Advanced Transportation*, *1*, 1–12. Advance online publication. DOI: 10.1155/2021/6634960

Jo, K., Choi, J., Kim, K., Kim, S., Nguyen, D.-L., Vo, X.-T., Priadana, A., & Tran, T.-D. (2025). *Artificial Behavior Intelligence: Technology, Challenges, and Future Directions*. ArXiv.org. https://arxiv.org/abs/2505.03315 DOI: 10.1109/HSI66212.2025.11142417

Kalagas, A. (2018). Co-op City: Zurich's experiment with nonprofit housing. *Medium.* https://medium.com/@alexiskalagas/coop-city-z%C3%BCrichs-experiment-with-non-profit-housing-64207f60cafc

Kamel Boulos, M. N., & Al-Shorbaji, N. M. (2014). On the Internet of Things, smart cities and the WHO Healthy Cities. *International Journal of Health Geographics*, *13*(1), 1–6. DOI: 10.1186/1476-072X-13-10 PMID: 24669838

Kamil, M., Muhammad, R., Roziqin, A., Sari, A. E., & Kismartini, . (2025). Examining Practice Smart City in Local Government : A Smart Governance Perspective. *International Journal of Sustainable Development and Planning*, *20*(2), 635–644. DOI: 10.18280/ijsdp.200214

Kamrowska-Załuska, D. (2021). Impact of AI-based tools and urban big data analytics on the design and planning of cities. *Land (Basel)*, *10*(11), 1209. DOI: 10.3390/land10111209

Kannammal, A., & Chandia, S. (2023). Applications of AI and IoT for smart cities. In *Research Trends in Artificial Intelligence: Internet of Things* (pp. 186–202). Bentham Science Publishers. DOI: 10.2174/9789815136449123010014

Kaza, S., Yao, L., Bhada-Tata, P., & Van Woerden, F. (2018). *What a Waste 2.0: A Global Snapshot of Solid Waste Management to 2050*. World Bank Publications. DOI: 10.1596/978-1-4648-1329-0

Kellow, P., Joel, J. P. C. R., Ousmane, D., Ashok, K. D., Victor, H. C. A., & Sergei, A. K. (2020). A Smart Waste Management Solution Geared towards Citizens. *Sensors (Basel)*, *20*(8), 2380. DOI: 10.3390/s20082380 PMID: 32331464

Kennedy, H. (2018). Living with data: Aligning data studies and data activism through a focus on everyday experiences of datafication. *Krisis: Journal for Contemporary Philosophy*, *2018*(1), 18–30. DOI: 10.21827/krisis.38.1.37184

Kessler, S. H., & Schmidt-Weitmann, S. (2021). Diseases and Emotions: An Automated Content Analysis of Health Narratives in Inquiries to an Online Health Consultation Service. *HealthCommunication.* 36 (2*): 226-235.* https://doi.org/. PMID 36112901. S2CID 210568504.DOI: 10.1080/10410236.2019.1673950

Khalil, U., Malik, O. A., Uddin, M., & Chen, C. (2022). A Comparative Analysis on Blockchain versus Centralized Authentication Architectures for IoT-Enabled Smart Devices in Smart Cities: A Comprehensive Review, Recent Advances, and Future Research Directions [Review of A Comparative Analysis on Blockchain versus Centralized Authentication Architectures for IoT-Enabled Smart Devices in Smart Cities: A Comprehensive Review, Recent Advances, and Future Research Directions]. Sensors, 22(14), 5168. Multidisciplinary Digital Publishing Institute. DOI: 10.3390/s22145168

Khan, L. U., Saad, W., Han, Z., & Hossain Ekram and Hong, C. S. (2020). Federated learning for Internet of Things: Recent advances, taxonomy, and open challenges.

Khan, A., Khan, S., Hassan, B., & Zheng, Z. (2022). CNN-Based Smoker Classification and Detection in Smart City Application. *Sensors (Basel)*, *22*(3), 892. DOI: 10.3390/s22030892 PMID: 35161637

Khan, W. Z., Rehman, M. H. U., Zangoti, H. M., Afzal, M. K., Armi, N., & Salah, K. (2020). Industrial Internet of Things: Recent advances, enabling technologies and open challenges. *Computers & Electrical Engineering*, *81*, 106522. DOI: 10.1016/j.compeleceng.2019.106522

Khelifi, F., Bradai, A., & Ksentini, A. (2021). Edge intelligence in smart city applications. *IEEE Communications Magazine*, *59*(10), 26–32. DOI: 10.1109/MCOM.001.2001081

Kilicay-Ergin, N., Barb, A., & Chaudhary, N. (2023). Knowledge elicitation methodology for evaluation of Internet of Things privacy characteristics in smart cities. *Systems Engineering*.

Kingma, D. P., & Welling, M. (2013). Auto-encoding variational Bayes. *arXiv preprint arXiv:1312.6114*.

Kirchherr, J., Reike, D., & Hekkert, M. (2017). Conceptualizing the circular economy: An analysis of 114 definitions. *Resources, Conservation and Recycling*, *127*, 221–232. DOI: 10.1016/j.resconrec.2017.09.005

Kirimtat, A., Krejcar, O., Kertész, A., & Taşgetiren, M. F. (2020). Future Trends and Current State of Smart City Concepts: A Survey. *IEEE Access : Practical Innovations, Open Solutions*, *8*, 86448–86467. DOI: 10.1109/ACCESS.2020.2992441

Kitchin, R. (2014). The real-time city? Big data and smart urbanism. *GeoJournal*, *79*(1), 1–14. DOI: 10.1007/s10708-013-9516-8

Ko, S., Kim, W., Shin, S.-C., & Shin, J. (2020). The economic value of sustainable recycling and waste management policies: The case of a waste management crisis in South Korea. *Waste Management, 104*, 220–227. https://doi.org/https://doi.org/10.1016/j.wasman.2020.01.020

Konstantia, M., Freiderikos, R., George, L., Stavros, O., Maria, P., Michalis, Z., & Michail, M. (2022). Intelligent robotic system for urban waste recycling. IEEE International Conference on Imaging Systems and Techniques, Proceedings, ISBN (Electronic) 9781665481021.

Koops, B. J. (2021). The concept of function creep. *Law, Innovation and Technology*, *13*(1), 29–56. DOI: 10.1080/17579961.2021.1898299

Kuang, Z., Su, J., Latifian, A., Eshraghi, S., & Ghafari, A. (2024). Utilizing Artificial neural networks (ANN) to regulate Smart cities for sustainable Urban Development and Safeguarding Citizen rights. *Scientific Reports*, *14*(1), 31592. Advance online publication. DOI: 10.1038/s41598-024-76964-z PMID: 39738236

Kuguoglu, B. K., van der Voort, H., & Janssen, M. (2021). The Giant Leap for Smart Cities: Scaling Up Smart City Artificial Intelligence of Things (AIoT) Initiatives. *Sustainability (Basel)*, *13*(21), 12295. DOI: 10.3390/su132112295

Kumari, R. (2024). Why Switzerland is next big opportunity for skilled tech talent in 2025? https://content.techgig.com/career-advice/why-switzerland-is-next-big-opportunity-for-skilled-tech-talent-in-2025/articleshow/115947673.cms

Kumar, N. M., & Mallick, P. K. (2018). The Internet of Things: Insights into the building blocks, component interactions, and architecture layers. *Procedia Computer Science*, *132*, 109–117. DOI: 10.1016/j.procs.2018.05.170

Kumar, N., Mallick, P. K., & Rodrigues, J. J. P. C. (2020). AI and IoT-based street lighting in smart cities: A review. *Future Generation Computer Systems*, *108*, 1076–1091. DOI: 10.1016/j.future.2020.03.019

Kumar, P., & Joshi, A. (2021). IoT-enabled child safety monitoring in smart homes. *International Journal of Smart Technology and Learning*, *2*(1), 44–53.

Kumar, R. (2022). Automated infant care systems using IoT technology. *Journal of Embedded Systems and Applications*, *9*(4), 101–108.

Kumar, R., Jain, V., Leong, H. W., & Teyarachakul, S. (2023). Convergence of IoT, Blockchain, and Computational Intelligence in Smart Cities. In *CRC Press eBooks*. Informa., DOI: 10.1201/9781003353034

Kumar, S., & Goyal, M. K. (2025). Water policy review: Ensuring sustainable water management for India. *Journal of Environmental Management*, *388*, 125823. Advance online publication. DOI: 10.1016/j.jenvman.2025.125823 PMID: 40440942

Kurupathi, S. R., & Maass, W. (2020, September). Survey on federated learning towards privacy preserving AI. Computer Science & Information Technology (CS & IT).

Larson, J., Mattu, S., Kirchner, L., & Angwin, J. (2016, May 23). *How We Analyzed the COMPAS Recidivism Algorithm*. ProPublica. https://www.propublica.org/article/how-we-analyzed-the-compas-recidivism-algorithm

Laski, P. A., Takosoglu, J. E., & Blasiak, S. (2015). Design of a 3-DOF tripod electro-pneumatic parallel manipulator. *Robotics and Autonomous Systems*, *72*, 59–70. DOI: 10.1016/j.robot.2015.04.009

Lee, C. S. (2019). Datafication, dataveillance, and the social credit system as China's new normal. *Online Information Review*, *43*(6), 952–970. DOI: 10.1108/OIR-08-2018-0231

Lee, H., & Lee, U. (2022). Toward dynamic consent for privacy-aware pervasive health and well-being: A scoping review and research directions. *IEEE Pervasive Computing*, *21*(4), 25–32. DOI: 10.1109/MPRV.2022.3210747

Lee, I., & Lee, K. (2015). The Internet of Things (IoT): Applications, investments, and challenges for enterprises. *Business Horizons*, *58*(4), 431–440. DOI: 10.1016/j.bushor.2015.03.008

Lee, J. H., Phaal, R., & Lee, S. H. (2013). An integrated service-device-technology roadmap for smart city development. *Technological Forecasting and Social Change*, *80*(2), 286–306. DOI: 10.1016/j.techfore.2012.09.020

Lee, M. (2021). Advanced smart cradle systems for modern childcare. *International Journal of Smart Home*, *15*(2), 55–63.

Lestari, M. D., Kantun, S., Hartanto, W., Suharso, P., & Widodo, J. (2020). Analysis of the financial literacy level of Micro, Small and Medium Enterprises (MSMEs) in Jember, East Java, Indonesia. In Sumardi (Ed.), *2nd International Conference on Environmental Geography and Geography Education, ICEGE 2019* (Vol. 485, Issue 1). Institute of Physics Publishing. DOI: 10.1088/1755-1315/485/1/012128

Li, H., Zhang, Q., & Lu, K. (2015). Integrating mobile sensing and social network for personalized health-care application. In *Proceedings of the ACM Symposium on Applied Computing* (Vol. 13-17-Apri, pp. 527–534). DOI: 10.1145/2695664.2695767

Li, Y.-H., Li, Y., Wei, M.-Y., & Li, G. (2024). Innovation and challenges of artificial intelligence technology in personalized healthcare [Review of Innovation and challenges of artificial intelligence technology in personalized healthcare]. Scientific Reports, 14(1). Nature Portfolio. DOI: 10.1038/s41598-024-70073-7

Lia, O. (2024). Top 7 smart cities in the world in 2024. https://earth.org/top-7-smart-cities-in-the-world/?

Liew, M. E. (2022). How AR and AI are making Zurich a smarter and safer city. https://govinsider.asia/intl-en/article/how-ar-and-ai-are-making-zurich-a-smarter-and-safer-city-david-weber-zurich?

Li, J., Li, X., Chen, X., Lee, P. P., & Lou, W. (2018). A hybrid cloud approach for secure authorized deduplication. *IEEE Transactions on Parallel and Distributed Systems*, *26*(5), 1206–1216. DOI: 10.1109/TPDS.2014.2318320

Limami, F., Hdioud, B., & Thami, R. O. H. (2024). Contextual emotion detection in images using deep learning. *Frontiers in Artificial Intelligence*, *7*, 1386753. Advance online publication. DOI: 10.3389/frai.2024.1386753 PMID: 38952408

Lin, J., Yu, W., Zhang, N., Yang, X., Zhang, H., & Zhao, W. (2017). A survey on Internet of Things: Architecture, enabling technologies, security and privacy, and applications. *IEEE Internet of Things Journal*, *4*(5), 1125–1142. DOI: 10.1109/JIOT.2017.2683200

Li, S., Xu, L. D., & Zhao, S. (2015). The internet of things: A survey. *Information Systems Frontiers*, *17*(2), 243–259. DOI: 10.1007/s10796-014-9492-7

Liu, C.-L., Huang, H.-H., & Yen, D. C. (2022). The effects of navigation cues and involvement of highly innovative consumers on attitudes toward mobile application stores. *Journal of Electronic Commerce in Organizations*, *20*(1), 1–20. Advance online publication. DOI: 10.4018/JECO.305736

Liu, H., Zhang, B., & Zhao, H. (2020). Edge computing for smart cities: A survey. *IEEE Internet of Things Journal*, *7*(6), 5855–5875. DOI: 10.1109/JIOT.2020.2975055

Liu, P., Zhang, Y., & Biljecki, F. (2024). Explainable spatially explicit geospatial artificial intelligence in urban analytics. *Environment and Planning. B, Urban Analytics and City Science*, *51*(5), 1104–1123. DOI: 10.1177/23998083231204689

Liu, Y., Wang, J., Liu, Q., Gheisari, M., Xu, W. L., Jiang, Z., & Zhang, J. (2023). FedTC: A personalized federated learning method with two classifiers. *Computers, Materials & Continua*, *76*(3), 3013–3027. DOI: 10.32604/cmc.2023.039452

Lnenicka, M., & Saxena, S. (2021). Re-defining open government data standards for smart cities' websites: A case study of selected cities. *Digital Policy. Regulation & Governance*, *23*(4), 398–411. DOI: 10.1108/DPRG-12-2020-0174

Longo, A., Zappatore, M., Bochicchio, M., & Navathe, S. B. (2017). Crowd-sourced data collection for urban monitoring via mobile sensors. *ACM Transactions on Internet Technology*, *18*(1), 1–21. DOI: 10.1145/3093895

Lu, J., & Merritt, J. (2024). *How equitable data practices can help shape urban planning*. [online] World Economic Forum. Available at: https://www.weforum.org/stories/2024/10/equitable-data-practices-urban-planning/

Lubongo, C., Bin Daej, M. A., & Alexandridis, P. (2024). Recent developments in technology for sorting plastic for recycling: The emergence of artificial intelligence and the rise of the robots. *Recycling*, *9*(4), 59. DOI: 10.3390/recycling9040059

Luo, L., Hou, X., Cai, W., & Guo, B. (2020). Incremental route inference from low-sampling GPS data: An opportunistic approach to online map matching. *Information Sciences*, *512*, 1407–1423. DOI: 10.1016/j.ins.2019.10.060

Luo, Y., Zhang, W., & Cao, J. (2019). Ambient intelligence in smart street lighting. *Journal of Ambient Intelligence and Smart Environments, 11*(5), 415–432. DOI: 10.3233/AIS-190532

Maalouf, A., & Mavropoulos, A. (2022). Re-assessing global municipal solid waste generation. *Waste Management & Research, 41*(4), 936–947. DOI: 10.1177/0734242X221074116 PMID: 35075952

MacArthur, E. (2013). Towards the circular economy. *Journal of Industrial Ecology, 2*, 23–44.

Maisyarah, R. A., Nugroho, R. A., & Te, A. Y. C. (2024). Examining of E-readiness in the Local Government toward Smart City: A Study from Balangan Regency. *Jurnal Ilmu Sosial, 23*(1), 1–26. https://doi.org/https://doi.org/10.14710/jis.23.1.2024.37-63

Maria, K., Fredy, R., George, P., Nikitas, M., & Michail, M. (2021). Robotic Waste Sorting Technology: Toward a Vision-Based Categorization System for the Industrial Robotic Separation of Recyclable Waste. *IEEE Robotics & Automation Magazine, 28*(2), 50–60. DOI: 10.1109/MRA.2021.3066040

Maria, V., Priyanka, P., Oluwaseyi, F., Katherine, H. I., Liang, Z., Hossain, S., & Sheikh, I. A. (2022). Development of a Noninvasive Blood Glucose Monitoring System Prototype: Pilot Study. *JMIR Formative Research, 6*, 8. PMID: 36018623

Martinez, R., & Masron, I. N. (2020). Jakarta: A city of cities. *Cities (London, England), 106*, 102868. Advance online publication. DOI: 10.1016/j.cities.2020.102868 PMID: 32863521

Material Economics. (2018). *The Circular Economy—A Powerful Force for Climate Mitigation*. Material Economics Sverige AB.

Matuszak, J. (2023). Swiss cities pioneering smart tech. https://knowhow.distrelec.com/internet-of-things/swiss-cities-pioneering-smart-tech

Mayors_User. (2024). *What makes Zurich a Smart City role model: five questions for Mr David Weber, Head of Smart City Zurich. Mayors of Europe.* https://mayorsofeurope.eu/news/what-makes-zurich-a-smart-city-role-model-five-questions-for-mr-david-weber-head-of-smart-city-zurich

Mazzarello, M., Duarte, F., Mora, S., Heine, C., & Ratti, C. (2025). Data Slots: Trade-offs between privacy concerns and benefits of data-driven solutions. *Humanities & Social Sciences Communications, 12*(1), 643. Advance online publication. DOI: 10.1057/s41599-025-04776-1

Mcmahon, J. (2020). From open air composting to efficient Anaerobic Digestion, https://biomassmagazine.com/articles/from-open-air-composting-to-efficient-anaerobic-digestion-17001?

Mehmood, R., Ahmad, F., Yaqoob, I., Imran, M., Guizani, M., & Alturki, B. (2020). Internet of Things (IoT)-based smart cities: Recent advances, requirements, and challenges. *IEEE Communications Surveys and Tutorials*, *22*(3), 1656–1686.

Mehta, R. and Rao, P. 2022, "IoT-enabled neonatal monitoring and smart cradle architecture," *Journal of Healthcare Engineering*, vol. 2022, no. 3, pp. 1–10.

Meijer, A., & Bolívar, M. P. R. (2016). Governing the smart city: A review of the literature on smart urban governance. *International Review of Administrative Sciences*, *82*(2), 392–408. DOI: 10.1177/0020852314564308

Menendez, M., & Ambühl, L. (2022). Implementing Design and Operational Measures for Sustainable Mobility: Lessons from Zurich. *Sustainability (Basel)*, *14*(2), 625. DOI: 10.3390/su14020625

Mensah, I. K. (2020). Impact of Government Capacity and E-Government Performance on the Adoption of E-Government Services. *International Journal of Public Administration*, *43*(4), 303–311. DOI: 10.1080/01900692.2019.1628059

Merlec, M. M., Lee, Y. K., Hong, S. P., & In, H. P. (2021). A smart contract-based dynamic consent management system for personal data usage under GDPR. *Sensors (Basel)*, *21*(23), 7994. DOI: 10.3390/s21237994 PMID: 34883997

Methuku, V. (2025). Decentralized machine learning for disease outbreak prediction: Enhancing data privacy with federated learning. *Int. J. Sci. Res. Arch.*, *14*(3), 1–8. DOI: 10.30574/ijsra.2025.14.3.0590

Miles, M. B., & Huberman, M. J. S. (2019). *Qualitative Data Analysis: A Methods Sourcebook* (4th ed.).

Mitteilung (2024). Zurich: Growth Engine for Digital Health in Switzerland, https://www.zh.ch/de/news-uebersicht/mitteilungen/2024/wirtschaft-arbeit/growth-engnge-digital-health.html

Miyazaki, K., Murayama, T., Uchiba, T., An, J., & Kwak, H. (2024). Public perception of generative AI on Twitter: An empirical study based on occupation and usage. *EPJ Data Science*, *13*(1), 2. Advance online publication. DOI: 10.1140/epjds/s13688-023-00445-y

Mkhonto, M., & Zuva, T. (2024). In Silhavy, R., & Silhavy, P. (Eds.), *Readiness for Smart City in Municipalities in Mbabane, Eswatini BT - Data Analytics in System Engineering* (pp. 402–410). Springer Nature Switzerland.

Mohanty, S. P., Choppali, U., & Kougianos, E. (2016). Everything you wanted to know about smart cities: The Internet of things is the backbone. *IEEE Consumer Electronics Magazine*, *5*(3), 60–70. DOI: 10.1109/MCE.2016.2556879

Monge, F., Barns, S., Kattel, R., & Bria, F. (2022). A new data deal: the case of Barcelona - UCL Discovery. *Ucl.ac.uk*. https://discovery.ucl.ac.uk/id/eprint/10196041/1/Kattel_new_data_deal_barcelona_fernando_barns_kattel_and_bria.pdf

Monteiro, C. M., Mendes, A. M., & Santos, C. (2023). Green roofs as an urban NbS strategy for rainwater retention: Influencing factors—a review. *Water (Basel)*, *15*(15), 2787. DOI: 10.3390/w15152787

Moy, L. (2021). Facing injustice: How face recognition technology may increase the incidence of misidentifications and wrongful convictions. *SSRN*, *30*, 337. DOI: 10.2139/ssrn.4101826

Mulhern, J. L. (2023). Safe Drinking water: Monitoring Quality control in Schwyz, Switzerland. https://www.ayyeka.com/blog/ensuring-safe-drinking-water-ayyekas-solution-for-water-quality-monitoring-in-schwyz-switzerland?

Murciano-Goroff, R., Zhuo, R., & Greenstein, S. (2024). *Navigating software vulnerabilities: Eighteen years of evidence from medium and large us organizations* (No. w32696). National Bureau of Economic Research.

Murroni, M., Anedda, M., Fadda, M., Ruiu, P., Popescu, V., Zaharia, C., & Giusto, D. (2023). 6G—Enabling the New Smart City: A Survey [Review of 6G—Enabling the New Smart City: A Survey]. Sensors, 23(17), 7528. Multidisciplinary Digital Publishing Institute. DOI: 10.3390/s23177528

Nam, H., Nam, T., Oh, M., & Choi, S. (2022). An Efficiency Measurement of E-Government Performance for Network Readiness : Non-Parametric Frontier Approach. *Journal of Open Innovation*, *8*(1), 10. DOI: 10.3390/joitmc8010010

Nam, T., & Pardo, T. A. (2011, June). Conceptualizing smart city with dimensions of technology, people, and institutions. In *Proceedings of the 12th annual international digital government research conference: digital government innovation in challenging times* (pp. 282-291). DOI: 10.1145/2037556.2037602

Neirotti, P., De Marco, A., Cagliano, A. C., Mangano, G., & Scorrano, F. (2014). Current trends in Smart City initiatives: Some stylised facts. *Cities (London, England)*, *38*, 25–36. DOI: 10.1016/j.cities.2013.12.010

Nguyen, D. C., Pathirana, P. N., Ding, M., Seneviratne, A., Li, J., & Niyato, D. (2021). Blockchain for 5G and beyond networks: A state-of-the-art survey. *Journal of Network and Computer Applications*, *174*, 102890.

Nguyen, T., & Park, S. (2023). Sensor-integrated smart cribs for neonatal monitoring. *International Journal of Medical Informatics*, *172*(5), 104–113.

Nidhya, R., Manish, K., Renjith, V. R., & Deepak, V. (2020). Enhanced Route Selection (ERS) algorithm for IoT enabled smart waste management system. *Environmental Technology & Innovation*, *20*, 101116. DOI: 10.1016/j.eti.2020.101116

Niranjan, A. (2024). Europe beats the US for walkable, livable cities, study shows, https://www.theguardian.com/environment/2024/sep/16/europe-beats-the-us-for-walkable-livable-cities-study-shows?

Nugroho, R. A., & Purbokusumo, Y. (2020). E-Government Readiness: Main Actor Readiness Assessment for E-Government Application in Indonesia. *Jurnal Ilmu Pengetahuan Dan Teknologi Komunikasi*, *22*(1), 1–17. DOI: 10.33164/iptekkom.22.1.2020.1-17

Obracht-Prondzyńska, H., Duda, E., Anacka, H., & Kowal, J. (2022). Greencoin as an AI-Based Solution Shaping Climate Awareness [Review of Greencoin as an AI-Based Solution Shaping Climate Awareness]. International Journal of Environmental Research and Public Health, 19(18), 11183. Multidisciplinary Digital Publishing Institute. DOI: 10.3390/ijerph191811183

OECD. (2024). Anti-Corruption and Integrity Outlook: Country Fact Sheet 2024, Anti-Corruption and Integrity Outlook 2024 – Country Notes: Switzerland | OECD.

Okoli, N. J., & Kabaso, B. (2024). Building a Smart Water City: IoT Smart Water Technologies, Applications, and Future Directions. In *Water (Switzerland)* (Vol. 16, Issue 4). DOI: 10.3390/w16040557

Olawade, D. B., Wada, O. Z., David-Olawade, A. C., Fapohunda, O., Ige, A. O., & Ling, J. (2024). Artificial intelligence potential for net zero sustainability: Current evidence and prospects. *Next Sustainability*, *4*, 100041. DOI: 10.1016/j.nxsust.2024.100041

Osakwe, Z. T., Ikhapoh, I., Arora, B. K., & Bubu, O. M. (2020). Identifying public concerns and reactions during the COVID-19 pandemic on Twitter: A text-mining analysis. *Public Health Nursing (Boston, Mass.)*, *38*(2), 145–151. DOI: 10.1111/phn.12843 PMID: 33258149

Ouallane, A. A., Bahnasse, A., Bakali, A., & Talea, M. (2021). Overview of Road Traffic Management Solutions based on IoT and AI. *Procedia Computer Science, 198*, 518–523. DOI: 10.1016/j.procs.2021.12.279

Padhiary, M., Roy, P., & Roy, D. (2025). The future of urban connectivity: AI and IoT in smart cities. In *Sustainable Smart Cities and the Future of Urban Development* (pp. 33–66). IGI Global.

Pajooh, H. H., Demidenko, S., Aslam, S., & Harris, M. (2022). Blockchain and 6G-Enabled IoT. *Inventions (Basel, Switzerland), 7*(4), 109. DOI: 10.3390/inventions7040109

Park, J., & Yoo, S. (2023). Evolution of the smart city: Three extensions to governance, sustainability, and decent urbanisation from an ICT-based urban solution. *International Journal of Urban Sciences, 27*(sup1, S1), 10–28. DOI: 10.1080/12265934.2022.2110143

Park, S., Choi, Y. T., & Bok, H. S. (2013). Does better e-readiness induce more use of e-government? Evidence from the Korean central e-government. *International Review of Administrative Sciences, 79*(4), 767–789. DOI: 10.1177/0020852313501247

PARQEX. (2025). Dynamic and predictive AI pricing for parking. https://parqex.com/dynamic-and-predictive-ai-pricing-for-parking

Partner, C. C. A. C. (n.d.). https://www.ccacoalition.org/partners/switzerland?

Pascu, L. (2018). *At $17 million, Atlanta network recovery six times more expensive than estimated.* Hot for Security. https://www.bitdefender.com/en-us/blog/hotforsecurity/at-17-million-atlanta-network-recovery-six-times-more-expensive-than-estimated

Patel, A., & Sharma, R. (2023). Embedded sensor-based intelligent baby cradle. *International Journal of Electronics and Communication Engineering, 12*(1), 77–84.

Patric, M., Diogo, M., Eduardo, D., Jonatan, C., Rodrigo, C., Juergen, R., Edison, P., & Rafael, K. (2019). An IoT-based smart cities infrastructure architecture applied to a waste management scenario. *Ad Hoc Networks, 87*(1), 200–208.

Potnis, D. D., & Pardo, T. A. (2011). Mapping the evolution of e-Readiness assessments. *Transforming Government: People. Transforming Government, 5*(4), 345–363. DOI: 10.1108/17506161111173595

Pratama, A. B., & Imawan, S. A. (2019). A scale for measuring perceived bureaucratic readiness for smart cities in Indonesia. *Public Administration and Policy, 22*(1), 25–39. DOI: 10.1108/PAP-01-2019-0001

Qin, Z., Li, F. Y., Li, G. Y., McCann, J. A., & Ni, Q. (2019). Low-Power Wide-Area Networks for Sustainable IoT. *IEEE Wireless Communications, 26*(3), 140–145. DOI: 10.1109/MWC.2018.1800264

Quality of Living City Ranking. (2024). https://www.mercer.com/insights/total-rewards/talent-mobility-insights/quality-of-living-city-ranking/?

Rachmawati, R., & Rohmah, A. A. (2025). Navigating Diversity of the Smart City Initiatives in Indonesia. *The Indonesian Journal of Geography, 57*(1), 191–199. DOI: 10.22146/ijg.97840

Raji, I. D., & Buolamwini, J. (2019). Actionable auditing: Investigating the impact of publicly naming biased performance results of commercial AI products. *AAAI/ACM Conference on AI, Ethics, and Society*. DOI: 10.1145/3306618.3314244

Ramani, D. R., Sujitha, B. B., & Tangade, S. (2025). Smart Environmental Monitoring Systems: IoT and Sensor-Based Advancements. *Environmental Monitoring Using Artificial Intelligence*, 45-60.

Ramson, S. R. J., & Moni, D. J. (2017). Wireless sensor networks based smart bin. *Computers & Electrical Engineering, 64*, 337–353. DOI: 10.1016/j.compeleceng.2016.11.030

Rashid, M. T., Wei, N., & Wang, D. (2023). A survey on social-physical sensing: An emerging sensing paradigm that explores the collective intelligence of humans and machines. *Collective Intelligence, 2*(2). Advance online publication. DOI: 10.1177/26339137231170825

Ratcliffe, A. (2024). How technology, including AI, gives claims investigators an edge, https://www.zurich.com/media/magazine/2024/how-technology-including-ai-gives-claims-investigators-an-edge?

Rathod, T., Jadav, N. K., Tanwar, S., Polkowski, Z., Yamsani, N., Sharma, R., Alqahtani, F., & Gafar, A. (2023). AI and Blockchain-Based Secure Data Dissemination Architecture for IoT-Enabled Critical Infrastructure. *Sensors (Basel), 23*(21), 8928. Advance online publication. DOI: 10.3390/s23218928 PMID: 37960626

Reshetova, E., & Hartley, K. (2025). Strengthening governance accountability in smart cities: The case of urban energy transition in Jakarta, Indonesia. In *Smart Cities to Smart Societies: Moving beyond Technology* (pp. 99–130). https://doi.org/ DOI: 10.4324/9781003439325_9

Restuccia, F., Das, S. K., & Payton, J. (2016). Incentive mechanisms for participatory sensing: Survey and research challenges. *ACM Transactions on Sensor Networks, 12*(2), 1–40. Advance online publication. DOI: 10.1145/2888398

Richardson, R., Schultz, J. M., & Crawford, K. (2019). Dirty data, bad predictions: How civil rights violations impact police data, predictive policing systems, and justice. *NYUL Rev. Online*, *94*, 15.

Rieke, N., Hancox, J., Li, W., Milletarì, F., Roth, H. R., Albarqouni, S., Bakas, S., Galtier, M. N., Landman, B. A., Maier-Hein, K., Ourselin, S., Sheller, M., Summers, R. M., Trask, A., Xu, D., Baust, M., & Cardoso, M. J. (2020). The future of digital health with federated learning. *NPJ Digital Medicine*, *3*(1), 119. DOI: 10.1038/s41746-020-00323-1 PMID: 33015372

Roy, P., Ghosh, S., Sarkar, S. K., Podder, A., & Paul, S. (2025). AI-Driven Ambient Intelligence Systems for Mental Health Monitoring and Proactive Intervention. *Cuestiones de Fisioterapia*, *54*(3), 2743–2752.

Roziqin, A., Kamil, M., Romadhan, A. A., & Zakaria, I. (2024). The dark side of local public sector innovation: insights from SAMPADE innovation. *Journal of Science and Technology Policy Management*. DOI: 10.1108/JSTPM-01-2024-0001

Rubinstein, I. S., & Petkova, B. (2019). Governing privacy in the datafied city. *The Fordham Urban Law Journal*, *47*, 755.

Rui, H., Aaron, U., Dian, J., Jiang, Z., & Jae-Won, C. (2022). Inductive proximity sensors within a ceramic package manufactured by material extrusion of binder-coated zirconia. *Sensors and Actuators. A, Physical*, *338*, 113497. DOI: 10.1016/j.sna.2022.113497

Salmi, A.-K., & Hökars, F. (2023). Smart urban traffic zones - demonstration of three smart zones in Sweden based on geofencing and sensor data. In *Transportation Research Procedia* (Vol. 72, pp. 1459–1466). DOI: 10.1016/j.trpro.2023.11.611

Sánchez, L., Elicegui, I., Cuesta, J., Muñoz, L., & Lanza, J. (2013). Integration of utilities infrastructures in a future internet enabled smart city framework. *Sensors (Basel)*, *13*(11), 14438–14465. DOI: 10.3390/s131114438 PMID: 24233072

Sanchez, T. W., Shumway, H., Gordner, T., & Lim, T. (2023). The prospects of artificial intelligence in urban planning. *International Journal of Urban Sciences*, *27*(2), 179–194. DOI: 10.1080/12265934.2022.2102538

Sangeetha, S., Kumari, K. A., Shrinika, M., Sujaybharath, P., Varsini, S. M., & Kumar, K. A. (2023). Ensuring Location Privacy in Crowdsensing System Using Blockchain. *Lecture Notes in Electrical Engineering, 995 LNEE*, 35–47. DOI: 10.1007/978-981-19-9748-8_4

Santoso, A. D., Aryansah, J. E., & Nasyaya, A. (2024). Writing about smart cities in Indonesia: A bibliometric analysis. *Journal of Regional and City Planning*, *35*(1), 69–89. DOI: 10.5614/jpwk.2024.35.1.4

Sathish, P. G., Subrata, H., & Atul, T. (2017). A review on automated sorting of source-separated municipal solid waste for recycling. *Waste Management (New York, N.Y.)*, *60*, 56–74. DOI: 10.1016/j.wasman.2016.09.015 PMID: 27663707

Satyanarayanan, M. (2017). The emergence of edge computing. *Computer*, *50*(1), 30–39. DOI: 10.1109/MC.2017.9

Schmidtke, N., Behrendt, F., Gerpott, F. T., & Wagner, M. (2022). Integration of New Business Models in Smart Logistics Zones. *International Journal of Supply and Operations Management*, *9*(1), 15–33. DOI: 10.22034/ijsom.2021.109063.2160

Schulte, K., Grieve, A., Barratt, B., Baker, T., Coonjobeeharry, H., & Mead, M. (2025). Advancing participatory sensing and knowledge production methods for city air quality governance: Applying the Breathe London Community Programme model. *Environmental Science & Policy*, *170*, 104092. Advance online publication. DOI: 10.1016/j.envsci.2025.104092

Schwartz, M. J. (2025, August 6). *Atlanta's Reported Ransomware Bill: Up to $17 Million.* https://www.bankinfosecurity.com/atlantas-reported-ransomware-bill-up-to-17-million-a-11281

Sebastian, G. (2022). Could incorporating cybersecurity reporting into SOX have prevented most data breaches at US publicly traded companies? An exploratory study. *International Cybersecurity Law Review*, *3*(2), 367–383. DOI: 10.1365/s43439-022-00062-x

Shafique, K., Khawaja, B. A., Sabir, F., Qazi, S., & Mustaqim, M. (2020). Internet of Things (IoT) for next-generation smart systems: A review of current challenges, future trends and prospects for emerging 5G-IoT scenarios. *IEEE Access : Practical Innovations, Open Solutions*, *8*, 23022–23040. DOI: 10.1109/ACCESS.2020.2970118

Shahid, K., Hassan, M., Husnain, A., & Mukhtar, S. (2022). Information and communication technologies for new generation of sustainable smart cities.In *Sustainable Smart Cities-A Vision for Tomorrow*.IntechOpen, 1(1), 1-10.

Shah, S. K., Tariq, Z., Lee, J., & Lee, Y. (2021). Event-Driven Deep Learning for Edge Intelligence (EDL-EI). *Sensors (Basel)*, *21*(18), 6023. DOI: 10.3390/s21186023 PMID: 34577228

Shahzadi, S., Iqbal, M. Z., Naqvi, S. A. R., & Amin, M. (2021). Role of AI in smart cities: A review. *International Journal of Distributed Sensor Networks*, *17*(5). Advance online publication. DOI: 10.1177/15501477211011040

Shan-Huen, H., & Pei-Chun, L. (2015). Vehicle routing–scheduling for municipal waste collection system under the "Keep Trash off the Ground" policy. *Omega*, *55*, 24–37. DOI: 10.1016/j.omega.2015.02.004

Shapiro, A. (2019). Predictive policing for reform? Indeterminacy and intervention in big data policing. *Surveillance & Society*, *17*(3/4), 456–472. DOI: 10.24908/ss.v17i3/4.10410

Sharma, R., & Gupta, P. (2022). Predictive analytics for child safety monitoring using AI. *Journal of Artificial Intelligence Research*, *14*(2), 88–97.

Shi, W., Cao, J., Zhang, Q., Li, Y., & Xu, L. (2016). Edge computing: Vision and challenges. *IEEE Internet of Things Journal*, *3*(5), 637–646. DOI: 10.1109/JIOT.2016.2579198

Shi, Y., Yang, K., Jiang, T., Zhang, J., & Letaief, K. B. (2020). Communication-Efficient Edge AI: Algorithms and Systems. *IEEE Communications Surveys and Tutorials*, *22*(4), 2167–2191. DOI: 10.1109/COMST.2020.3007787

Shraddha, M. K., Sameer, U., Amogh, R. M., Shreyas, S. A., & Meenaxi, M. R. (2025). Grading and Classification of Tomatoes Using ESP32-CAM and Impulse Cloud based IoT Platform for Sustainability. Seventh International Conference on Recent Trends in Image Processing and Pattern Recognition (RTIP2R-2024). *Procedia Computer Science*, *260*, 938–946. DOI: 10.1016/j.procs.2025.03.277

Shreelatha, R. P., & Kumar, V. (2017). Wireless baby monitoring system using mobile application. *International Journal of Computer Applications*, *162*(4), 18–23.

Silva, B. N., Khan, M., & Han, K. (2018). Towards sustainable smart cities: A review of trends, architectures, components, and open challenges in smart cities. *Sustainable Cities and Society*, *38*, 697–713. DOI: 10.1016/j.scs.2018.01.053

Simoiu, C. (2020). *Secure by Default: A Behavioral Approach to Cyber Security*. Stanford University.

Singh, M. (2019). Solid waste management in urban India: Imperatives for improvement. *Journal of Contemporary Issues in Business and Government*, *25*(1), 87–92.

Singh, R., & Thomas, J. (2020). Remote baby monitoring using wireless sensor networks. *International Journal of Engineering Research & Technology (Ahmedabad)*, *13*(5), 89–96.

Sivakumar, C. L. V., Mone, V., & Abdumukhtor, R. (2024). Addressing privacy concerns with wearable health monitoring technology. *Wiley Interdisciplinary Reviews. Data Mining and Knowledge Discovery*, *14*(3), e1535. DOI: 10.1002/widm.1535

Smart Cities World. (2025). Switzerland to launch public transport autonomous driving project,https://www.smartcitiesworld.net/connected-and-autonomous-vehicles/switzerland-to-launch-public-transport-autonomous-driving-project-11081

Smyre, B. (2018, March 27). *City of Atlanta 2018 Ransomware Hack: What We Know and What You Can Learn From It | Raxis*. Raxis.com. https://raxis.com/blog/city-of-atlanta-2018-ransomware-hack-what-we-know-and-what-you-can-learn-from-it/

Sofeska, E., & Sofeski, E. (2022). Developing Projects for Realizing of the Program "Skopje 2020 Smart Strategy" by Enhancing Citizen Approach, Engineering, Digitalization, and Sensing of the City District Toward Smarter Sustainability Urban Potential in the Small Ring of Skopje. In *Advances in Science, Technology and Innovation* (pp. 69–84). DOI: 10.1007/978-3-030-86499-6_7

Son, T. H., Weedon, Z., Yiğitcanlar, T., Sanchez, T. W., Corchado, J. M., & Mehmood, R. (2023). Algorithmic urban planning for smart and sustainable development: Systematic review of the literature. *Sustainable Cities and Society*, *94*, 104562. DOI: 10.1016/j.scs.2023.104562

Soriano-Gonzalez, R., Martin, X. A., Perez-Bernabeu, E., & Carracedo, P. (2024). Modeling and Optimization of NO2 Stations in the Smart City of Barcelona. *Applied Sciences (Basel, Switzerland)*, *14*(22), 10355. Advance online publication. DOI: 10.3390/app142210355

Soumyabrata, S., Rituparna C. (2023). IoT based smart waste management system in aspect of COVID-19. Journal of Open Innovation: Technology, Market, and Complexity, 9, 2, 100048.

Spicer, Z., Goodman, N., & Wolfe, D. A. (2023). How 'smart' are smart cities? Resident attitudes towards smart city design. *Cities (London, England)*, *141*(July), 104442. DOI: 10.1016/j.cities.2023.104442

Stadt Zurich. (n.d.). Open Government Data, https://www.stadt-zuerich.ch/opendata

StraitsTimes. (2024). Singapore is world's 5th smartest city. Retrieved April 3, 2025, from https://www.straitstimes.com/singapore/singapore-is-5th-smartest-city-in-the-world-top-in-asia-global-index

Sulistyaningsih, T., Loilatu, M. J., & Roziqin, A. (2024). Research trends on smart urban governance in Asia: A bibliometric analysis. *Journal of Science and Technology Policy Management*, *15*(5), 997–1105. DOI: 10.1108/JSTPM-03-2022-0045

Swiss information. (2018). Switzerland's first municipal blockchain vote hailed a success, https://www.swissinfo.ch/eng/business/crypto-valley-_-switzerland-s-first-municipal-blockchain-vote-hailed-a-success/44230928?

Syahmi, H. (2025). Privacy and ethical implications of big data utilization in public transportation surveillance. *International Journal of Advanced Cybersecurity Systems, Technologies, and Applications*, *9*(1), 1–10.

Symon, A. F., Hasan, M., & Hossain, T. (2017). Design and implementation of a smart baby monitoring system using IoT. *IEEE Internet of Things Journal*, *4*(6), 2157–2165.

Tabassum, K., & Ibrahim, A. (2019). A secure and privacy-aware framework for future smart cities. *Int. J. Comput. Netw. Technol.*, *07*(01), 16–20. DOI: 10.12785/ijcnt/070103

Taddese, A. A., Addis, A. C., & Tam, B. T. (2025). Data stewardship and curation practices in AI-based genomics and automated microscopy image analysis for high-throughput screening studies: Promoting robust and ethical AI applications. *Human Genomics*, *19*(1), 16. Advance online publication. DOI: 10.1186/s40246-025-00716-x PMID: 39988670

Taher, T., Ahmed, R. U., Haider, M. A., Das, S., Yasmin, M. N., & Mamun, N. (2018). Accident prevention smart zone sensing system. *5th IEEE Region 10 Humanitarian Technology Conference 2017, R10-HTC 2017, 2018-Janua*, 67–70. DOI: 10.1109/R10-HTC.2017.8288908

Tang, C., Qiu, C., & Xiao, X. (2023). Machine-learning and water energy harvesting based wireless water consumption sensing system in buildings. *E-Prime - Advances in Electrical Engineering. Electronics and Energy*, *5*, 100260. Advance online publication. DOI: 10.1016/j.prime.2023.100260

Tee, W. Z., Dave, R., Seliya, N., & Vanamala, M. (2022). A Close Look into Human Activity Recognition Models using Deep Learning. arXiv (Cornell University). / arXiv.2204.13589DOI: 10.1109/CNIOT55862.2022.00043

Tekinerdogan, B., Köksal, Ö., & Çelik, T. (2023). System Architecture Design of IoT-Based Smart Cities. *Applied Sciences (Basel, Switzerland)*, *13*(7), 4173. Advance online publication. DOI: 10.3390/app13074173

TEKTELIC. (2023). Top 3 smart cities in 2021: How IOT technologies help building smart cities. https://tektelic.com/expertise/best-smart-cities/?

Telepatil, A. R., Patil, P. P., Yaja, S. S., & Jadhav, S. R. (2021). IoT based smart baby cradle system for infant monitoring. *International Journal of Advanced Research in Computer and Communication Engineering*, *10*(5), 112–117.

Texas Instruments Incorporated. (2021). Sensor Design for Inductive Sensing Applications Using LDC, SNOA930C – March 2015 – Revised May 2021.

The Telegraph. (2021). From Auckland to Zurich: The cities with the most green space. https://www.telegraph.co.uk/travel/galleries/cities-most-green-space/?

Tianchen, J., Huaiying, F., Rencheng, Z., Jianhong, Y., Lulu, F., Yangyang, H., & Zhengxing, C. (2023). Low-value recyclable waste identification based on NIR feature analysis and RGB-NIR fusion. *Infrared Physics & Technology*, *131*, 104693. DOI: 10.1016/j.infrared.2023.104693

Tiffin, N., George, A., & LeFevre, A. E. (2019). How to use relevant data for maximal benefit with minimal risk: Digital health data governance to protect vulnerable populations in low-income and middle-income countries. *BMJ Global Health*, *4*(2), e001395. DOI: 10.1136/bmjgh-2019-001395 PMID: 31139457

Toh, C. K. (2020). Security for smart cities. *IET Smart Cities*, *2*(2), 95–104. https://doi.org/https://doi.org/10.1049/iet-smc.2020.0001. DOI: 10.1049/iet-smc.2020.0001

Toppeta, D. (2010). *The Smart City Vision: How Innovation and ICT Can Build Smart, "Livable"*. Sustainable Cities. The Innovation Knowledge Foundation.

Torrens, P. M. (2022). Smart and sentient retail high streets. *Smart Cities*, *5*(4), 1670–1720. DOI: 10.3390/smartcities5040085

Torres-Sospedra, J., & Ometov, A. (2021). *Data from Smartphones and Wearables. Data*. MDPI AG., DOI: 10.3390/data6050045

Townsend, A. M. (2000). Life in the real-time city: Mobile telephones and urban metabolism. *Journal of Urban Technology*, *7*(2), 85–104. DOI: 10.1080/713684114

Townsend, A. M. (2013). *Smart cities: Big data, civic hackers, and the quest for a new utopia*. WW Norton & Company.

Trading Economics. (2024). Switzerland employment rate, https://tradingeconomics.com/switzerland/employment-rate

Tzounis, A., Katsoulas, N., Bartzanas, T., & Kittas, C. (2017). Internet of Things in agriculture, recent advances and future challenges. *Biosystems Engineering*, *164*, 31–48. DOI: 10.1016/j.biosystemseng.2017.09.007

U.S. Attorney's Office. (2018, December 5). *Atlanta U.S. Attorney Charges Iranian nationals for City Of Atlanta ransomware attack*. https://www.justice.gov/usao-ndga/pr/atlanta-us-attorney-charges-iranian-nationals-city-atlanta-ransomware-attack

UAE. (2023, November 7). *Dubai Metaverse Strategy - The Official Portal of the UAE Government*. https://u.ae/en/about-the-uae/strategies-initiatives-and-awards/strategies-plans-and-visions/government-services-and-digital-transformation/dubai-metaverse-strategy

Ullah, A., Anwar, S. M., Li, J., Nadeem, L., Mahmood, T., Rehman, A., & Saba, T. (2024). Smart cities: The role of Internet of Things and machine learning in realizing a data-centric smart environment. *Complex & Intelligent Systems*, *10*(1), 1607–1637. DOI: 10.1007/s40747-023-01175-4

Ullo, S. L., & Sinha, G. R. (2020). Advances in smart environment monitoring systems using iot and sensors. In *Sensors (Switzerland)* (Vol. 20, Issue 11). DOI: 10.3390/s20113113

UN Environment Programme. International SWA. (2024). Global Waste Management Outlook 2024—Beyond an Age of Waste: Turning Rubbish into a Resource. Technical Report, Available online: https://wedocs.unep.org/20.500.11822/44939

UN-DESA. (2024). World Population Prospects 2024. Retrieved May 27, 2025, from https://www.un.org/development/desa/pd/world-population-prospects-2024

United Nations. (2018). 68% of world population is projected to line in Urban areas by 2050, https://www.un.org/uk/desa/68-world-population-projected-live-urban-areas-2050-says-un

University of Zurich. (n.d.). The Digital Society Initiative Community Mobility, https://mobility.dsi.uzh.ch/about/?

Vera, C. H., Karl, F. D., Richard, F. H., & Daniele, V. (2013). Models and Algorithms for the Integrated Planning of Bin Allocation and Vehicle Routing in Solid Waste Management. *Transportation Science*, *48*, 1.

Verma, A., Sahay, K., Prakash, S., Kumar, A., Kumar, A., Lahza, H. F. M., & Lahza Husam and Albahar, M. A. (2023). A systematic review on machine learning fundamentals for smart home: Towards smart city solution. In Preprints.

Verma, S., & Kaabouch, N. (2017). Privacy preservation in big data: A survey. *Journal of Big Data*, *4*(1), 1–18. DOI: 10.1186/s40537-017-0086-7

Vicentini, F., Giusti, A., Rovetta, A., Fan, X., He, Q., Zhu, M., & Liu, B. (2009). Sensorized waste collection container for content estimation and collection optimization. *Waste Management (New York, N.Y.)*, *29*(5), 1467–1472. DOI: 10.1016/j.wasman.2008.10.017 PMID: 19103480

Visvizi, A., & Lytras, M. D. (2018). Rescaling and refocusing smart cities research: From mega cities to smart villages. *J. Sci. Technol. Policy Manag.*, *9*(2), 134–145. DOI: 10.1108/JSTPM-02-2018-0020

Vujko, A., Knežević, M., & Arsić, M. (2025). The Future Is in Sustainable Urban Tourism: Technological Innovations, Emerging Mobility Systems and Their Role in Shaping Smart Cities. *Urban Science (Basel, Switzerland)*, *9*(5), 169. Advance online publication. DOI: 10.3390/urbansci9050169

W4 Marketing AG. (2018). *ERZ Mobile App development Clean Zurich*. https://www.w-4.ch/en/about-us/references/erz-app?

Walker, A. (2023, May 4). *Singapore's digital twin – from science fiction to hi-tech reality*. Infrastructure Global. https://infra.global/singapores-digital-twin-from-science-fiction-to-hi-tech-reality

Wang, D., Abdelzaher, T., & Kaplan, L. (2015). Social Sensing: Building Reliable Systems on Unreliable Data. In *Social Sensing: Building Reliable Systems on Unreliable Data*. DOI: 10.1016/C2013-0-18808-3

Wang, H., Yurochkin, M., Sun, Y., Papailiopoulos, D., & Khazaeni, Y. (2020). Federated learning with matched averaging.

Wang, D., Szymanski, B. K., Abdelzaher, T., Ji, H., & Kaplan, L. (2019). The age of social sensing. *Computer*, *52*(1), 36–45. DOI: 10.1109/MC.2018.2890173

Wang, K., Zhao, Y., Gangadhari, R. K., & Li, Z. (2021). Analyzing the adoption challenges of the Internet of Things (IoT) and artificial intelligence (AI) for smart cities in China. *Sustainability (Basel)*, *13*(19), 10983. DOI: 10.3390/su131910983

Wang, S., Zhang, Y., Zhang, Y., Wang, Y., Yang, J., & Wang, Y. (2020). A survey on mobile edge networks: Convergence of computing, caching and communications. *IEEE Access : Practical Innovations, Open Solutions*, *5*, 6757–6779. DOI: 10.1109/ACCESS.2017.2685434

Wang, W. Y., & Wang, A. I. (2021). The role of federated learning in smart cities: Privacy-aware AI for decentralized systems. *IEEE Communications Magazine*, *59*(9), 42–48.

Wang, W., Mirjafari, S., Harari, G., Ben-Zeev, D., Brian, R., Choudhury, T., & Campbell, A. (2020). Social Sensing: Assessing Social Functioning of Patients Living with Schizophrenia using Mobile Phone Sensing. In *Conference on Human Factors in Computing Systems - Proceedings*. DOI: 10.1145/3313831.3376855

Wang, X., & Zhang, Y. (2023). AI-assisted baby care and monitoring systems. *Sensors (Basel)*, *23*(4), 1450–1462.

Wang, Z., Ma, J., Wang, X., Hu, J., Qin, Z., & Ren, K. (2022). Threats to training: A survey of poisoning attacks and defenses on machine learning systems. *ACM Computing Surveys*, *55*(7), 1–36. DOI: 10.1145/3538707

Washburn, D., Sindhu, U., Balaouras, S., Dines, R. A., Hayes, N. M., & Nelson, L. E. (2010). *Helping CIOs Understand "Smart City" Initiatives: Defining the nSmart City, Its Drivers, and the Role of the CIO*. Forrester Research, Inc.

Wen-Tsai, S., Indra, G. T. I., & Sung-Jung, H. (2023). An IoT-Based Aquaculture Monitoring System Using Firebase. *Computers, Materials & Continua*, *76*(2), 2179–2200. DOI: 10.32604/cmc.2023.041022

Wibowo, T. S., Endroyono, & Pratomo, I. (2020). Analysis of Malang City Readiness in Realizing Smart Tourism with New Integrated E-Readiness Model. *2020 International Conference on Smart Technology and Applications (ICoSTA)*, 1–6. DOI: 10.1109/ICoSTA48221.2020.1570616406

Wijayaningsih, N., Arifah, M. N., & Izzati, M. N. (2020). Smart zone: Application improving zero waste lifestyle based on community development. *Jurnal Ilmiah Penalaran Dan Penelitian Mahasiswa*, *4*(2), 102–117.

Wikipedia contributors. (n.d.). 2000-watt society. Wikipedia. https://en.wikipedia.org/wiki/2000-watt_society

Wikipedia Contributors. (n.d.). Green infrastructure. https://en.wikipedia.org/wiki/Green_infrastructure

Wikipedia contributors. (n.d.). ServiceHunter. https://en.wikipedia.org/wiki/ServiceHunter

Wilts, H., Garcia, B. R., Garlito, R. G., Gómez, L. S., & Prieto, E. G. (2021). Artificial intelligence in the sorting of municipal waste as an enabler of the circular economy. *Resources*, *10*(4), 28. DOI: 10.3390/resources10040028

Wissenstransfer. (2012). Traffic management in the inner city of Zurich. https://www.nsl.ethz.ch/en/traffic-management-in-the-inner-city-of-zurich

Wolniak, R., Jonek-Kowalska, I., & Grebski, W. (2023). Environmental monitoring in smart city–smartphone applications aspects. *Zeszyty Naukowe Politechniki Śląskiej. Organizacja i Zarządzanie*, *185*, 605–615.

World Economic forum (2020). Exploring Blockchain Technology for Government Transparency: Blockchain-Based Public Procurement to Reduce Corruption. https://www3.weforum.org/docs/WEF_Blockchain_Government_Transparency_Report_Supplementary%20Research.pdf?

Worldbank. (2018). What a Waste 2.0: A Global Snapshot of Solid Waste Management to 2050.

Wu, Q., He, K., & Chen, X. (2020b). Personalized federated learning for intelligent IoT applications: A cloud-edge based framework.

Wu, J., Luo, K., Wang, Y., & Wang, Z. (2021). Urban road greenbelt configuration: The perspective of PM2.5 removal and air quality regulation. *Environment International*, *157*, 106786. DOI: 10.1016/j.envint.2021.106786 PMID: 34314980

Wu, Q., He, K., & Chen, X. (2020a). Personalized federated learning for intelligent IoT applications: A cloud-edge based framework. *IEEE Computer Graphics and Applications*, *1*, 35–44. PMID: 32396074

Xiaoyu, W., Jia, L., Linpeng, Y., & Zhenming, X. (2020). Auto-sorting commonly recovered plastics from waste household appliances and electronics using near-infrared spectroscopy. *Journal of Cleaner Production*, *246*, 118732. DOI: 10.1016/j.jclepro.2019.118732

Xie, L., Su, Z., Chen, N., & Li, R. (2023). Differential Privacy-Based Incentive Scheme for App-Assisted Mobile Edge Crowdsensing. *Proceedings - IEEE Global Communications Conference, GLOBECOM*, 5518–5523. DOI: 10.1109/GLOBECOM54140.2023.10437797

Xinxing, C., Huaiyang, H., Yuxuan, L., Jiqing, L., & Ming, L. (2022). Robot for automatic waste sorting on construction sites. *Automation in Construction*, *141*, 104387. DOI: 10.1016/j.autcon.2022.104387

Xu, A., Biehl, J., & Lee, A. (2024). Getting it Just Right: Towards Balanced Utility, Privacy, and Equity in Shared Space Sensing. *ACM Transactions on the Internet of Things*, *5*(2), 1–26. Advance online publication. DOI: 10.1145/3648479

Xu, X., Rahimi, R., & Zhou, J. (2019). A survey on edge computing for the Internet of Things. *IEEE Access : Practical Innovations, Open Solutions*, *7*, 156974–157001. DOI: 10.1109/ACCESS.2019.2950427

Xu, Z., Mei, L., Choo, K. K. R., Lv, Z., Hu, C., Luo, X., & Liu, Y. (2018). Mobile crowd sensing of human-like intelligence using social sensors: A survey. *Neurocomputing*, *279*, 3–10. DOI: 10.1016/j.neucom.2017.01.127

Yang, D., Xue, G., Fang, X., & Tang, J. (2012). Crowdsourcing to smartphones: Incentive mechanism design for mobile phone sensing. *Proceedings of the Annual International Conference on Mobile Computing and Networking, MOBICOM*, 173–184. DOI: 10.1145/2348543.2348567

Yang, F., & Xu, J. (2018). Privacy concerns in China's smart city campaign: The deficit of China's Cybersecurity Law. *Asia & the Pacific Policy Studies*, *5*(3), 533–543. DOI: 10.1002/app5.246

Yann, G., Arnab, S., & Paul, C. (2015). A Smart Waste Management with Self-Describing Complex Objects. *International Journal On Advances in Intelligent Systems*, *8*, 1–16.

Yassine, A., Singh, S., Hossain, M. S., & Muhammad, G. (2019). IoT big data analytics for smart homes with fog computing. *Future Generation Computer Systems*, *91*, 563–573. DOI: 10.1016/j.future.2018.08.040

Yelbek, U., Alexandr, N., Timoth, M., Zhanbolat, S., Sungat, A., Alizhan, K., Assel, T.M., Aigul, K.K. (2024). Advancing sanitary surveillance: Innovating a live-feed sewer monitoring framework for effective water level and chamber cover detections. Heliyon, 10, 6, e27395.

Yen, C. P., & Hung, T. W. (2021). Achieving equity with predictive policing algorithms: A social safety net perspective. *Science and Engineering Ethics*, *27*(3), 36. DOI: 10.1007/s11948-021-00312-x PMID: 34075448

Yigitcanlar, T., Degirmenci, K., Butler, L., & Desouza, K. C. (2022). What are the key factors affecting smart city transformation readiness? Evidence from Australian cities. *Cities, 120*(September 2020), 103434. DOI: 10.1016/j.cities.2021.103434

Yigitcanlar, T. (2015). Smart cities: An effective urban development and management model? *Australian Plants*, *52*(1), 27–34. DOI: 10.1080/07293682.2015.1019752

Yin, X., Qiu, H., Wu, X., & Zhang, X. (2024). An efficient attribute-based participant selecting scheme with blockchain for federated learning in smart cities. *Computers*, *13*(5), 118. DOI: 10.3390/computers13050118

Yu, W., Mao, Q., Yang, S., Zhang, S., & Rong, Y. (2017). Social Sensing: The Necessary Component of Planning Support System for Smart City in the Era of Big Data. *Lecture Notes in Geoinformation and Cartography*, 231–244. DOI: 10.1007/978-3-319-57819-4_13

Yucer, S., Tektas, F., Al Moubayed, N., & Breckon, T. (2024). Racial bias within face recognition: A survey. *ACM Computing Surveys*, *57*(4), 1–39. DOI: 10.1145/3705295

Yu, D. (2025). Toward Integrated Urban Observatories: Synthesizing Remote and Social Sensing in Urban Science. *Remote Sensing (Basel)*, *17*(12), 2041. Advance online publication. DOI: 10.3390/rs17122041

Yunex Traffic. (2024). Large scale project successfully completed in Zurich: Yunex Traffic modernizes over 200 traffic signals. https://www.yunextraffic.com/newsroom/large-scale-project-zuerich

Yurdem, B., Kuzlu, M., Güllü, M. K., Çatak, F. Ö., & Tabassum, M. (2024). Federated Learning: Overview, Strategies, Applications, Tools and Future Directions [Review of Federated Learning: Overview, Strategies, Applications, Tools and Future Directions]. Heliyon, 10(19). Elsevier BV. DOI: 10.1016/j.heliyon.2024.e38137

Zanella, A., Bui, N., Castellani, A., Vangelista, L., & Zorzi, M. (2014). Internet of Things for smart cities. *IEEE Internet of Things Journal*, *1*(1), 22–32. DOI: 10.1109/JIOT.2014.2306328

Zatsu, V., Shine, A. E., Tharakan, J. M., Peter, D., Ranganathan, T. V., Alotaibi, S. S., Mugabi, R., Muhsinah, A. B., Waseem, M., & Nayik, G. A. (2024). Revolutionizing the food industry: The transformative power of artificial intelligence-a review [Review of Revolutionizing the food industry: The transformative power of artificial intelligence-a review]. Food Chemistry X, 24, 101867. Elsevier BV. DOI: 10.1016/j.fochx.2024.101867

Zeli, W., Heng, L., & Xintao, Y. (2020). Vision-based robotic system for on-site construction and demolition waste sorting and recycling. *Journal of Building Engineering*, *32*, 101769. DOI: 10.1016/j.jobe.2020.101769

Zeng, F., Pang, C., & Tang, H. (2023). Sensors on the Internet of Things Systems for Urban Disaster Management: A Systematic Literature Review. In *Sensors* (Vol. 23, Issue 17). DOI: 10.3390/s23177475

Zhang, H., Ibrahim, A., Parsia, B., Poliakoff, E., & Harper, S. (2023). Passive social sensing with smartphones: A systematic review. *Computing*, *105*(1), 29–51. DOI: 10.1007/s00607-022-01112-2

Zhang, K., Ni, J., Yang, K., Liang, X., Ren, J., & Shen, X. S. (2017). Security and privacy in smart city applications: Challenges and solutions. *IEEE Communications Magazine*, *55*(1), 122–129. DOI: 10.1109/MCOM.2017.1600267CM

Zhang, Y., Kwan, M.-P., Yu, B., Liu, Y., Song, L., & Chen, N. (2025). Quantitative Identification of Mixed Urban Functions: A Probabilistic Approach Based on Physical and Social Sensing Data. *Transactions in GIS*, *29*(1), e13272. Advance online publication. DOI: 10.1111/tgis.13272

Zhao, Y., Zhang, X., Wang, J., Zhang, K., & Ordonez de Pablos, P. (2020). How do features of social media influence knowledge sharing? An ambient awareness perspective. *Journal of Knowledge Management*, *24*(2), 439–462. DOI: 10.1108/JKM-10-2019-0543

Zheng, L., Kwok, W. M., Aquaro, V., & Qi, X. (2019). *Digital Government.* Smart Cities and Sustainable Development., DOI: 10.1145/3326365.3326403

Ziemba, E. (2019). The Contribution of ICT Adoption to the Sustainable Information Society. *Journal of Computer Information Systems*, *59*(2), 116–126. DOI: 10.1080/08874417.2017.1312635

Zurich Financial Centre. (2023). Zurich Financial Centre 2023/2024. https://www.zh.ch/content/dam/zhweb/bilder-dokumente/themen/wirtschaft-arbeit/wirtschaftsstandort/dokumente/englisch/zurich_financial_centre_2023-24_executive_summary.pdf

Zurich Insurance Group. (n.d.). We care about business integrity. *Zurich Insurance Group.* https://www.zurich.com/about-us/corporate-governance/code-of-conduct/we-care-about-business-integrity

Zurich, E. T. H. (2025). The Chair of Air Quality and Practice Technology. https://ie.ifu.ethz.ch

About the Contributors

Hari Mohan Rai is a distinguished researcher and academician with expertise in Artificial Intelligence (AI), Internet of Things (IoT), biomedical signal and image processing, and cybersecurity. He holds a Ph.D. in Electrical and Computer Engineering from the Indian Institute of Technology (ISM), Dhanbad, and a Master of Engineering in Control Systems from Government Engineering College, Jabalpur, where he was awarded a Gold Medal. Dr. Rai currently serves as an Assistant Professor in the Department of Computer Science at Nazarbayev University, Astana, Kazakhstan. He has also worked remotely as an Associate Professor in the Department of Cyber Security at the International IT University (IITU), Almaty, Kazakhstan. His earlier positions include Research Assistant Professor at the School of Computing, Gachon University, South Korea, along with teaching and research roles at international institutions such as Samarkand State University (Uzbekistan) and several universities in India. Across more than a decade of academic experience, he has contributed extensively to both pedagogy and research, with over 50 published papers (including more than 30 in SCI journals) and a citation record approaching 2100. His research contributions include cutting-edge models such as LightweightUNet, GAN-SkipNet, and U-MaskNet, focused on enhancing medical diagnostics through AI. Among his notable works is the co-edited volume Optimization Tools and Techniques for Enhanced Computational Efficiency (IGI Global, 2025), which explores advanced computational methodologies for engineering and data science. In addition, he has authored multiple book chapters and contributed as an editor for various volumes in artificial intelligence, image analysis, and smart technologies. Beyond research, Dr. Rai is skilled in tools such as Python (TensorFlow, PyTorch), MATLAB, and cybersecurity frameworks like Metasploit and Frida. His interdisciplinary expertise allows him to bridge gaps between technology and real-world medical applications. A committed educator and innovator, Dr. Rai continues to drive forward-thinking projects with international collaborators to advance healthcare, digital systems, and smart technologies.

Razaque Abdul is a renowned researcher and educator in Computer Science & Engineering, specializing in cybersecurity, machine learning, and quantum computing. He holds a PhD from the University of Bridgeport, USA, and completed his Post Doctorate at the University of Illinois, USA. With over 170 international academic publications, including journals, conferences, and book chapters, Dr. Razaque has significantly advanced knowledge in cybersecurity, biometric authentication, computer networks, and mobile malware detection. He has authored four academic books and served as Editor-in-Chief for the International Journal for Engineering and Technology (IJET), Singapore, from 2012 to 2015. He holds certifications including CCNA, MCP, and CISM, and has received awards such as the Google PhD Fellowship and Best Paper Awards at prestigious conferences. Dr. Razaque's contributions to research and education have garnered international recognition and acclaim.

Syed Atif Moqurrab is a Lecturer in the School of Computer Science and Technology at the University of Bedfordshire. He holds a PhD in Computer Science from COMSATS Institute of Information Technology, Pakistan, and a master's degree from the National University of Computer and Emerging Sciences, Pakistan. His research focuses on privacy, security, and advanced data analytics, with significant contributions to Artificial Intelligence and privacy-preserving techniques. Prior to joining Bedfordshire, Dr. Moqurrab served as a Research Fellow at the University of Southampton and as a Research Assistant Professor at Gachon University, South Korea. He also held an Assistant Professorship at Air University, Pakistan, where he taught subjects ranging from machine learning to data science and led various academic committees. Dr. Moqurrab's extensive publication record showcases innovative approaches in deep learning, data privacy, and cybersecurity. His extensive publication record features innovative solutions for preserving privacy, advancing data analytics, IoT, and deep learning models.

***

Nathaniel Akande is a cybersecurity and data governance professional with over 8 years of experience driving secure digital transformation across sectors. He holds a BSc in Electrical Engineering, an MSc in Cybersecurity, and an MBA and is certified as an ISO/IEC 27001 Lead Implementer. Nathaniel has led enterprise projects involving data governance and AI risk management, with a focus on aligning technology, compliance, and business value. With an emphasis on balancing technology, compliance, and business value, Nathaniel has led enterprise projects involving data governance and AI risk management. He has also authored several articles in the Daily Newspaper. Nathaniel had also facilitated speaking and engagement sessions at student- and religious-based conferences, where insights are shared on cybersecurity, ethical leadership, and digital responsibility.

Rasulov Avazbek is a motivated and emerging scholar currently pursuing his Master's studies at Tashkent University of Information Technologiesi, Tashkent, Uzbekistan. He is developing a strong academic foundation in information technologies, with a growing interest in research, innovation, and the practical application of digital solutions to real-world challenges. His academic journey reflects dedication to continuous learning, analytical thinking, and professional growth in the rapidly evolving field of technology. As a postgraduate student, he is actively engaged in expanding his knowledge in computing, information systems, and modern technological advancements. He is committed to strengthening both theoretical understanding and practical skills that support research and industry-oriented development. Through academic participation, project involvement, and scholarly curiosity, Rasulov Avazbek demonstrates promising potential as a future contributor to the field of information technology. His enthusiasm for learning and innovation supports his ongoing development as a young researcher and academic professional.

B. Arthi is currently working as an Associate Professor in the Department of Computing Technologies, School of Computing, SRM Institute of Science and Technology, Kattankulathur. She holds a Ph.D degree in the field of Computer Science and Engineering from Anna University. She has 19+ years of experience in teaching. She has published several articles in various reputed journals and conferences. She has also published book chapters in Taylor & Francis, CRC Press and Springer Book series. Her areas of interest include AI & ML, Computer Vision and Software Engineering.

Rajni Bala is presently working as an Associate Professor in Chitkara Business School, Chitkara University, Punjab, India. She is having 17 years of teaching experience. She has 35 research publications in her account indexed in Scopus and UGC care. She has presented a number of papers in various national and international conferences. She has 8 patents and 5 copyrights in her account. She has delivered workshops on Writing Literature Review, Statistical Tools on Excel & Innovative technologies in education etc.

Shannaz Mutiara Deniar is an Assistant Professor in the Department of International Relations at the University of Muhammadiyah Malang, Indonesia. She works on paradiplomacy, international development, and East Asian Studies.

J. Viji Gripsy is an Associate Professor in the Department of Computer Science with Cognitive Systems at PSGR Krishnammal College for Women, Coimbatore. She earned her Ph.D. in Computer Science from Bharathiar University in 2015 and completed a postdoctoral fellowship at Srinivas University in 2024. With over a decade of academic experience, her research focuses on mobile ad-hoc networks, wireless sensor networks, cybersecurity, and AI-driven healthcare systems. Dr. Gripsy has authored more than 60 publications, including journal articles, conference papers, and book chapters, and holds an h-index of 6 with over 100 citations. Her notable projects include developing Tamil speech-enabled mobile interfaces for children with learning disabilities, funded by the Indian Council of Social Science Research and GRG Trust. She actively contributes to academic governance, serving on Boards of Studies for multiple institutions, and is recognized in the AD Scientific Index 2025 for her contributions to mobile ad-hoc and wireless sensor networks.

Lokesh Jain is currently serving as an Associate Professor in the Department of Artificial Intelligence and Data Science at Vivekananda Institute of Professional Studies - Technical Campus, New Delhi. He obtained his Ph.D. in Computer Science from Delhi Technological University, Delhi, and completed his M.Tech. in Information Systems from Netaji Subhas Institute of Technology, Delhi. He earned his undergraduate degree with honors from the Uttar Pradesh Technical University and has also qualified for the UGC-NET and GATE in Computer Science. With over 18 years of teaching and research experience, Dr. Jain has published research papers in reputable SCI, IEEE, and SCOPUS-indexed international journals, conferences, and transactions, and also serves as a reviewer for SCOPUS-indexed journals. He has been recognized with several prestigious awards, including the Premium Research Award in 2024 and the Commendable Research Awards in 2020 and 2021 by Delhi Technological University, as well as the Global Youth Icon Award in 2024 by ICERT, India. He has also chaired sessions at numerous national and international conferences. He is actively associated with several reputed international journals and conferences as a reviewer. His areas of research interest include Machine Learning, Data Science, Deep learning, Database Systems, Data Mining, Online Social Networking, Computational Intelligence, and Object-Oriented Techniques.

K. Kavitha received her Bachelor's degree in Electronics and Communication Engineering from Madurai Kamaraj University, Madurai, India and her Master's degree in Microwave & Optical Engineering from Madurai Kamaraj University, Madurai, India in She completed her doctoral degree at Anna University, Chennai, specializing in Hyperspectral Image Classification. Currently, she holds the position of Professor and Head in the Department of Electronics and Communication Engineering at Velammal College of Engineering and Technology in Madurai,

Tamil Nadu. Her career spans over 23 years, during which she has made significant contributions in the realms of teaching and research. Her research interests encompass diverse areas, including Antenna Design, Hyperspectral Image Classification and Optical System Design. She has published and presented over 100 research articles in renowned international journals and international and national conferences. Presently, she is supervising the research works of ten Ph.D. scholars in various areas of her expertise. In recognition of her outstanding contributions to education, Dr.K.Kavitha was honored with Dr.A.P.K Abdul Kalam Teachers Excellence Award.

Muhammad Kamil is a Lecturer in the Department of Government Studies, Universitas Muhammadiyah Malang, Indonesia. He holds a Master of Politic and Government from Universitas Gadjah Mada. He has research interests in urban and sustainability governance.

Kazi Kutubuddin Sayyad Liyakat has completed his B.E., M.E., Ph.D. in E&TC Engineering and Post Doctorate in "IoT in Healthcare Applications", and is nowadays working as a Professor & Head of Department, E&TC Engineering Department and was Dean R&D. His area of Interest is IoT, IoRT, IoBT, AI, ML, and AIIoT. He had suggested KSK Approach, KSK1 Approach, KK Approach, KVS Approach and DL Approach in the arena of IoT and its security. He has published more than 220+ papers, including Book chapters with various renowned publishers like IGI Global, CRC, NOVA publishers. Also he had received various 50 Patents from Indian Patent House (Utility, Design and Copyright), UK Design Grant Patents, South African Design Grant Patents and Canadian Copyright Patents. He received funds for various projects on IoT applications. He published 35 books under different National and International publishers. He worked as Editor for 2 book. He worked as a Reviewer for Scopus Conferences and Journal and for IGI Global. Also, he works as Editorial Board Member for 6 Journals. He got 4 Best Researcher Awards, 1 Best Faculty Award, a appreciation letter form Nxtwave, 3 Best HoD Award and 2-Appreciation Letter from MoE's Innovation Cell, Government of India.

M. Aruna is currently working as an Associate Professor in the Department of Computing Technologies, SRM Institute of Science and Technology, Kattankulathur. She holds Ph.D. degree from Anna University, Chennai, India. She has 19+ years of experience in Teaching. Her area of interest includes Computer Vison, AR/VR, Artificial Intelligence, Machine Learning and IOT. She has published several articles in various reputed Journals indexed in SCI, WOS, SCOPUS. She has also published book chapters in Taylor & Francis (CRC Press), Springer Book Series. She was granted with an Australian Patent and published five Indian Patents. She is an active reviewer of various Journals and acted as a TPC member and Reviewer of reputed International Conference. She has been involved in various Industry Projects and Consultancy works.

M. Praneesh is an Assistant Professor in the Department of Computer Science with Data Analytics at Sri Ramakrishna College of Arts and Science, Coimbatore, where he contributes to teaching and research in advanced computing and data-driven technologies. He holds an M.Sc., M.Phil., PGDNLP, and Ph.D., and his scholarly work spans areas such as satellite image processing, deep and machine learning, natural language processing, and computational intelligence, with numerous publications and research projects to his credit. Dr. Praneesh's academic profile reflects a strong commitment to both education and research in core computer science and data analytics domains.

John T. Mesia Dhas is an Associate Professor in the Department of Computer Science and Engineering at T John Institute of Technology, Bangalore. He holds an M.E. and a Ph.D. in Computer Science and Engineering, with over 15 years of teaching experience. His research interests include blockchain applications in supply chain transparency and software effort estimation for educational platforms. Dr. Dhas has authored over 35 publications and has been actively involved in placement training, GATE coaching, and academic administration, including work with Anna University, AICTE, and NAAC.

Rifki Muhammad Bintang is a master student in the Master's Program of Public Administration at Universitas Brawijaya, Indonesia. He holds a Bachelor's degree in Government Studies from Universitas Muhammadiyah Malang. His academic interests focus on public policy, governance, community empowerment, and sustainable development. He has experience as a research assistant and is actively involved in policy-oriented research, particularly in the areas of public sector management and data-driven policy analysis. His work aims to contribute to evidence-based policymaking and the advancement of effective and inclusive public governance.

N Meenakshi, completed her Bachelor of Technology Information Technology from Anna University and M.E Embedded Systems from Sastra University, in 2006 and 2009. She completed Ph.D. in Computer science and engineering and Embedded systems in M.S university 2019. Currently she is working as Assistant Professor in the Department of Computational Intelligence (School of Computing), SRMIST, Kattankulathur. She has over 13 years of full time teaching experience in reputed engineering colleges in Chennai. She has published 19 and communicated 8 technical papers in reputed International Journals indexed in Scopus, Springer,

WOS and ESCI. Her main areas of interest include Cyber Security, Digital Forensics, Artificial Intelligence, IOT and Wireless Sensor Networks.

G. Padmapriya is currently working as an Associate Professor in the Department of Computing Technologies, School of Computing, SRM Institute of Science and Technology, Kattankulathur. She received her M.E. degree in Computer Science and Engineering from Anna University, Chennai in the year 2008 and was awarded Ph.D. in Information and Communication Engineering from Anna University, Chennai, India in the year 2016. she is having 19 years teaching experience. Her research interests include Data Mining, Machine Learning, Natural Language Processing and Data Science.

Damodharan Palaniappan works as an Associate Professor & Head in the Department of Information Technology, Marwadi University, Rajkot, Gujarat. He completed his Ph.D. at Anna University, Chennai, M.E. in Computer Science and Engineering at C.I.E.T, Coimbatore and B.E. in Computer Science and Engineering at CSI College of Engineering, Ketty, Nilgiri. He has 19 years' experience of teaching and research. Currently, he guides 6 Ph.D. research scholars. He has authored 8 book chapters, and he has published research papers in reputed journals. His interested areas of research are Data Mining, Image Processing, Software, Machine Learning and Deep Learning.

Ravikumar R. Natarajan is an experienced academician with over 12+ years of teaching and research expertise across institutions in India, the UAE, and the Maldives. He currently serves as an Assistant Professor at Marwadi University, Rajkot, specializing in Machine Learning, Recommender Systems, and AI-driven applications in healthcare and security. He holds a Ph.D. in Machine Learning and Recommender Systems from Amity University, Jaipur, a Master of Engineering in CSE from VMRF University, Salem, and a Bachelor of Engineering in CSE from Anna University. He also has a MiniMBA from the Swiss eLearning Institute. Dr. Ravikumar has authored more than 8+ multiple SCIE/Scopus-indexed journal articles, 25+ IEEE/Springer conference papers, and co-authored a book on Artificial Intelligence. He has also published 9 patents and contributed 10+ book chapters in emerging domains such as Federated Learning, Blockchain, and Digital Twins for IGI Global, Wiley and Springer publisher. He actively mentors students through initiatives like the Research Activities Club (RAC) and Student Research Internship Training (SRIT), promoting innovation and research culture.

Ali Roziqin is a Lecturer in the Department of Government Studies, Universitas Muhammadiyah Malang, Indonesia. He obtained his bachelor's degree from the Department of Public Administration at Universitas Diponegoro, Indonesia.

He holds a Master of Public Administration from the Universitas Gadjah Mada, Indonesia. His research interests are in bureaucracy, public policy and administration, and public sector innovation. He has published papers in the reputable international journals like Sustainability (MDPI), Local Environment (Taylor and Francis), Tourism Critiques, Journal of Science and Technology Policy Management, Journal of the Knowledge Economy, Frontiers in Communication and Public Administration and Policy: An Asia-Pacific Journal, as well as in the IGI Global

Aarthi S is a dedicated Assistant Professor in Computer Science with over six years of teaching experience. She is currently pursuing her Ph.D. at Marwadi University, Gujarat, and holds an M.Phil. and MCA from Bharathiar University, Tamil Nadu. Throughout her academic career, Aarthi has specialized in programming languages such as C++, Java, and C, and contributed to research in recommender systems and AI-related data privacy. She has published journal articles, contributed to book chapters, and actively participated in technical workshops. Known for her student mentorship and engagement in academic events, Aarthi remains committed to fostering innovation and excellence in computer science education.

S. Raj Kumar received his bachelor's degree in Electronics and Communication Engineering from Anna University, Chennai, and his master's degree in VLSI Design from the same institution. He completed his doctoral degree at Anna University, Chennai, specializing in VLSI device modeling. Currently, he serves as an Assistant Professor in the Department of Electronics Engineering (VLSI Design and Technology) at Velammal College of Engineering and Technology, Madurai, Tamil Nadu, India. His career spans over 11 years, during which he has made significant contributions to teaching and research. His research interests include diverse areas such as modeling and simulation of analog VLSI. He has published over 10 research papers in international journals and presented 20 papers at international conferences. He has also received the award for "Best Outcome in Academic Performance".

Sridhar S. is an Assistant Professor in the Department of Computing Technologies at SRM Institute of Science and Technology, where he has been since June 2025. From February 2023 to June 2025, he served as Assistant Professor in the Department of Computer Science and Business Systems at Rajalakshmi Institute of Technology, Anna University. From April 2018 to January 2023, he served as Assistant Professor in the Department of Computer Science at Saveetha School of Engineering, Saveetha University. From January 2016 to January 2018, he served as Assistant Professor in the Department of Computer Science at Indra Gandhi Women's Engineering College. From 2015 to 2016 he served as server administrator in AGS Pvt., Ltd. He received a B.E. degree from Anna University in

2013 and an M.Tech. degree from B.S. Abdur Rahman University, Chennai, in 2015. He was awarded a Ph.D. degree in "A SECURED SOCIAL NETWORKING RECOMMENDATION SYSTEM USING SOFT COMPUTING TECHNIQUES" by Saveetha School of Engineering, Saveetha Institute of Medical and Technical Sciences, Chennai, in 2022. His research interests span both cloud computing and the Internet of Things. Much of his work has been on improving the understanding, design, and performance of security in cloud computing towards social media. In the cloud computing arena, he has worked on characterizing the Internet and the World Wide Web. He has explored the presence and implications of self-similarity and heavy-tailed distributions towards trustworthy people suggestions in social media. In addition, he has made numerous contributions to Internet measurement and modelling; and he has examined the impact of trustworthy people in social media. Professor Sridhar has published more than 48 papers in his career. Out of those, 20 papers are published in SCI journals with high impact factors, and he is the author of 20+ papers on networking and computer systems, cloud computing, and IoT. He holds five patents deriving from his research and 1 registered trademark. As of 2025, Google Scholar reports over 635 citations to his work, and Scopus reports over 285 citations to his work.

Valisher Sapayev is a faculty member in the Department of General Professional Subjects at Mamun University, Khiva, Uzbekistan. He is actively engaged in undergraduate teaching and academic development, contributing to curriculum delivery and interdisciplinary learning across foundational professional subjects. His academic interests include higher education pedagogy, professional skill development, and the integration of contemporary instructional approaches to enhance student learning outcomes. Dr. Sapayev is involved in institutional academic activities and collaborations aimed at strengthening teaching quality and fostering a learner-centered educational environment.

Yusril Saputra is a university student with a strong interest in technology and smart city development. Actively involved in academic and research activities, Yusril is known as a diligent and analytical individual with a high enthusiasm for finding innovative solutions to social issues.

Binastya Anggara Sekti has been a lecturer at the Faculty of Computer Science, Esa Unggul University, Jakarta, since 2016. He is actively involved in the Artificial Intelligence and Information Security Study Group. He holds a Bachelor's degree in Informatics Engineering from the Bandung Institute of Technology and a Master's degree in Information Systems Management from Trisakti University, Jakarta. Prior to entering academia, he accumulated over 25 years of professional experience as an IT practitioner in various companies across Jakarta. His work focused on developing data centers for internet service providers and delivering IT solutions for the manufacturing and oil & gas sectors. His research interests include artificial intelligence, cybersecurity, and the development of IT infrastructure.

Turobova Sevara is a faculty member in the Department of Social Sciences Education at Termez University of Economics and Service, Termez, Uzbekistan. She is engaged in teaching and academic activities related to social sciences education, with a focus on strengthening students' conceptual understanding, critical thinking, and civic awareness. Her academic interests include social science pedagogy, educational development, and the application of innovative teaching methodologies to improve learning outcomes in higher education. She actively contributes to curriculum implementation and scholarly initiatives that support quality education and interdisciplinary learning.

Premavathi T is an Assistant Professor in the Department of Computer Engineering – Artificial Intelligence at Marwadi University, Rajkot, Gujarat. She earned her M.E. in Software Engineering with Distinction from Anna University and is currently pursuing her Ph.D. at Marwadi University. With over 15 years of academic experience, Prof. Premavathi specializes in Machine Learning, Deep Learning, and core programming subjects. She actively mentors students, supervises academic projects, and plays a key role in guiding research and innovation. Her academic portfolio includes 23 book chapters, 5 patents, and 5 research articles, including 1 publication in an SCI-indexed journal. She has also presented papers at 5 national and international conferences.

Darshan Tank, Ph.D. Senior Lecturer – Government Polytechnic, Rajkot, Gujarat, India Dr. Darshan Tank is a dedicated academic and researcher specializing in cloud computing and cybersecurity. Currently a Senior Lecturer in Information Technology at Government Polytechnic Rajkot, he holds a Ph.D. and brings over a decade of expertise in virtualization security, malware detection, and real-time analytics

Khayrulla Urozboev serves as Chief Specialist in the Department of International Scientific Journals and Ratings at Alfraganus University, Tashkent, Uzbekistan. With a strong focus on academic publishing and research evaluation systems, he plays a key role in supporting faculty and researchers in enhancing the visibility and impact of their scholarly work. His expertise includes journal indexing processes, international ranking frameworks, bibliometric analysis, and compliance with global publication standards such as Scopus and Web of Science. He actively contributes to improving institutional research performance by guiding publication strategies, promoting high-quality submissions, and facilitating collaborations with international journals and indexing bodies. Khayrulla Urozboev is also involved in capacity-building initiatives, workshops, and training programs aimed at strengthening research culture and publication ethics among academic communities. His professional contributions support the advancement of global academic standards and the positioning of Alfraganus University within international research and ranking ecosystems.

Index

A

B

C

D

E

F

G

I

L

M

O

P

Q

R

S

T

U

W

Z

www.ingramcontent.com/pod-product-compliance
Lightning Source LLC
LaVergne TN
LVHW061528290926
849162LV00004B/703
* 9 7 9 8 3 3 7 3 4 9 6 2 6 *